Advances in ensuring the microbiological safety of fresh produce

It is widely recognised that agriculture is a significant contributor to global warming and climate change. Agriculture needs to reduce its environmental impact and adapt to current climate change whilst still feeding a growing population, i.e. become more 'climate-smart'. Burleigh Dodds Science Publishing is playing its part in achieving this by bringing together key research on making the production of the world's most important crops and livestock products more sustainable.

Based on extensive research, our publications specifically target the challenge of climate-smart agriculture. In this way we are using 'smart publishing' to help achieve climate-smart agriculture.

Burleigh Dodds Science Publishing is an independent and innovative publisher delivering high quality customer-focused agricultural science content in both print and online formats for the academic and research communities. Our aim is to build a foundation of knowledge on which researchers can build to meet the challenge of climate-smart agriculture.

For more information about Burleigh Dodds Science Publishing simply call us on +44 (0) 1223 839365, email info@bdspublishing.com or alternatively please visit our website at www.bdspublishing.com.

Related titles:

Advances in postharvest management of horticultural produce
Print (ISBN 978-1-78676-288-7); Online (ISBN 978-1-78676-291-7, 978-1-78676-290-0)

Achieving sustainable cultivation of vegetables
Print (ISBN 978-1-78676-236-8); Online (ISBN 978-1-78676-239-9, 978-1-78676-238-2)

Developing smart agri-food supply chains: Using technology to improve safety and quality
Print (ISBN 978-1-78676-749-3); Online (ISBN 978-1-78676-751-6, 978-1-78676-752-3)

Chapters are available individually from our online bookshop:
https://shop.bdspublishing.com

climate•SMART•publishing
IN AGRICULTURAL SCIENCE

BURLEIGH DODDS SERIES IN AGRICULTURAL SCIENCE
NUMBER 136

Advances in ensuring the microbiological safety of fresh produce

Edited by Professor Karl R. Matthews, Rutgers University, USA

burleigh dodds
SCIENCE PUBLISHING

Published by Burleigh Dodds Science Publishing Limited
82 High Street, Sawston, Cambridge CB22 3HJ, UK
www.bdspublishing.com

Burleigh Dodds Science Publishing, 1518 Walnut Street, Suite 900, Philadelphia, PA 19102-3406, USA

First published 2023 by Burleigh Dodds Science Publishing Limited

Library of Congress Control Number: 2023940904

British Library Cataloguing in Publication Data
A catalogue record for this book is available from the British Library

ISBN 978-1-80146-268-6 (print)
ISBN 978-1-80146-270-9 (PDF)
ISBN 978-1-80146-269-3 (ePub)
ISSN 2059-6936 (print)
ISSN 2059-6944 (online)

DOI: 10.19103/AS.2023.0121

Typeset by Deanta Global Publishing Services, Dublin, Ireland

Contents

Series list

Title	Series number
Achieving sustainable cultivation of maize - Vol 1 From improved varieties to local applications Edited by: Dr Dave Watson, CGIAR Maize Research Program Manager, CIMMYT, Mexico	001
Achieving sustainable cultivation of maize - Vol 2 Cultivation techniques, pest and disease control Edited by: Dr Dave Watson, CGIAR Maize Research Program Manager, CIMMYT, Mexico	002
Achieving sustainable cultivation of rice - Vol 1 Breeding for higher yield and quality Edited by: Prof. Takuji Sasaki, Tokyo University of Agriculture, Japan	003
Achieving sustainable cultivation of rice - Vol 2 Cultivation, pest and disease management Edited by: Prof. Takuji Sasaki, Tokyo University of Agriculture, Japan	004
Achieving sustainable cultivation of wheat - Vol 1 Breeding, quality traits, pests and diseases Edited by: Prof. Peter Langridge, The University of Adelaide, Australia	005
Achieving sustainable cultivation of wheat - Vol 2 Cultivation techniques Edited by: Prof. Peter Langridge, The University of Adelaide, Australia	006
Achieving sustainable cultivation of tomatoes Edited by: Dr Autar Mattoo, USDA-ARS, USA and Prof. Avtar Handa, Purdue University, USA	007
Achieving sustainable production of milk - Vol 1 Milk composition, genetics and breeding Edited by: Dr Nico van Belzen, International Dairy Federation (IDF), Belgium	008
Achieving sustainable production of milk - Vol 2 Safety, quality and sustainability Edited by: Dr Nico van Belzen, International Dairy Federation (IDF), Belgium	009
Achieving sustainable production of milk - Vol 3 Dairy herd management and welfare Edited by: Prof. John Webster, University of Bristol, UK	010
Ensuring safety and quality in the production of beef - Vol 1 Safety Edited by: Prof. Gary Acuff, Texas A&M University, USA and Prof. James Dickson, Iowa State University, USA	011
Ensuring safety and quality in the production of beef - Vol 2 Quality Edited by: Prof. Michael Dikeman, Kansas State University, USA	012
Achieving sustainable production of poultry meat - Vol 1 Safety, quality and sustainability Edited by: Prof. Steven C. Ricke, University of Arkansas, USA	013
Achieving sustainable production of poultry meat - Vol 2 Breeding and nutrition Edited by: Prof. Todd Applegate, University of Georgia, USA	014

Introduction

Increasing consumer demand for low-input cultivation and minimal processing has significantly increased the risk of microbiological contamination of fresh produce. This both presents a health risk to consumers and undermines trust in the food supply chain from farm to fork.

This volume reviews our current understanding of the main pathogenic risks to fresh produce, including their epidemiology, genetics and behaviour. The book is split into three parts: Part 1 chapters focus on understanding current pathogenic risks in fresh produce, specifically focusing on *Salmonella*, *Listeria* and *Escherichia coli*. Chapters in Part 2 examine developments in detection and risk assessment of pathogens, such as rapid detection/high throughput screening techniques, modelling pathogen behaviour and microbiological risk assessment for pathogens in fresh produce. Part 3 chapters address improving food safety along the value chain, specifically discussing sources of pathogen contamination, the role of good agricultural practices in preventing contamination, as well as developments in sanitising and packaging techniques. Chapters also discuss the role of good manufacturing practice, hazard analysis and critical control point systems to maintain fresh produce safety and improving safe consumer handling of fresh produce.

Part 1 Pathogenic risks

The book opens with a chapter that focuses on understanding contamination of fresh produce by *Salmonella*. Chapter 1 begins by first reviewing the various plant traits and agricultural practices that aid *Salmonella* association with fruit and vegetable crops. It also discusses the genetic strategies that *Salmonella* employs to compete in the plant niche. The chapter moves on to examine *Salmonella* in various types of produce, such as sprouts and microgreens, root, bulbous and stalk vegetables and leafy vegetables and fresh herbs. It also examines *Salmonella* in fruit, tree nuts and seeds. A section on *Salmonella* strategies in the plant niche is also provided.

The subject of Chapter 2 is advances in understanding and presenting contamination of fresh produce by *Listeria monocytogenes*. The chapter begins by first describing the pathogen's pathogenicity and gene regulation, followed by an overview of recent *L. monocytogenes* outbreaks linked to fresh produce contamination. A section on the sources of fresh produce supply chain contamination is also included, focusing specifically on pre-harvest, harvest and postharvest contamination. The chapter includes a case study on *L. monocytogenes* contamination in cantaloupes, which is then followed by an overview of current strategies used to control *L. monocytogenes* in the fresh produce supply chain. Sections on *L. monocytogenes* gene regulation, survival mechanisms and its adaptation to various forms of stress are also included. The chapter examines biocontrol of *L. monocytogenes* and also describes the novel processing technologies to control the pathogen.

The final chapter of Part 1 examines advances in understanding contamination of fresh produce by pathogenic *Escherichia coli*. Chapter 3 first describes the pathogenicity and virulence of *Escherichia coli*, then moves on to analyse produce contamination with pathogenic *E. coli*. The chapter then describes the interaction of pathogenic *E. coli* with fresh produce and highlights various produce outbreaks linked to pathogenic *E. coli*. A section on food safety regulations and future actions for *E. coli* contamination is also provided, focusing specifically on the research describing the pathogen's growth and survival in soil and plants and current implications for pathogen control.

Part 2 Detection and risk assessment

Part 2 begins with a chapter that focuses on developments in rapid detection/high throughput screening techniques for identifying pathogens in food. Chapter 4 highlights current conventional methods in place for detecting foodborne microorganisms, focusing on techniques such as culture-based assay and immunoassays. The chapter goes on to review molecular methods such as polymerase chain reaction, loop-mediated isothermal detection, nucleic acid sequence-based amplification, recombinase polymerase amplification, DNA microarray and whole-genome sequencing. This is then followed by an overview of current spectroscopic methods, specifically Raman spectroscopy, Fourier-transform infrared spectroscopy, near-infrared spectroscopy and hyperspectral imaging spectroscopy. The chapter describes mass spectrometry methods such as matrix-assisted laser desorption/ionisation-time-of-flight mass spectrometry, optical phenotyping methods such as forward light scattering and flow cytometry and biosensor methods. Sections on the detection of parasites and the advantages and limitations of microbial detection methods are also included.

The focus of Chapter 5 is advances in modelling pathogen behaviour in fresh produce. Chapter 5 begins by first describing methods for modelling spoilage of fresh produce, which is then followed by an overview of current modelling methods for behaviour in fresh produce, focusing specifically on growth modelling and pathogen inactivation. A section on modelling transfer of pathogens during fresh produce processing is also included, highlighting how microbial transfer can be modelled between processing equipment and foods.

The next chapter draws specific attention to advances in quantitative microbiological risk assessment for pathogens in fresh produce. Chapter 6 takes a microorganism and historical perspective to review advances in quantitative microbial risk assessment of fresh produce. The chapter reviews microbial risk assessment research over the past 25 years, specifically focused on *Listeria monocytogenes*, *Salmonella*, *Escherichia coli* O157:H7 and Norovirus as important pathogens that have all caused outbreaks linked to fresh produce.

Part 3 Improving safety along the value chain

The first chapter of Part 3 reviews advances in understanding sources of pathogenic contamination of fresh produce, focusing specifically on soil and soil amendments. Chapter 7 first discusses pathogen survival in soils containing biological amendments, which is then followed by an examination of *E. coli* and *Salmonella* survival in manure dust. A section on the prevalence of antibiotic resistance genes in manure-amended soils is also provided, followed by that draw attention to outbreaks in fruit and vegetables in 2018 and 2020. The chapter also provides an overview of recommendations for the use of soil and soil amendments alongside the potential risk.

Chapter 8 draws attention to the role of Good Agricultural Practices (GAPs) in preventing pathogenic microbial contamination of fresh produce. The chapter begins by describing the evolution of GAPs from voluntary to regulation, then moves on to highlight the reasons for adopting these GAPs in fresh produce safety. A section on why GAPs remain the foundation for fresh produce is also provided, followed by an overview of commitment and training involved in the implementation of GAPs. The chapter reviews the scalability of GAPs and the feasibility of implementing them. A section on ways of assessing risks is also included, focusing specifically on practices such as worker training, preharvest, harvest and postharvest water, soil amendments, wildlife and domesticated animals, sanitation and storage and transportation. The chapter moves on to examine how GAPs can be improved, how produce safety can add to farming stress, as well as the relevance of GAPs for the industry in the future.

The subject of Chapter 9 is advances in sanitising techniques and their assessment for assuring the safety of fresh produce. The chapter first reviews the importance of postharvest decontamination, drawing attention to the limitations of postharvest washing. This is then followed by an analysis of the advances in postharvest washing processes such as monitoring chlorine concentration, predicting chlorine demand of wash waters, pre-oxidation of wash water and advances in alternative sanitisers. The chapter moves on to review non-aqueous decontamination methods such as electron beams, ultraviolet light, gas-phase decontamination, application of gas plasma and the gas-phase hydroxyl-radical process. Considerations for developing a standard validation method for fresh produce decontamination methods are also reviewed.

Chapter 10 draws specific attention to developments in packaging techniques and their assessment for assuring the safety of fresh produce. The chapter begins by describing the current packaging methods in use for physical protection. It moves on to analyse the use of modified atmosphere packaging, focusing on the history and scope of the modified atmosphere packing concept, highlights the different types and packaging materials used, how a modified atmosphere is created and also the use of modified humidity packaging. A section on the use of edible coatings and films is also provided, followed by a section on active packaging methods.

The next chapter focuses on the role of good manufacturing practice (GMP) and hazard analysis and critical control point (HACCP) systems in maintaining the safety of minimally-processed fresh produce. Chapter 11 first examines the key hazards in minimally-processed fresh produce and lessons from previous outbreaks and incidents. It then goes on to review the developments in GMPs for application in minimally-processed fresh produce. A section on the application of HACCP system in minimally-processed fresh produce is also included. The chapter also considers current GMP and HACCP system practices to assure the safety of minimally-processed fresh produce.

The final chapter of the book focuses on improving safe consumer handling of fresh produce. Chapter 12 begins by first discussing consumer produce consumption, drawing specific attention to the consumption of high risk produce items. The chapter then moves on to review pathogens isolated from produce samples, which is followed by an analysis of outbreaks and illnesses caused by produce. A section on organic produce is also provided, focusing specifically on microbial quality of conventionally grown versus organic produce and foodborne outbreaks associated with organic foods. The chapter also provides an overview of farmers markets, comparing the quality of produce items from these markets to traditional retail and also identifies the foodborne illnesses that can occur due to farmers markets. Consumer knowledge and behaviours around safe produce handling and consumption is also discussed, followed by an overview of produce safety and vulnerable and immunocompromised consumers. The chapter includes an overview of fresh produce safety handling recommendations and resources, as well as a discussion on the need to include behaviour theories and formative research in the development of new and targeted education materials around produce safety.

Acknowledgement

We wish to acknowledge the following for their help in reviewing a particular chapter:

- Chapter 3: Dr Li Ma, Oklahoma State University, USA, Emeritus Professor Larry Beuchat, University of Georgia, USA and Dr Alejandro Castillo, Texas A&M University, USA

Part 1

Pathogenic risks

Chapter 1

Advances in understanding contamination of fresh produce by *Salmonella*

Shirley A. Micallef, University of Maryland, USA

1 Introduction

Salmonella enterica is a member of the Enterobacteriaceae, a bacterial family that was last emended in 2016 based on genome- and multi-gene-based phylogenetic trees (Adeolu et al., 2016). The family contains other taxa of importance to public health (e.g. *Escherichia coli, Klebsiella, Enterobacter, Shigella* and *Citrobacter*). *S. enterica* is accepted as the type species of the genus *Salmonella* and comprises six subspecies, *S. enterica* subspp. *enterica* (I), *salamae* (II), *arizonae* (IIIa), *diarizona* (IIIb), *houtenae* (IV) and *indica* (VI) (Tindall et al., 2005). Food safety is primarily concerned with *S. enterica* subsp. *enterica* which is subtyped based on surface antigens yielding over 1500 serovars (Grimont and Weill, 2007) and can be divided into typhoidal (e.g. *S. enterica* serovar Typhi) and non-typhoidal (e.g. *S. enterica* serovar Typhimurium) types. Further strain differentiation within serovars has been resolved by DNA fingerprinting methods such as pulsed-field gel electrophoresis (PFPG) and multiple-locus variable numbers of tandem repeats analysis (MLVA), the most widely adopted techniques in source tracking before the advent of whole genome sequencing (WGS) (Tang et al., 2019). Typhoidal *Salmonella* has declined remarkably in many parts of the world although it continues to be a burden in some countries in Asia, Africa and Oceania (Stanaway et al., 2019). On the other hand, food safety concerns with non-typhoidal salmonellae,

http://dx.doi.org/10.19103/AS.2023.0121.01

which exhibit an astounding capacity to persist in various foods and in food production areas, constitute a primary concern on a global scale (Majowicz et al., 2010).

Non-typhoidal *S. enterica* subsp. *enterica* (hereafter referred to as *Salmonella*) is a diverse and adaptable taxon. Although best studied for its colonization and infection of vertebrate gastrointestinal tracts, *Salmonella* also colonizes insects such as flies and cockroaches (Nasirian, 2019; Wales et al., 2010) and can infect *Caenorhabditis elegans* (Aballay et al., 2000). Its implication in foodborne illness outbreaks involving fruit and vegetables also points to the ability of *Salmonella* to associate with plants (Callejón et al., 2015), supporting the notion that plants can serve as transitional hosts for this pathogen before returning to an enteric lifestyle, following ingestion of the colonised plant material by herbivores (Fletcher et al., 2013). In addition, *Salmonella* can persist in various environments including surface water, reclaimed wastewater, sediments and manure-amended soil (Callahan et al., 2019; Haley et al., 2009; Pornsukarom and Thakur, 2016; Sharma et al., 2020; Walters et al., 2013). Surveys of produce production areas where multiple samples types were collected for bacterial isolation suggest that surface water may present a more favourable habitat for *Salmonella* than soil (Bell et al., 2015; Gorski et al., 2011; Micallef et al., 2012). Mammals, birds, amphibians and reptiles are likely sources of environmental *Salmonella* as related strains have been recovered from wildlife and environmental samples (Gorski et al., 2011; Gruszynski et al., 2014). Once introduced in an environmental medium, *Salmonella* populations may be able to persist for prolonged periods independently from recurrent episodes of introduction (Čučak et al., 2018; Topalcengiz et al., 2020). Moreover, several studies have reported the co-existence of multiple *Salmonella* serovars (Callahan et al., 2019; Gorski et al., 2022), demonstrating the high diversity of *Salmonella* in the agro-environment, which probably increases the likelihood of contamination events. The risk of contamination of plant-based foods during production is linked to this diverse genetic pool in the environment, enhancing the probability of possessing traits that give various *Salmonella* strains the genetic mechanisms needed to successfully associate with plants.

2 The diversity of the *Salmonella*-plant association

Salmonellosis outbreaks have been linked to an impressive diversity of vegetable, fruit, nut, herb and spice crops. Examples of plant-based foods that have been contaminated with *Salmonella* leading to illnesses are given here and have included edible parts of plant members of the Solanaceae (tomatoes, peppers and paprika) (Barton Behravesh et al., 2011; Bennett et al., 2015; Lehmacher et al., 1995), the Cucurbitaceae (melons and cucumber) (Angelo et al., 2015; Laughlin et al., 2019; Walsh et al., 2014), the Rosaceae (nuts such as almonds and stone fruit such as peaches) (CDC, 2020; Isaacs et al., 2005), the Fabaceae (alfalfa sprouts and peanuts) (CDC, 2009; Harfield et al., 2019; Kirk et al., 2004; Sheth et al., 2011), the Asteraceae (lettuce) (Lienemann et al., 2011), the Brassicaceae (arugula/rocket) (Nygård et al., 2008), the Caricaceae (papaya) (Hassan et al., 2019), the Alliaceae (onions) (McCormic et al., 2022), the Apiaceae (herbs such as cilantro and parsley) (Campbell et al., 2001), the Anacardiaceae (mango and pistachio nuts) (CDC, 2016, 2012) and others.

The United States Interagency Food Safety Analytics Collaboration (IFSAC) publishes source attribution estimates for human salmonellosis infections from all foods that occurred in the United States between 1998 and 2019 (IFSAC, 2021). *Salmonella* was the cause of 61.6% of outbreaks from all foods, based on a method used to assess outbreaks caused by *Salmonella, E. coli* O157:H7, *Listeria monocytogenes* and *Campylobacter* spp., with additional weight given to outbreaks occurring in the most recent 5 years (IFSAC, 2021). IFSAC categorizes plant-based foods as vegetable row crops (leafy greens), seeded vegetables (tomatoes, cucumber, pepper), fruit, other produce (such as nuts), sprouts and grains/beans (IFSAC, 2021). Plant-based foods accounted for 42.7% of all *Salmonella* illnesses from all foods, with 13.5% of infections attributed to fruit, 12.6% to seeded vegetables, 7.3% to other produce, 4.2% to vegetable row crops, 4.2% to sprouts and 0.9% to grains and beans. Fruit was ranked the second-highest category for a number of cases after chicken, and seeded vegetables and other produce were ranked fourth and fifth, respectively, with the top five categories making up 63% of all illnesses. Despite proactive measures to minimize food safety problems through grower and industry guidance and government regulations, *Salmonella* in fresh produce remains a concern of the highest priority for food safety.

Certain serovars are repeatedly associated with produce commodities (Jackson et al., 2013). Using the IFSAC categorization and according to the United States National Outbreak Reporting System (NORS) data for outbreaks occurring between 1999 and 2000, the serovars associated with the most outbreaks from plant-based food were primarily *Salmonella* Newport, followed by *Salmonella* Javiana, Enteritidis, Typhimurium and Saintpaul (CDC, 2022). *Salmonella* Newport accounted for 33% of seeded vegetable outbreaks, 25% of underground-grown vegetable outbreaks, 18% of herb and 16% of fruit-associated outbreaks. *Salmonella* Javiana is responsible for several outbreaks linked to vegetables grown underground and *Salmonella* Enteritidis caused 26% of row crop, 21% of sprout, 18% of herb and 14% of seed/nut outbreaks (CDC, 2022). These serovars may be considered generalists as they also cause several meat, poultry, dairy or egg-related outbreaks. However, some serovars may recur with the same commodity, such that serovar-commodity pairs may not only be a factor of the geographical prevalence of a particular serovar but also a consequence of serovar and commodity characteristics that make the interaction particularly favourable. Examples of these recurring associations include *Salmonella* Poona on cantaloupe melons and more recently cucumber (Laughlin et al., 2019; Walsh et al., 2014), and *Salmonella* Newport on tomato and more recently cucumber and onion (Angelo et al., 2015; Bennett et al., 2015; Greene et al., 2008; McCormic et al., 2022). Other serovars less specifically associated with one commodity but frequently involved in fresh produce contamination issues include *Salmonella* serovars Saintpaul, Braenderup, Enteritidis and Javiana (Barton Behravesh et al., 2011; Bennett et al., 2015; CDC, 2020; Isaacs et al., 2005). Sprouts are one exception to this idea of specific pathogen-commodity pairs, as the multitude of sprout outbreaks that have been reported in the last decades have involved an astounding range of *Salmonella* serovars (Miyahira and Antunes, 2021), placing sprouts in a food safety category of their own.

3 *Salmonella* contamination of fruit and vegetables

Since *Salmonella* is prevalent in the environment, multiple routes of transmission of *Salmonella* to fresh produce crops have been identified. Water is of particular concern as it can act as a reservoir and a vehicle for the dissemination of *Salmonella* to crops. Moreover, water is used in multiple processes in both the pre- and post-harvest settings, providing many opportunities for *Salmonella* to come in contact with fresh produce during production, harvest and post-harvest handling of fruit, vegetables and nuts. During production, water is used primarily for irrigation. However, many other practices make use of copious amounts of water that come directly in contact with the crop, such as pesticide mixing and application, frost protection and evapotranspiration for cooling harvests. Experimental evidence linking agricultural practices to *Salmonella* dissemination to crops points to irrigation and rain splash as factors that increase pathogen transmission risk in the field (Cevallos-Cevallos et al., 2012; Gu et al., 2018). Other identified risk factors have implicated soil amendment use, mulching and wildlife (Cevallos-Cevallos et al., 2012; Jay-Russell, 2013; Strawn et al., 2013).

Despite the potential widespread prevalence of *Salmonella* in the environment, tracing back an outbreak to an environmental source has proven to be challenging. A few salmonellosis outbreaks implicating crops have tentatively source-tracked contamination back to irrigation water (Barton Behravesh et al., 2011; Greene et al., 2008; Voelker, 2021), and *Salmonella* isolates recovered from surface water have been reported to match illness outbreak strains in some cases (Li et al., 2014). However, most outbreak investigations fail to establish an environmental source for the contamination, even when the production area or facility is identified. This could be due to the sporadic and ephemeral nature of *Salmonella* presence in the agro-environment. Its detection could be restricted to a narrow window during which several factors converge to provide favourable conditions for *Salmonella* populations to flourish, increasing the contamination risk. Such a notion is backed by studies that report not only a year-long presence of *Salmonella* in surface waters (Acheamfour et al., 2021; Callahan et al., 2019; Sharma et al., 2020) but only sporadic presence in production areas, possibly connected with climatic events (Marine et al., 2015). The latter study entailed an environmental survey of 32 conventional and organic leafy greens farms in three mid-Atlantic states of the United States over two growing seasons in fall and spring. Out of 577 samples including water, sediment, soil, compost and leafy crop samples, *Salmonella* was only isolated in October. Remarkably, positive samples included a mixture of crops (spinach, bok choy and kale) and serovars (*Salmonella* serovars Mbandaka, Braenderup and Newport) from both conventional and organic farms in all three states. The authors attributed this narrow interval of detection to climatic conditions in the region during that period (Marine et al., 2015).

4 The metabolite landscape in the plant niche

To associate with produce, *Salmonella* must adapt to a variety of plant niches, since various parts of plants of a multitude of crop species may be consumed. As such, the association of *Salmonella* with fruit and vegetables cannot be considered monotypical but dependent on the crop species and edible plant structure. Different produce types provide unique ecological niches for bacteria to colonise as a result of their characteristic topology and physicochemical properties of plant organs (roots, leaves and fruit), offering a range of attachment sites and nutritional landscapes. Moreover, bacteria must also cope with a suite of abiotic factors and varying levels of microbial competition in the rhizosphere and phyllosphere.

The rhizosphere, the portion of soil under the influence of roots and their exudates, represents a rich habitat for microbes where competition for resources is high. As seeds germinate and the resulting seedlings develop below and above ground systems and initiate photosynthesis, plant primary metabolites (sugars, amino acids and organic acids) are released into the soil in the form of root exudates (Badri and Vivanco, 2009; Narasimhan et al., 2003). The release of primary metabolites appears to be a passive process with diffusion facilitated by concentration gradients between the root and the soil that are maintained as microorganisms utilize these metabolites (Canarini et al., 2019). Plant primary metabolites serve as carbon and nitrogen sources for many microorganisms in the rhizosphere (Lopes et al., 2016), which could include commensals, beneficial microbes, phytopathogens and enteric pathogens. The former groups are considered well adapted to the rhizosphere, possessing traits that enhance their competitiveness in this niche, such as chemotaxis towards organic acids released by roots, perception of flavonoid-signalling molecules of plant origin, antibiotic production that confers a competitive advantage over other microbes, and plant hormone production to modify plant growth and modulate the plant interaction (Backer et al., 2018; Bag et al., 2022; Feng et al., 2021; Lopes et al., 2021).

Although *Salmonella* can persist in soil and associate with edible roots (Islam et al., 2004), the ability of *Salmonella* to compete for a niche in the rhizosphere has not been well established. When artificially applied in high concentration to tomato transplants grown in steam-sterilized soil, *Salmonella* did become established in the tomato rhizosphere and even translocated to aerial tissues (Zheng et al., 2013). However, under more natural conditions where microbial diversity is high and *Salmonella* populations are sporadically introduced at relatively low abundance, *Salmonella* may be less successful in this niche. Such a scenario might be envisaged when *Salmonella* is introduced to the soil via drip irrigation. Levels in surface water have been reported to be less than 0.2 MPN/100 mL for sites across the United States (Acheamfour et al., 2021; Antaki et al., 2016; Sharma et al., 2020), although certain regions may naturally harbour higher populations (McEgan et al., 2013). At these relatively low concentrations, *Salmonella* may not be able to overcome the habitat transition from water to soil under high competitive pressure.

The phyllosphere, comprised of all aerial plant surfaces including leaves, fruit, stems and flowers, is often described as a harsh environment for microbial epiphytes, characterized by fluctuating conditions and a variety of stresses including desiccation, high ultraviolet (UV) radiation and nutrient limitations. Yet, many microbes are adapted to inhabit the phyllosphere, whose microbiome is diverse and has been shown to differ among food crop species (Leff and Fierer, 2013), while also exhibiting specificity to plant anatomical structure (Ottesen et al., 2013). Concurrently, season and weather play a role in driving the phyllosphere microbiome of fruit and vegetables (Allard et al., 2020; Ibekwe et al., 2021). At first glance, the phyllosphere might appear hostile to a taxon that is adapted to the enteric niche. However, the physicochemical traits of a plant surface, which determine bacterial colonization patterns, are varied (Schlechter et al., 2019) and microsites with more favourable conditions can provide a safe haven for attachment. Leaves, for instance, constitute a heterogenous surface whereby the epidermal cell layer covered by the cuticle is interspersed with trichomes, guard cells and stomata, hydathodes at the leaf margins and veins that create ridges and troughs. Many of these structures have been identified as preferential attachment sites where bacterial biofilms or aggregates can form (Kusstatscher et al., 2020; Monier and Lindow, 2004; Morris et al., 1997). Specifically, *Salmonella* has been reported to accumulate at the base of type I (non-glandular) trichomes on tomato leaves (Barak et al., 2011), sites characterized by high transpiration rates on the leaf surface (Fich et al., 2020). On lettuce leaves, stomatal openings were reported to induce a chemotaxis response in *Salmonella* under light conditions, when leaves would be actively photosynthesizing (Kroupitski et al., 2009). Therefore, despite the potentially hostile habitat presented by the phyllosphere, microsites with adequate levels of primary metabolites and water can serve as oases, allowing *Salmonella* to colonize various plant anatomical structures.

While *Salmonella* is supported by metabolites that can serve as carbon and nitrogen sources, less is known about *Salmonella* interaction with plant-specialized metabolites that may have an inhibitory effect (George and Brandl, 2021). Sugars and organic acids in root, leaf and fruit washes of tomato correlated positively with *Salmonella* growth in those washes, while saturated fatty acids appeared to have an inhibitory effect (Dev Kumar and Micallef, 2017; Han and Micallef, 2016). Specialized metabolites include a wide range of compounds that have important roles in defence against plant pathogens and herbivores, attractants for beneficial insect or microbe interactions and managing biotic and abiotic stress (Garagounis et al., 2021). Roots actively exude a wide range of specialized metabolites (Badri and Vivanco, 2009; Narasimhan et al., 2003), including phenolic compounds, flavonoids, anthocyanins, glucosinolates and terpenoids that can shape the rhizosphere microbiome (Massalha et al., 2017). While several *in vitro* and *in planta* studies have investigated the antimicrobial effect of fruit extracts and plant-derived bioactive compounds on *Salmonella* (Alvarado-Martinez et al., 2020; Byun et al., 2022; Guo et al., 2020; Nohynek et al., 2006; Zhu et al., 2014), any action these compounds may have on *Salmonella* populations *in situ* - in the plants that synthesize them - remains largely unknown. *Salmonella* populations inoculated on leafy greens that were subjected to drought stress, which shifted the plant metabolome and enhanced phenolic and flavonoid leaf content, declined more rapidly than on optimally watered plants (Liu et al., 2023, 2022). However, bacterial populations could have been affected by other factors such as a lower leaf water potential or cuticular changes triggered by the plant response to drought. More research is needed in this area to identify what bioactive compounds *Salmonella* encounters on the surface or internal tissues of intact produce, or damaged tissues of chopped fruit and vegetables, as well as their potential anti-*Salmonella* effect.

5 *Salmonella* on produce

5.1 *Salmonella* in sprouts and microgreens

Sprouts are the germinated seeds of a variety of species including several species in the Fabaceae such as alfalfa (*Medicago sativa*), mung beans (*Vigna radiata*), clover (*Trifolium* spp.), and fenugreek (*Trigonella foenum-graecum*) and the Brassicaceae such as broccoli (*Brassica oleracea* var. *italica*) and radish (*Raphanus sativus*), to name a few. In sprouts, the emerging radicle, hypocotyl and the cotyledons are eaten. As the seedling develops and the first true leaves emerge, the plant is called a microgreen and the roots are discarded while the stems and leaves are eaten. Seedling roots exude substantial amounts of exudates composed of sugars, amino acids, organic acids and ions that can support microbial growth (Han and Micallef, 2016; Vančura and Hanzlíková, 1972; Yuan et al., 2015; Zhang et al., 2014). Indeed, the primary role of root exudation of this varied cocktail of molecules is thought to be for the recruitment of beneficial microbes (Feng et al., 2021). It is therefore not surprising that sprouts constitute such a favourable environment for enteric pathogens such as *Salmonella* to be competitive in this niche. Experimental assessment of *Salmonella* persistence on sprouts indicated it did not disrupt microbial metabolic profiles of the bacterial communities on alfalfa and Swiss chard (*Beta vulgaris* subp. *Vulgaris*) sprouts (Reed et al., 2018). The adaptation to this plant niche is further evidenced by the numerous outbreaks caused by a multitude of *Salmonella* serovars in the past decades (Dechet et al., 2014). The seed is frequently the source of contamination and inadequate seed disinfection before sprouting has been identified as a possible failure in sprout salmonellosis outbreaks (Dechet et al., 2014; Harfield et al., 2019; Winthrop et al., 2003). The high food safety risks associated with this commodity have necessitated strong interventions and have resulted in strict guidelines for sprout growers (Sikin et al., 2013).

5.2 *Salmonella* in root, bulbous and stalk vegetables

Root, bulbous and stalk vegetables are grown fully or partially below the soil level and hence provide a unique scenario for contamination due to the direct contact with surface and sub-surface soil that may have been amended with manure or compost, water, as well as water percolating through soil. Carrots (*Daucus carota*) and radishes are the commonest root vegetables to be eaten raw. In these vegetables, the tap root forms the edible portion of the crop. A variety of serovars have been isolated from market carrots and radishes, including *Salmonella* serovars Anatum, Newport, Saintpaul, Virchow and Weltevreden (Singh et al., 2007). The persistence of *Salmonella* Typhimurium on carrots and radishes was reported to extend to 203 and 84 days, respectively, under experimental transmission through contaminated soil (Islam et al., 2004). Despite these perceived risks, few incidents involving root vegetables have been reported in the United States, including one outbreak in 2005 caused by *Salmonella* Typhimurium on carrots which sickened eight people (CDC, 2022) and a 2021 recall on carrots (FDA, 2021).

Being roots, these vegetables provide carbohydrates that could support *Salmonella* growth. However, it is also plausible to hypothesize that when cut or juiced, bacteria are also exposed to specialized metabolites which may have an inhibitory effect on susceptible bacteria. The dynamics of *Salmonella*

Typhimurium populations in black carrot juice showed an initial proliferation of *Salmonella* in the first 24 hours of incubation, followed by a decline in the subsequent 24 hours, with no detection after seven days (Degirmenci et al., 2012). Similar bacterial dynamics were reported for *Salmonella* Typhimurium during the fermentation of carrot juice, with initial growth not causing a concern, as an eight-day incubation resulted in no detectable *Salmonella* (Van Beeck et al., 2020). How much of this decline is due to nutrient depletion after 24 hours or to the action of inhibitory compounds has not been determined. Carrots contain high levels of phenolic compounds and carotenoids (Haq et al., 2016; Haq and Prasad, 2017), with the main phenolic compound being hydroxycinnamic acid, chlorogenic acid (Zhang and Hamauzu, 2004). Chlorogenic acid has been shown to have a bactericidal effect on *Salmonella* as a result of cell wall damage, and a minimum inhibitory concentration (MIC) of 6.25 mg/L for a five-log culture of *Salmonella* Typhimurium has been reported (Sun et al., 2020; Yang et al., 2022). Treating chilled chicken with chlorogenic acid reduced *Salmonella* Enteritidis by 3 logCFU/g, showing a direct bactericidal effect (Sun et al., 2020) that could play a role in cut or juiced carrot preparations. Radishes also contain a broad range of specialized metabolites including glucosinolates, phenolic compounds and anthocyanins, the most abundant of which are the pelargonidins (Park et al., 2016; Gutiérrez and Perez, 2004). Root extracts of radish had a direct inhibitory effect on *Salmonella* Typhi (Abdou et al., 1972), while a pelargonidin extracted from pomegranate inhibited the growth of *Salmonella* Typhi and Paratyhi A and B (Naz et al., 2007).

Stalk vegetables include Florence fennel (*Foeniculum vulgare* var. *dulce*) and celery (*Apium graveolens*) in which the base of the stalks growing out of the soil is commonly consumed, along with other aboveground parts of the plant. In both these vegetables, the edible stalks that are partially grown underground are in fact the leaf petioles, and in the case of Florence fennel, the bulbous base of the petioles is used. The alternating petiole arrangement fanning open just above the soil surface provides a protected area that could capture soil particles and harbour bacteria. Bulbous vegetables such as onions (*Allium cepa*) also provide a comparable niche where the fleshy scale leaves emerge from the bulb. Fennel and celery have not been implicated in salmonellosis outbreaks; however, these vegetables can be colonized with *Salmonella*. An Italian study conducted several decades ago demonstrates the ease with which *Salmonella* can associate with field-grown fennel, with over 70% of samples found to be positive, harbouring higher average levels than lettuce from the same region, and yielding six serovars, namely *Salmonella* Typhimurium, Schottmuelleri, Anatum, Dublin, Thompson and Typhi (Ercolani, 1976). Experimentally, cut celery was able to maintain a 3 logCFU/g population at 12°C and support 2 log of growth at 22°C of a five-serovar *Salmonella* mix at room temperature (Vandamm et al., 2013).

Onions have a more notorious history with *Salmonella*. The NORS includes five outbreaks with onions as the sole ingredient occurring between 2009 and 2020, specifically onions (three outbreaks; serovar Javiana), scallions (one outbreak; serovar Javiana) and red onions (1 outbreak; serovar Newport) (CDC, 2022). The latter involved a large outbreak of over 1120 cases and an intensive investigation that traced back red onions to a grower in California, but despite thorough environmental testing, failed to identify a contamination source within the operation (McCormic et al., 2022). Interestingly, a comparison of seven *A. cepa* cultivars revealed a relatively high stomatal density for all varieties, ranging from 118 and 203 stomata/mm^2 (da Silva et al., 2014), suggesting that internalization of *Salmonella* into onions might be possible. Indeed, Ge et al. demonstrated that *Salmonella* applied to the leaves of Welsh onion (*Allium fistulosum*) followed by root irrigation resulted in the internalization of *Salmonella* into the green leaves and migration to the underground portions within 48 hours (Ge et al., 2013). Stomatal density of this species, however, was reported to be 60 stomata/mm^2 (Yang et al., 2017). Although Ge et al. did not determine whether *Salmonella* remained in the apoplast or moved to the phloem to be translocated to the bulb, application of *Salmonella* to the rhizosphere in a separate study did result in root uptake and translocation to the leaves in this species (Ge et al., 2012).

5.3 *Salmonella* in leafy vegetables and fresh herbs

The selection of leafy greens available to consumers has exploded in the past decade to include a multitude of varieties of lettuce (*Lactuca sativa*), spinach (*Spinacia oleracea*), kale and cabbage (*Brassica oleracea*), arugula/rocket (*Eruca sativa*) and others. Herbaceous vegetables that are consumed uncooked also include several herbs, with basil (*Ocimum basilicum*), cilantro/coriander (*Coriandrum sativum*) and parsley (*Petroselinum crispum*) being the types previously implicated in salmonellosis cases (CDC, 2022). Market surveys have reported the detection of *Salmonella* from leafy greens and herbs, marking these crops as potentially at high risk for salmonellosis (Plessis et al., 2017; Quiroz-Santiago et al., 2009; Sant'Ana et al., 2011). However, the incidence of contamination is generally low although informal market channels can elevate risk, suggesting that strict measures can be effective in minimizing *Salmonella* on leafy vegetables (Arthur et al., 2007; Desiree et al., 2021; Van Pelt et al., 2018). *Salmonella* contamination of leafy greens and herbs, however, has led to outbreaks (Elviss et al., 2009; Herman et al., 2015; Nygård et al., 2008) that has prompted research aimed at elucidating how *Salmonella* interacts with leafy vegetables.

As previously discussed, leaves present a potentially hostile niche for *Salmonella*. *Salmonella* attaches more readily to cut compared to intact lettuce and cabbage leaves, suggesting that leaked nutrients from injured plant tissue give *Salmonella* an advantage (Kroupitski et al., 2009; Patel and Sharma, 2010). The abaxial side and the petiole end of leaves also appear to be more favourable sites along the leaf (Kroupitski et al., 2011; Van der Linden et al., 2016), possibly as a result of lower UV radiation, higher humidity and high stomatal density. Additionally, the patchy availability of water and nutrients on the leaf surface means that specific microsites are preferentially colonized by *Salmonella*. *Salmonella* has been shown to aggregate at the base of trichomes (Barak et al., 2011) and around stomata, even migrating to the substomatal space and the apoplast in actively photosynthesizing iceberg lettuce but not in lettuce kept in the dark, indicating that *Salmonella* responded to photosynthate diffusing from the leaf apertures (Kroupitski et al., 2009). The ability to penetrate stomata appears to be plant-specific, and in a comparative study, arugula and iceberg lettuce experienced higher rates of entry than Romaine lettuce or basil, while little internal *Salmonella* was detected in parsley (Golberg et al., 2011). Environmental factors can play a significant role in determining *Salmonella* establishment on or entry into leaves. Night spray-irrigation of parsley with *Salmonella*-contaminated water yielded higher populations than day irrigation (Kisluk and Yaron, 2012) and internalization occurred at a higher frequency in iceberg lettuce grown during the summer months (Golberg et al., 2011). The genetic variability that led to differential entry rates of *Salmonella* Typhimurium 14028s into lettuce leaves was linked to stomatal width and aperture area but not stomatal density (Jacob and Melotto, 2019).

Despite the ability of *Salmonella* to attach and enter substomatal and intercellular spaces, its ability to persist in this niche appears to constitute another challenge. Several leafy greens studies report *Salmonella* population declines over time, especially if *Salmonella* have been subjected to prior stress or injury, but the pathogen can still persist for several days both epiphytically and endophytically (Chen and Meng, 2021; Jacob and Melotto, 2019). Plants recognize (Backer et al., 2018) *Salmonella* microbe-associated molecular patterns (MAMPs) (Garcia et al., 2014) that can induce a basal defense response in plants that could include bursts of reactive oxygen (ROS) and nitrogen species (RNS), stomatal closure, callose deposition and induction of salicylic acid-dependent gene expression (Ferelli et al., 2020; Iniguez et al., 2005; Jacob and Melotto, 2019; Roy et al., 2013; Shirron and Yaron, 2011). In a variety of lettuce cultivars, the decline of both epiphytic and apoplastic *Salmonella* populations was lettuce genotype-dependent and inversely correlated with the magnitude of the induced ROS burst and callose deposition (Jacob and Melotto, 2019). Plant immunity responses triggered by *Salmonella* recognition were shown to have a restrictive effect on *Salmonella* populations (Ferelli et al., 2020; Iniguez et al., 2005; Meng et al., 2013), thereby raising the possibility of exploiting the plants' own defense mechanisms to devise novel food safety strategies (Melotto et al., 2020). *Salmonella* colonization success was also linked to leaf age in lettuce and plant developmental stage in kale, with juvenile leaves providing a more favourable niche (Brandl and Amundson, 2008; Liu et al., 2022). This has implications for the consumption of 'baby leafy greens' which often include mixtures of multiple crops. On the other hand, plant responses to abiotic stresses might bolster plant defense against *Salmonella*, as was demonstrated for kale and lettuce subjected to water restriction, indicating that the physiological responses to this stress modified the leaf habitat in a way that hampered *Salmonella* association (Liu et al., 2023, 2022). Understanding the plant-mediated mechanisms modulating *Salmonella* interactions can lead to novel strategies that benefit the food safety of leafy greens (George and Brandl, 2021; Liu et al., 2022).

5.4 *Salmonella* on fruit

Edible fruit is a diverse category of produce, including many types such as pomme (e.g. apple and pear), stone (e.g. peach, plum), citrus (e.g. orange and grapefruit) and tropical (e.g. mango and avocado) fruit, fruit that grow on herbaceous plants (e.g. melons, tomato, peppers, cucumber and strawberries) and berries (e.g. raspberries, blackberries, and blueberries). Fruit arises from fertilized ovule-containing flowers of angiosperms, inside which the seeds develop. The edible fruit consists of a flesh structure called the pericarp. The exocarp is the outermost layer of the pericarp making up the fruit skin and the mesocarp and endocarp are the inner layers that typically surround the seed. Exceptions include the achenes of strawberries contained in a fleshy receptacle that is eaten. *Salmonella* contamination of fruit is a major food safety concern but interactions vary by plant growth habit, cropping system, agricultural practices and crop species, the latter driven by specific morphological and physicochemical characteristics of the fruit. The firmness of the fruit and the thickness and topography of the exocarp are factors that determine whether a type of fruit is at higher risk for food safety issues, as a result of either *Salmonella* attachment capacity or the ability to reduce loads during post-harvest washing. As examples, one might think comparatively of smooth firm apple peels, rugged, porous rinds of cantaloupe melons, and soft, delicate skins of raspberries to envisage how form and structure may affect bacterial contamination of fruit surfaces and post-harvest handling practices that can enhance food safety. Another risk factor unique to fruit (and nuts) is a mode of flower fertilization (insect pollinators, wind pollination and self-pollination) that may place the fruit at higher risk of being exposed to enteric pathogens that become entrapped in the developing fruit.

Salmonella can inhabit the fruit exocarp as an epiphyte or be pulled into the interior layers of a fruit after internalization via natural openings, flower fertilization or fruit slicing during post-harvest preparation (Burris et al., 2020; Soto et al., 2007; Ukuku and Sapers, 2001). In general, the pulp or flesh of a fruit will support higher populations and possibly the growth of *Salmonella* than the peel (Knudsen et al., 2001; Rezende et al., 2016; Ukuku, 2004). Physicochemical factors that might modulate survival include moisture content, organic acid and fatty acid levels, pH, sugar content and secondary metabolites. Many types of fruit have high levels of the sugars glucose and fructose, and organic acid contents of malic and citric acid, with modulation of these and other sugars, organic acids and phenolics occurring during the complex process of fruit ontogeny and ripening (Batista-Silva et al., 2018). *Salmonella* can metabolize these compounds and can also tolerate acidic environments (Foster, 1995; Gutnick et al., 1969). As a result, genetic variation in physicochemical properties among cultivars of a given fruit can lead to a differential propensity to support *Salmonella*. For example, *Salmonella* Newport population levels in fruit washes on tomatoes correlated positively and negatively with proportions of sugars and fatty acids, respectively (Han and Micallef, 2016), and *Salmonella* decline was greatest on low pH apple cultivars such as Granny Smith (Gurtler et al., 2020). Unripe tomato fruit surfaces were also found to be less favourable for *Salmonella* than ripe fruit (Micallef et al., 2022). Features of fruit exocarp that may aid the 'trapping' of *Salmonella*, and hence their susceptibility to surface attachment include macroscopic surface irregularities (e.g. cantaloupe melon), which can greatly reduce the effectiveness of post-harvest washing steps (Wang et al., 2009). However, even smooth-skinned fruit can allow for *Salmonella* to attach and persist due to surface roughness at the microscopic scale, with even short contact times resulting in pathogen attachment (Macarisin et al., 2010; Mathew et al., 2018). Some fruit possesses natural apertures on the surface, including stomata (e.g. cucumber) and lenticels (e.g. apples) that can serve as natural openings for *Salmonella* during fruit-washing steps. The practice of hydrocooling can draw pathogens in through these paths of least resistance, as was demonstrated with *Salmonella* drawn into warm mango via lenticels when immersed in water at cooler temperatures (Soto et al., 2007).

5.5 *Salmonella* in tree nuts and seeds

What we culinarily refer to as nuts includes a variety of seeds that come from true nuts (such as hazelnut), fruit (or drupes such as almond, walnut and coconut), pine cones (pine nuts) and legumes (peanut). Peanuts are in fact the seeds of the peanut plant that are formed in underground pods. *Salmonella* illnesses have been associated with whole nuts, seeds, nut butters and shredded coconut (CDC, 2022). Microbiological assessments from around the world have detected a multitude of *Salmonella* serovars on raw almonds, cashews, hazelnuts, pecans, walnuts, macadamia nuts, pine nuts, peanuts, sesame seeds and related products (Aguilar-Vázquez et al., 2022; Brar et al., 2016; Brockmann et al., 2004; Danyluk et al., 2007; Zhang et al., 2017). Nuts and seeds differ from fresh produce in having low moisture levels, and present an environment that should impair *Salmonella* proliferation. However, *Salmonella* is well adapted to persist in low-water activity foods and longitudinal studies following experimental inoculation of various nuts have detected viable *Salmonella* after several months and even years of storage (Kataoka et al., 2014; Uesugi et al., 2006). Moreover, *Salmonella* may be introduced to nuts during harvesting and wet conditions during harvesting can lead to *Salmonella* growth (Danyluk et al., 2008). Interventions to improve the food safety of nuts and seeds should be at the level of controlling animal reservoirs in nut orchards and improving harvesting and storage practices (FAO, 2012; Munck et al., 2020).

6 *Salmonella* strategies in the plant niche

The recognition of plants as transitional hosts for *Salmonella* has spurred a series of studies on the genetic mechanisms that come into play when *Salmonella* is inhabiting a plant niche. To persist on this host, *Salmonella* must attach to the plant surface, avert the plant immune response, migrate to a supportive microniche, compete for space, acquire nutrients and replicate. Transcriptomic studies have revealed a series of processes that provide clues as to the strategies invoked by *Salmonella* to persevere in or on plant hosts. *Salmonella* establishment and viability on leaves appears to be facilitated by the environmental modifications moulded by resident epiphytes unless an epiphyte exerts direct antibiosis against it (Poza-Carrion et al., 2013). Different strategies may be at play in this cooperation, such as via the secretion of extracellular polymeric substance (EPS) that reduces desiccation stress for the EPS-producing and the migrant bacteria or as a result of plant immunity suppression by resident phytopathogens (Potnis et al., 2014; Poza-Carrion et al., 2013). Regulation of the *Salmonella* aggregative behaviour known as the rdar morphotype, which involves the synthesis of thin aggregative fimbriae (Tafi or curli), cellulose, other polysaccharides and protein to form an extracellular matrix, has been shown to be needed for *Salmonella* environmental persistence (White et al., 2006), including adhesion to alfalfa sprouts (Barak et al., 2007, 2005). The ensuing biofilms have also been shown to exhibit enhanced resistance to chlorine (Scher et al., 2005).

Once attached to plant surfaces, *Salmonella* must contend with basal plant defense responses, such as ROS and RNS bursts and expression of pathogenesis-related (PR) genes, which are triggered in response to *Salmonella* MAMPs such as flagellin (Ferelli et al., 2020; Garcia et al., 2014; Meng et al., 2013; Schikora et al., 2008). The guard cells of the plant model *Arabidopsis thaliana* release nitric oxide to regulate the closure of stomata in response to bacterial MAMPs such as flagellin and lipopolysaccharide (Melotto et al., 2006). To counter these plant-triggered responses, *Salmonella* upregulates the expression of ROS and RNS detoxification genes and xenobiotic efflux (*mar*) genes (Ferelli et al., 2020; Han et al., 2020). In lettuce, *Salmonella* can induce only transient closure of stomatal apertures which could be a consequence of weak pattern-triggered immunity (PTI), (Roy et al., 2013) or an attenuation of the signals that govern this response (Ferelli et al., 2020; Johnson et al., 2020). Genetic analyses have indicated that, once established, *Salmonella* persists in or on various plant tissues by employing a number of physiological processes such as *de novo* amino acid synthesis, sulphur metabolism, anaerobic respiration, propanediol metabolism, cobalamin biosynthesis and iron acquisition (de Moraes et al., 2018, 2017; Goudeau et al., 2013; Han et al., 2020). A number of stress, virulence and Type III secretion system (T3SS) genes have also been identified as facilitating *Salmonella* plant colonization (Han et al., 2020; Johnson et al., 2020; Neumann et al., 2014). Effector proteins, in particular, elicit interest as they may play a role in quenching plant immunity. Although the effector proteins synthesized in plant-colonizing *Salmonella* are shared with those on animal hosts, the expression patterns differ (Han et al., 2020; Schikora et al., 2011). The function of some effectors is being elucidated, and a role in subverting lettuce stomata closure in response to pathogen recognition has been described (Johnson et al., 2020).

7 Conclusion

Despite the gaps in our knowledge, *Salmonella* has sufficient capabilities in its repertoire to allow it to utilize unusual substrates, synthesize essential metabolites and navigate various abiotic conditions within and on the surface of plant hosts to allow it to exploit the plant niche during its spells outside animal hosts. It can colonize any plant structure both when the plant is alive during cultivation and after it has been harvested while also having the capacity to tolerate desiccation and disinfection. Despite strict measures to control *Salmonella* at every step in the food production chain, this highly adaptable pathogen is still challenging us to develop novel interventions. For fresh produce, the future of this may lie in better elucidating the *Salmonella*-plant association such that we may fully tap into natural plant defenses and traits that render the plant niche unhospitable for *Salmonella*.

8 References

Aballay, A., Yorgey, P. and Ausubel, F. M. 2000. *Salmonella* Typhimurium proliferates and establishes a persistent infection in the intestine of *Caenorhabditis elegans*. *Current Biology* 10(23), 1539–1542. https://doi.org/10.1016/S0960-9822(00)00830-7.

Abdou, I. A., Abou-Zeid, A. A., El-Sherbeeny, M. R. and Abou-El-Gheat, Z. H. 1972. Antimicrobial activities of *Allium sativum*, *Allium cepa*, *Raphanus sativus*, *Capsicum frutescens, Eruca sativa, Allium kurrat* on bacteria. *Qualitas Plantarum et Materiae Vegetabiles* 22(1), 29–35. https://doi.org/10.1007/BF01099735.

Acheamfour, C. L., Parveen, S., Hashem, F., Sharma, M., Gerdes, M. E., May, E. B., Rogers, K., Haymaker, J., Duncan, R., Foust, D., Taabodi, M., Handy, E. T., East, C., Bradshaw, R., Kim, S., Micallef, S. A., Callahan, M. T., Allard, S., Anderson-Coughlin, B., Craighead, S., Gartley, S., Vanore, A., Kniel, K. E., Solaiman, S., Bui, A., Murray, R., Craddock, H. A., Kulkarni, P., Rosenberg Goldstein, R. E. and Sapkota, A. R. 2021. Levels of *Salmonella enterica* and *Listeria monocytogenes* in alternative irrigation water vary based on water source on the eastern shore of Maryland. *Microbiology Spectrum* 9(2), e0066921. https://doi.org/10.1128/Spectrum.00669-21.

Adeolu, M., Alnajar, S., Naushad, S. and Gupta, R. 2016. Genome-based phylogeny and taxonomy of the 'Enterobacteriales': Proposal for Enterobacterales ord. nov. divided into the families Enterobacteriaceae, Erwiniaceae fam. nov., Pectobacteriaceae fam. nov., Yersiniaceae fam. nov., Hafniaceae fam. nov., Morganellaceae fam. nov., and Budviciaceae fam. nov. *International Journal of Systematic and Evolutionary Microbiology* 66(12), 5575–5599. https://doi.org/10.1099/ijsem.0.001485.

Aguilar-Vázquez, J. C., Godínez-Oviedo, A., Lucero-Mejía, J. E., D'Souza, D. H., Palacios-Marmolejo, A. and Hernández-Iturriaga, M. 2022. Microbiological profile, prevalence, and characterization of *Salmonella enterica* in peanuts, pecans, raisins, sun-dried tomatoes, and chocolate sprinkles sold in bulk in markets in Querétaro, Mexico. *Journal of Food Protection* 85(9), 1265–1272. https://doi.org/10.4315/JFP-21-432.

Allard, S. M., Ottesen, A. R. and Micallef, S. A. 2020. Rain induces temporary shifts in epiphytic bacterial communities of cucumber and tomato fruit. *Scientific Reports* 10(1), 1765. https://doi.org/10.1038/s41598-020-58671-7.

Alvarado-Martinez, Z., Bravo, P., Kennedy, N. F., Krishna, M., Hussain, S., Young, A. C. and Biswas, D. 2020. Antimicrobial and antivirulence impacts of phenolics on *Salmonella enterica* serovar Typhimurium. *Antibiotics* 9(10), 668. https://doi.org/10.3390/antibiotics9100668.

Angelo, K. M., Chu, A., Anand, M., Nguyen, T. A., Bottichio, L., Wise, M., Williams, I., Seelman, S., Bell, R., Fatica, M., Lance, S., Baldwin, D., Shannon, K., Lee, H., Trees, E., Strain, E., Gieraltowski, L. and Centers for Disease Control and Prevention (CDC) 2015. Outbreak of *Salmonella* Newport infections linked to cucumbers–United States, 2014. *MMWR. Morbidity and Mortality Weekly Report* 64(6), 144–147.

Antaki, E. M., Vellidis, G., Harris, C., Aminabadi, P., Levy, K. and Jay-Russell, M. T. 2016. Low concentration of *Salmonella enterica* and generic *Escherichia coli* in farm ponds and irrigation distribution systems used for mixed produce production in Southern Georgia. *Foodborne Pathogens and Disease* 13(10), 551–558. https://doi.org/10.1089/fpd.2016.2117.

Arthur, L., Jones, S., Fabri, M. and Odumeru, J. 2007. Microbial survey of selected Ontario-grown fresh fruits and vegetables. *Journal of Food Protection* 70(12), 2864–2867. https://doi.org/10.4315/0362-028X-70.12.2864.

Backer, R., Rokem, J. S., Ilangumaran, G., Lamont, J., Praslickova, D., Ricci, E., Subramanian, S. and Smith, D. L. 2018. Plant growth-promoting rhizobacteria: Context, mechanisms of action, and roadmap to commercialization of biostimulants for sustainable agriculture. *Frontiers in Plant Science* 9, 1473. https://doi.org/10.3389/fpls.2018.01473.

Badri, D. V. and Vivanco, J. M. 2009. Regulation and function of root exudates. *Plant, Cell and Environment* 32(6), 666–681. https://doi.org/10.1111/j.1365-3040.2008.01926.x.

Bag, S., Mondal, A., Majumder, A., Mondal, S. K. and Banik, A. 2022. Flavonoid mediated selective cross-talk between plants and beneficial soil microbiome. *Phytochemistry Reviews: Proceedings of the Phytochemical Society of Europe* 21(5), 1739–1760. https://doi.org/10.1007/s11101-022-09806-3.

Barak, J. D., Gorski, L., Naraghi-Arani, P. and Charkowski, A. O. 2005. *Salmonella enterica* virulence genes are required for bacterial attachment to plant tissue. *Applied and Environmental Microbiology* 71(10), 5685–5691. https://doi.org/10.1128/AEM.71.10.5685-5691.2005.

Barak, J. D., Jahn, C. E., Gibson, D. L. and Charkowski, A. O. 2007. The role of cellulose and O-antigen capsule in the colonization of plants by *Salmonella enterica*. *Molecular Plant-Microbe Interactions* 20(9), 1083–1091. https://doi.org/10.1094/MPMI-20-9-1083.

Barak, J. D., Kramer, L. C. and Hao, L. Y. 2011. Colonization of tomato plants by *Salmonella enterica* is cultivar dependent, and Type 1 trichomes are preferred colonization sites. *Applied and Environmental Microbiology* 77(2), 498–504. https://doi.org/10.1128/AEM.01661-10.

Barton Behravesh, C., Mody, R. K., Jungk, J., Gaul, L., Redd, J. T., Chen, S., Cosgrove, S., Hedican, E., Sweat, D., Chávez-Hauser, L., Snow, S. L., Hanson, H., Nguyen, T. A., Sodha, S. V., Boore, A. L., Russo, E., Mikoleit, M., Theobald, L., Gerner-Smidt, P., Hoekstra, R. M., Angulo, F. J., Swerdlow, D. L., Tauxe, R. V., Griffin, P. M., Williams, I. T. and Salmonella Saintpaul Outbreak Investigation Team. 2011. 2008 outbreak of *Salmonella* Saintpaul infections associated with raw produce. *New England Journal of Medicine* 364(10), 918–927. https://doi.org/10.1056/NEJMoa1005741.

Batista-Silva, W., Nascimento, V. L., Medeiros, D. B., Nunes-Nesi, A., Ribeiro, D. M., Zsögön, A. and Araújo, W. L. 2018. Modifications in organic acid profiles during fruit development and ripening: Correlation or causation? *Frontiers in Plant Science* 9, 1689. https://doi.org/10.3389/fpls.2018.01689.

Bell, R. L., Zheng, J., Burrows, E., Allard, S., Wang, C. Y., Keys, C. E., Melka, D. C., Strain, E., Luo, Y., Allard, M. W., Rideout, S. and Brown, E. W. 2015. Ecological prevalence, genetic diversity, and epidemiological aspects of Salmonella isolated from tomato agricultural regions of the Virginia Eastern Shore. *Frontiers in Microbiology* 6, 415. https://doi.org/10.3389/fmicb.2015.00415.

Bennett, S. D., Littrell, K. W., Hill, T. A., Mahovic, M. and Behravesh, C. B. 2015. Multistate foodborne disease outbreaks associated with raw tomatoes, United States, 1990–2010: A recurring public health problem. *Epidemiology and Infection* 143(7), 1352–1359. https://doi.org/10.1017/S0950268814002167.

Brandl, M. T. and Amundson, R. 2008. Leaf age as a risk factor in contamination of lettuce with *Escherichia coli* O157:H7 and *Salmonella enterica*. *Applied and Environmental Microbiology* 74(8), 2298–2306. https://doi.org/10.1128/AEM.02459-07.

Brar, P. K., Strawn, L. K. and Danyluk, M. D. 2016. Prevalence, level, and types of *Salmonella* isolated from North American in-shell pecans over four harvest years. *Journal of Food Protection* 79(3), 352–360. https://doi.org/10.4315/0362-028X.JFP-15-365.

Brockmann, S. O., Piechotowski, I. and Kimmig, P. 2004. *Salmonella* in sesame seed products. *Journal of Food Protection* 67(1), 178–180. https://doi.org/10.4315/0362-028X-67.1.178.

Burris, K. P., Simmons, O. D., Webb, H. M., Deese, L. M., Moore, R. G., Jaykus, L. A., Zheng, J., Reed, E., Ferreira, C. M., Brown, E. W. and Bell, R. L. 2020. Colonization and internalization of *Salmonella enterica* and its prevalence in cucumber plants. *Frontiers in Microbiology* 11, 1135. https://doi.org/10.3389/fmicb.2020.01135.

Byun, S., Chen, C., Yin, H. and Patel, J. 2022. Antimicrobial effect of natural fruit extracts against *Salmonella* on whole and fresh-cut cucumbers. *Journal of Food Processing and Preservation* 46(4), e16437. https://doi.org/10.1111/jfpp.16437.

Callahan, M. T., Van Kessel, J. A. and Micallef, S. A. 2019. Salmonella enterica recovery from river waters of the Maryland Eastern Shore reveals high serotype diversity and some multidrug resistance. *Environmental Research* 168, 7–13. https://doi.org/10.1016/j.envres.2018.09.012.

Callejón, R. M., Rodríguez-Naranjo, M. I., Ubeda, C., Hornedo-Ortega, R., Garcia-Parrilla, M. C. and Troncoso, A. M. 2015. Reported foodborne outbreaks due to fresh produce in the United States and European union: Trends and causes. *Foodborne Pathogens and Disease* 12(1), 32–38. https://doi.org/10.1089/fpd.2014.1821.

Campbell, J. V., Mohle-Boetani, J., Reporter, R., Abbott, S., Farrar, J., Brandl, M., Mandrell, R., Werner, S. B. 2001. An outbreak of Salmonella serotype Thompson associated with fresh cilantro. *Journal of Infectious Diseases* 183, 984–987. https://doi.org/10.1086/319254.

Canarini, A., Kaiser, C., Merchant, A., Richter, A. and Wanek, W. 2019. Root exudation of primary metabolites: Mechanisms and their roles in plant responses to environmental stimuli. *Frontiers in Plant Science* 10, 157. https://doi.org/10.3389/fpls.2019.00157.

CDC. 2012. Multistate outbreak of *Salmonella* Braenderup infections associated with mangoes (final update). Available at: https://www.cdc.gov/salmonella/montevideo-03-16/index.html.

CDC. 2016. Multistate outbreak of *Salmonella* Montevideo and *Salmonella* Senftenberg infections linked to wonderful pistachios (final update). Available at: https://www.cdc.gov/salmonella/montevideo-03-16/index.html.

CDC. 2020. Outbreak of *Salmonella* enteritidis infections linked to peaches. Available at: https://www.cdc.gov/salmonella/enteritidis-08-20/index.html. Last accessed 23 February 2023.

CDC. 2022. National outbreak reporting system (NORS) 1999–2020. Available at: https://wwwn.cdc.gov/norsdashboard/.

Centers for Disease Control and Prevention (CDC). 2009. Outbreak of *Salmonella* serotype Saintpaul infections associated with eating alfalfa sprouts - United States, 2009. *MMWR. Morbidity and Mortality Weekly Report* 58(18), 500–503.

Cevallos-Cevallos, J. M., Danyluk, M. D., Gu, G., Vallad, G. E. and van Bruggen, A. H. C. 2012. Dispersal of *Salmonella* typhimurium by rain splash onto tomato plants. *Journal of Food Protection* 75(3), 472–479. https://doi.org/10.4315/0362-028X.JFP-11-399.

Chen, Z. and Meng, J. 2021. Persistence of *Salmonella enterica* and *Enterococcus faecium* NRRL B-2354 on baby spinach subjected to temperature abuse after exposure to sub-lethal stresses. *Foods* 10(9), 2141. https://doi.org/10.3390/foods10092141.

Čučak, D., Babić, O., Tamaš, I., Simeunović, J., Karaman, M., Kovač, D., Novaković, M., Markov, S., Knežević, P., Stojanov, I., Obradović, V. and Radnović, D. 2018. Prevalence, antibiotic resistance and diversity of *Salmonella* isolates from soils and sediments in Serbia. *International Journal of Environmental Research* 12(6), 829–841. https://doi.org/10.1007/s41742-018-0138-3.

da Silva, V. C. Pd., Bettoni, M. M., Bona, C. and Foerster, L. A. 2014. Morphological and chemical characteristics of onion plants (*Allium cepa* L.) associated with resistance to onion thrips. *Acta Scientiarum. Agronomy* 37(1), 85. https://doi.org/10.4025/actasciagron.v37i1.17436.

Danyluk, M. D., Brandl, M. T. and Harris, L. J. 2008. Migration of *Salmonella* Enteritidis phage Type 30 through almond hulls and shells. *Journal of Food Protection* 71(2), 397–401. https://doi.org/10.4315/0362-028X-71.2.397.

Danyluk, M. D., Jones, T. M., Abd, S. J., Schlitt-Dittrich, F., Jacobs, M. and Harris, L. J. 2007. Prevalence and amounts of *Salmonella* found on raw California almonds. *Journal of Food Protection* 70(4), 820–827. https://doi.org/10.4315/0362-028X-70.4.820.

de Moraes, M. H., Desai, P., Porwollik, S., Canals, R., Perez, D. R., Chu, W., McClelland, M. and Teplitski, M. 2017. Salmonella persistence in tomatoes requires a distinct set of metabolic functions identified by transposon insertion sequencing. *Applied and Environmental Microbiology* 83(5), e03028-16. https://doi.org/10.1128/AEM.03028-16.

de Moraes, M. H., Soto, E. B., Salas González, I., Desai, P., Chu, W., Porwollik, S., McClelland, M. and Teplitski, M. 2018. Genome-wide comparative functional analyses reveal adaptations of *Salmonella* sv. Newport to a plant colonization lifestyle. *Frontiers in Microbiology* 9, 877. https://doi.org/10.3389/fmicb.2018.00877.

Dechet, A. M., Herman, K. M., Chen Parker, C., Taormina, P., Johanson, J., Tauxe, R. V. and Mahon, B. E. 2014. Outbreaks caused by sprouts, United States, 1998–2010: Lessons learned and solutions needed. *Foodborne Pathogens and Disease* 11(8), 635–644. https://doi.org/10.1089/fpd.2013.1705.

Degirmenci, H., Karapinar, M. and Karabiyikli, S. 2012. The survival of *E. coli* O157:H7, *S.* Typhimurium and *L. monocytogenes* in black carrot (*Daucus carota*) juice. *International Journal of Food Microbiology* 153(1–2), 212–215. https://doi.org/10.1016/j.ijfoodmicro.2011.11.017.

Desiree, K., Schwan, C. L., Ly, V., Hok, L., Bello, N. M., Nwadike, L., Phebus, R. K. and Vipham, J. L. 2021. Investigating *Salmonella enterica, Escherichia coli*, and Coliforms on fresh vegetables sold in informal markets in Cambodia. *Journal of Food Protection* 84(5), 843–849. https://doi.org/10.4315/JFP-20-219.

Dev Kumar, G. and Micallef, S. A. 2017. Susceptibility of *Salmonella enterica* isolates from tomato farm environments to fatty acids naturally found on tomato fruit. *Foodborne Pathogens and Disease* 14(5), 293–301. https://doi.org/10.1089/fpd.2016.2239.

Elviss, N. C., Little, C. L., Hucklesby, L., Sagoo, S., Surman-Lee, S., de Pinna, E., Threlfall, E. J. and Food, Water and Environmental Surveillance Network 2009. Microbiological study of fresh herbs from retail premises uncovers an international outbreak of salmonellosis. *International Journal of Food Microbiology* 134(1–2), 83–88. https://doi.org/10.1016/j.ijfoodmicro.2009.01.015.

Ercolani, G. L. 1976. Bacteriological quality assessment of fresh marketed lettuce and fennel. *Applied and Environmental Microbiology* 31(6), 847–852. https://doi.org/10.1128/aem.31.6.847-852.1976.

FAO. 2012. Prevention and control of *Salmonella* and enterohemorrhagic *E. coli* in tree nuts. EMPRES food safety, lessons learned series. Available at: https://ucfoodsafety.ucdavis.edu/sites/g/files/dgvnsk7366/files/inline-files/163174.pdf.

FDA. 2021. Grimmway farms voluntarily recalls certain retail-packaged carrots due to potential *Salmonella* contamination. Grimmway farms voluntarily recalls. Available at: https://www.fda.gov/safety/recalls-market-withdrawals-safety-alerts/grimmway-farms-voluntarily-recalls-certain-retail-packaged-carrots-due-potential-salmonella.

Feng, H., Fu, R., Hou, X., Lv, Y., Zhang, N., Liu, Y., Xu, Z., Miao, Y., Krell, T., Shen, Q. and Zhang, R. 2021. Chemotaxis of beneficial rhizobacteria to root exudates: The first step towards root-microbe rhizosphere interactions. *International Journal of Molecular Sciences* 22(13), 6655. https://doi.org/10.3390/ijms22136655.

Ferelli, A. M. C., Bolten, S., Szczesny, B. and Micallef, S. A. 2020. *Salmonella enterica* elicits and is restricted by nitric oxide and reactive oxygen species on tomato. *Frontiers in Microbiology* 11, 391. https://doi.org/10.3389/fmicb.2020.00391.

Fich, E. A., Fisher, J., Zamir, D. and Rose, J. K. C. 2020. Transpiration from tomato fruit occurs primarily via trichome-associated transcuticular polar pores. *Plant Physiology* 184(4), 1840–1852. https://doi.org/10.1104/pp.20.01105.

Fletcher, J., Leach, J. E., Eversole, K. and Tauxe, R. 2013. Human pathogens on plants: Designing a multidisciplinary strategy for research. *Phytopathology* 103(4), 306–315. https://doi.org/10.1094/PHYTO-09-12-0236-IA.

Foster, J. W. 1995. Low pH adaptation and the acid tolerance response of *Salmonella typhimurium*. *Critical Reviews in Microbiology* 21(4), 215–237. https://doi.org/10.3109/10408419509113541.

Garagounis, C., Delkis, N. and Papadopoulou, K. K. 2021. Unraveling the roles of plant specialized metabolites: Using synthetic biology to design molecular biosensors. *New Phytologist* 231(4), 1338–1352. https://doi.org/10.1111/nph.17470.

Garcia, A. V., Charrier, A., Schikora, A., Bigeard, J., Pateyron, S., de Tauzia-Moreau, M. L., Evrard, A., Mithöfer, A., Martin-Magniette, M. L., Virlogeux-Payant, I. and Hirt, H. 2014. *Salmonella enterica* flagellin is recognized via FLS2 and activates PAMP-triggered immunity in *Arabidopsis thaliana*. *Molecular Plant* 7(4), 657–674. https://doi.org/10.1093/mp/sst145.

GBD 2017 Typhoid and Paratyphoid Collaborators, Reiner, R. C., Blacker, B. F., Goldberg, E. M., Khalil, I. A., Troeger, C. E., Andrews, J. R., Bhutta, Z. A., Crump, J. A., Im, J., Marks, F., Mintz, E., Park, S. E., Zaidi, A. K. M., Abebe, Z., Abejie, A. N., Adedeji, I. A., Ali, B. A., Amare, A. T., Atalay, H. T., Avokpaho, E. F. G. A., Bacha, U., Barac, A., Bedi, N., Berhane, A., Browne, A. J., Chirinos, J. L., Chitheer, A., Dolecek, C., El Sayed Zaki, M., Eshrati, B., Foreman, K. J., Gemechu, A., Gupta, R., Hailu, G. B., Henok, A., Hibstu, D. T., Hoang, C. L., Ilesanmi, O. S., Iyer, V. J., Kahsay, A., Kasaeian, A., Kassa, T. D., Khan, E. A., Khang, Y.-H., Magdy Abd El Razek, H., Melku, M., Mengistu, D. T., Mohammad, K. A., Mohammed, S., Mokdad, A. H., Nachega, J. B., Naheed, A., Nguyen, C. T., Nguyen, H. L. T., Nguyen, L. H., Nguyen, N. B., Nguyen, T. H., Nirayo, Y. L., Pangestu, T., Patton, G. C., Qorbani, M., Rai, R. K., Rana, S. M., Ranabhat, C. L., Roba, K. T., Roberts, N. L. S., Rubino, S., Safiri, S., Sartorius, B., Sawhney, M., Shiferaw, M. S., Smith, D. L., Sykes, B. L., Tran, B. X., Tran, T. T., Ukwaja, K. N., Vu, G. T., Vu, L. G., Weldegebreal, F., Yenit, M. K., Murray, C. J. L. and Hay, S. I. 2019. The global burden of typhoid and paratyphoid fevers: A systematic analysis for the global burden of disease study 2017. *The Lancet: Infectious Diseases* 19(4), 369–381. https://doi.org/10.1016/S1473-3099(18)30685-6.

Ge, C., Lee, C. and Lee, J. 2012. The impact of extreme weather events on *Salmonella* internalization in lettuce and green onion. *Food Research International* 45(2), 1118–1122. https://doi.org/10.1016/j.foodres.2011.06.054.

Ge, C., Lee, C. and Lee, J. 2013. Localization of viable *Salmonella* typhimurium internalized through the surface of green onion during preharvest. *Journal of Food Protection* 76(4), 568–574. https://doi.org/10.4315/0362-028X.JFP-12-374.

George, A. S. and Brandl, M. T. 2021. Plant bioactive compounds as an intrinsic and sustainable tool to enhance the microbial safety of crops. *Microorganisms* 9(12), 2485. https://doi.org/10.3390/microorganisms9122485.

Golberg, D., Kroupitski, Y., Belausov, E., Pinto, R. and Sela, S. 2011. *Salmonella* Typhimurium internalization is variable in leafy vegetables and fresh herbs. *International Journal of Food Microbiology* 145(1), 250–257. https://doi.org/10.1016/j.ijfoodmicro.2010.12.031.

Gorski, L., Liang, A. S., Walker, S., Carychao, D., Aviles Noriega, A., Mandrell, R. E. and Cooley, M. B. 2022. *Salmonella enterica* Serovar diversity, distribution, and prevalence in public-access waters from a central California coastal leafy green-growing region from 2011 to 2016. *Applied and Environmental Microbiology* 88(3), e0183421. https://doi.org/10.1128/AEM.01834-21.

Gorski, L., Parker, C. T., Liang, A., Cooley, M. B., Jay-Russell, M. T., Gordus, A. G., Atwill, E. R. and Mandrell, R. E. 2011. Prevalence, distribution, and diversity of *Salmonella enterica* in a major produce region of California. *Applied and Environmental Microbiology* 77(8), 2734–2748. https://doi.org/10.1128/AEM.02321-10.

Goudeau, D. M., Parker, C. T., Zhou, Y., Sela, S., Kroupitski, Y. and Brandl, M. T. 2013. The *Salmonella* transcriptome in lettuce and cilantro soft rot reveals a niche overlap with the animal host intestine. *Applied and Environmental Microbiology* 79(1), 250–262. https://doi.org/10.1128/AEM.02290-12.

Greene, S. K., Daly, E. R., Talbot, E. A., Demma, L. J., Holzbauer, S., Patel, N. J., Hill, T. A., Walderhaug, M. O., Hoekstra, R. M., Lynch, M. F. and Painter, J. A. 2008. Recurrent multistate outbreak of *Salmonella* Newport associated with tomatoes from contaminated fields, 2005. *Epidemiology and Infection* 136(2), 157–165. https://doi.org/10.1017/S095026880700859X.

Grimont, P. A. and Weill, F.-X. 2007. *Antigenic Formulae of the Salmonella Serovars* (9th edn.). Paris: World Health Organization, WHO Collaborating Centre for Reference and Research on Salmonella, 166.

Gruszynski, K., Pao, S., Kim, C., Toney, D., Wright, K., Ross, P. G., Colon, A. and Levine, S. 2014. Evaluating wildlife as a potential source of *Salmonella* serotype Newport (JJPX01.0061) contamination for tomatoes on the eastern shore of Virginia. *Zoonoses and Public Health* 61(3), 202–207. https://doi.org/10.1111/zph.12061.

Gu, G., Strawn, L. K., Oryang, D. O., Zheng, J., Reed, E. A., Ottesen, A. R., Bell, R. L., Chen, Y., Duret, S., Ingram, D. T., Reiter, M. S., Pfuntner, R., Brown, E. W. and Rideout, S. L. 2018. Agricultural practices influence salmonella contamination and survival in pre-harvest tomato production. *Frontiers in Microbiology* 9, 2451. https://doi.org/10.3389/fmicb.2018.02451.

Guo, L., Gong, S., Wang, Y., Sun, Q., Duo, K. and Fei, P. 2020. Antibacterial activity of olive oil polyphenol extract against *Salmonella* typhimurium and *Staphylococcus aureus*: Possible mechanisms. *Foodborne Pathogens and Disease* 17(6), 396–403. https://doi.org/10.1089/fpd.2019.2713.

Gurtler, J. B., Keller, S. E., Fan, X., Olanya, O. M., Jin, T. and Camp, M. J. 2020. Survival of *Salmonella* during apple dehydration as affected by apple cultivar and antimicrobial pretreatment. *Journal of Food Protection* 83(5), 902–909. https://doi.org/10.4315/JFP-19-475.

Gutiérrez, R. M. and Perez, R. L. 2004. *Raphanus sativus* (Radish): Their chemistry and biology. *The Scientific World Journal* 4, 811–837. https://doi.org/10.1100/tsw.2004.131.

Gutnick, D., Calvo, J. M., Klopotowski, T. and Ames, B. N. 1969. Compounds which serve as the sole source of carbon or nitrogen for *Salmonella typhimurium* LT-2. *Journal of Bacteriology* 100(1), 215–219. https://doi.org/10.1128/jb.100.1.215-219.1969.

Haley, B. J., Cole, D. J. and Lipp, E. K. 2009. Distribution, diversity, and seasonality of waterborne Salmonellae in a rural watershed. *Applied and Environmental Microbiology* 75(5), 1248–1255. https://doi.org/10.1128/AEM.01648-08.

Han, S., Ferelli, A. M. C., Lin, S. S. and Micallef, S. A. 2020. Stress response, amino acid biosynthesis and pathogenesis genes expressed in Salmonella enterica colonizing tomato shoot and root surfaces. *Heliyon* 6(9), e04952. https://doi.org/10.1016/j.heliyon.2020.e04952.

Han, S. and Micallef, S. A. 2016. Environmental metabolomics of the tomato plant surface provides insights on *Salmonella enterica* colonization. *Applied and Environmental Microbiology* 82(10), 3131–3142. https://doi.org/10.1128/AEM.00435-16.

Haq, R., Kumar, P. and Prasad, K. 2016. Physico-chemical, antioxidant and bioactive changes in cortex core sections of carrot (*Daucus carota* var. *Pusa rudhira*). *Journal of Food Measurement and Characterization* 10(3), 701–708. https://doi.org/10.1007/s11694-016-9354-8.

Haq, R. and Prasad, K. 2017. Antioxidant activity, phenolic, carotenoid and color changes in packaged fresh carrots stored under refrigeration temperature. *Journal of Food Measurement and Characterization* 11(4), 1542–1549. https://doi.org/10.1007/s11694-017-9533-2.

Harfield, S., Beazley, R., Denehy, E., Centofanti, A., Dowsett, P., Housen, T. and Flood, L. 2019. An outbreak and case-control study of *Salmonella* Havana linked to alfalfa sprouts in South Australia, 2018. *Communicable Diseases Intelligence* 43. https://doi.org/10.33321/cdi.2019.43.45.

Hassan, R., Whitney, B., Williams, D. L., Holloman, K., Grady, D., Thomas, D., Omoregie, E., Lamba, K., Leeper, M., Gieraltowski, L. and Outbreak Investigation Team 2019. Multistate outbreaks of *Salmonella* infections linked to imported Maradol papayas - United States, December 2016–September 2017. *Epidemiology and Infection* 147, e265. https://doi.org/10.1017/S0950268819001547.

Herman, K. M., Hall, A. J. and Gould, L. H. 2015. Outbreaks attributed to fresh leafy vegetables, United States, 1973–2012. *Epidemiology and Infection* 143(14), 3011-3021. https://doi.org/10.1017/S0950268815000047.

Ibekwe, A. M., Ors, S., Ferreira, J. F. S., Liu, X. and Suarez, D. L. 2021. Influence of seasonal changes and salinity on spinach phyllosphere bacterial functional assemblage. *PLoS ONE* 16(6), e0252242. https://doi.org/10.1371/journal.pone.0252242.

IFSAC. 2021. Foodborne illness source attribution estimates for 2019 for *Salmonella*, *Escherichia coli* O157, *Listeria monocytogenes* and *Campylobacter* using multi-year outbreak surveillance data, United States. GA and D. C. Interagency Food Safety Analytics Collaboration (IFSAC). United States Department of Health and Human Services' Centers for Disease Control and Prevention and U.S. Food and Drug Administration, U.S. Department of Agriculture's Food Safety and Inspection Service. Interagency Food Safety Analytics Collaboration.

Iniguez, A. L., Dong, Y., Carter, H. D., Ahmer, B. M. M., Stone, J. M. and Triplett, E. W. 2005. Regulation of enteric endophytic bacterial colonization by plant defenses. *Molecular Plant-Microbe Interactions* 18(2), 169–178. https://doi.org/10.1094/MPMI-18-0169.

Isaacs, S., Aramini, J., Ciebin, B., Farrar, J. A., Ahmed, R., Middleton, D., Chandran, A. U., Harris, L. J., Howes, M., Chan, E., Pichette, A. S., Campbell, K., Gupta, A., Lior, L. Y.,

Pearce, M., Clark, C., Rodgers, F., Jamieson, F., Brophy, I., Ellis, A. and Salmonella Enteritidis PT30 Outbreak Investigation Working Group 2005. An international outbreak of salmonellosis associated with raw almonds contaminated with a rare phage type of *Salmonella* enteritidis. *Journal of Food Protection* 68(1), 191–198. https://doi.org/10.4315/0362-028X-68.1.191.

Islam, M., Morgan, J., Doyle, M. P., Phatak, S. C., Millner, P. and Jiang, X. 2004. Fate of *Salmonella enterica* serovar Typhimurium on carrots and radishes grown in fields treated with contaminated manure composts or irrigation water. *Applied and Environmental Microbiology* 70(4), 2497–2502. https://doi.org/10.1128/AEM.70.4.2497-2502.2004.

Jackson, B. R., Griffin, P. M., Cole, D., Walsh, K. A. and Chai, S. J. 2013. Outbreak-associated *Salmonella enterica* serotypes and food commodities, United States, 1998–2008. *Emerging Infectious Diseases* 19(8), 1239–1244. https://doi.org/10.3201/eid1908.121511.

Jacob, C. and Melotto, M. 2019. Human pathogen colonization of lettuce dependent upon plant genotype and defense response activation. *Frontiers in Plant Science* 10, 1769. https://doi.org/10.3389/fpls.2019.01769.

Jay-Russell, M. T. 2013. What is the risk from wild animals in food-borne pathogen contamination of plants?. *CABI Reviews* 2013, 1–16. https://doi.org/10.1079/PAVSNNR20138040.

Johnson, N., Litt, P. K., Kniel, K. E. and Bais, H. 2020. Evasion of plant innate defense response by *Salmonella* on lettuce. *Frontiers in Microbiology* 11, 500. https://doi.org/10.3389/fmicb.2020.00500.

Kataoka, A., Enache, E., Black, D. G., Elliott, P. H., Napier, C. D., Podolak, R. and Hayman, M. M. 2014. Survival of *Salmonella* Tennessee, *Salmonella* typhimurium DT104, and *Enterococcus faecium* in peanut paste formulations at two different levels of water activity and fat. *Journal of Food Protection* 77(8), 1252–1259. https://doi.org/10.4315/0362-028X.JFP-13-553.

Kirk, M. D., Little, C. L., Lem, M., Fyfe, M., Genobile, D., Tan, A., Threlfall, J., Paccagnella, A., Lightfoot, D., Lyi, H., McINTYRE, L., Ward, L., Brown, D. J., Surnam, S. and Fisher, I. S. T. 2004. An outbreak due to peanuts in their shell caused by *Salmonella enterica* serotypes Stanley and Newport – Sharing molecular information to solve international outbreaks. *Epidemiology and Infection* 132(4), 571 577. https://doi.org/10.1017/S095026880400216X.

Kisluk, G. and Yaron, S. 2012. Presence and persistence of *Salmonella enterica* serotype Typhimurium in the phyllosphere and rhizosphere of spray-irrigated parsley. *Applied and Environmental Microbiology* 78(11), 4030–4036. https://doi.org/10.1128/AEM.00087-12.

Knudsen, D. M., Yamamoto, S. A. and Harris, L. J. 2001. Survival of *Salmonella* spp. and *Escherichia coli* O157:H7 on fresh and frozen strawberries. *Journal of Food Protection* 64(10), 1483–1488. https://doi.org/10.4315/0362-028X-64.10.1483.

Kroupitski, Y., Golberg, D., Belausov, E., Pinto, R., Swartzberg, D., Granot, D. and Sela, S. 2009. Internalization of *Salmonella enterica* in leaves is induced by light and involves chemotaxis and penetration through open stomata. *Applied and Environmental Microbiology* 75(19), 6076–6086. https://doi.org/10.1128/AEM.01084-09.

Kroupitski, Y., Pinto, R., Belausov, E. and Sela, S. 2011. Distribution of *Salmonella typhimurium* in romaine lettuce leaves. *Food Microbiology* 28(5), 990–997. https://doi.org/10.1016/j.fm.2011.01.007.

Kroupitski, Y., Pinto, R., Brandl, M. T., Belausov, E. and Sela, S. 2009. Interactions of *Salmonella enterica* with lettuce leaves. *Journal of Applied Microbiology* 106(6), 1876–1885. https://doi.org/10.1111/j.1365-2672.2009.04152.x.

Kusstatscher, P., Wicaksono, W. A., Bergna, A., Cernava, T., Bergau, N., Tissier, A., Hause, B. and Berg, G. 2020. Trichomes form genotype-specific microbial hotspots in the phyllosphere of tomato. *Environmental Microbiome* 15(1), 17. https://doi.org/10.1186/s40793-020-00364-9.

Laughlin, M., Bottichio, L., Weiss, J., Higa, J., McDonald, E., Sowadsky, R., Fejes, D., Saupe, A., Provo, G., Seelman, S., Concepción-Acevedo, J., Gieraltowski, L. and Outbreak Investigation Team. 2019. Multistate outbreak of *Salmonella* Poona infections associated with imported cucumbers, 2015–2016. *Epidemiology and Infection* 147, e270. https://doi.org/10.1017/S0950268819001596.

Leff, J. W. and Fierer, N. 2013. Bacterial communities associated with the surfaces of fresh fruits and vegetables. *PLoS ONE* 8(3), e59310. https://doi.org/10.1371/journal.pone.0059310.

Lehmacher, A., Bockemühl, J. and Aleksic, S. 1995. Nationwide outbreak of human salmonellosis in Germany due to contaminated paprika and paprika-powdered potato chips. *Epidemiology and Infection* 115(3), 501–511. https://doi.org/10.1017/S0950268800058660.

Li, B., Vellidis, G., Liu, H., Jay-Russell, M., Zhao, S., Hu, Z., Wright, A. and Elkins, C. A. 2014. Diversity and antimicrobial resistance of *Salmonella enterica* isolates from surface water in Southeastern United States. *Applied and Environmental Microbiology* 80(20), 6355–6365. https://doi.org/10.1128/AEM.02063-14.

Lienemann, T., Niskanen, T., Guedes, S., Siitonen, A., Kuusi, M. and Rimhanen-Finne, R. 2011. Iceberg lettuce as suggested source of a nationwide outbreak caused by two *Salmonella* serotypes, Newport and Reading, in Finland in 2008. *Journal of Food Protection* 74(6), 1035–1040. https://doi.org/10.4315/0362-028X.JFP-10-455.

Liu, X., Li, Y. and Micallef, S. A. 2022. Developmentally related and drought-induced shifts in the kale metabolome limited *Salmonella enterica* association, providing novel insights to enhance food safety. *Food Microbiology* 108, 104113. https://doi.org/10.1016/j.fm.2022.104113.

Liu, X., Li, Y. and Micallef, S. A. 2023. Natural variation and drought-induced differences in metabolite profiles of red oak-leaf and Romaine lettuce play a role in modulating the interaction with Salmonella enterica. *International Journal of Food Microbiology* 385, 109998. https://doi.org/10.1016/j.ijfoodmicro.2022.109998.

Lopes, L. D., Pereira e Silva, M. de C. and Andreote, F. D. 2016. Bacterial abilities and adaptation toward the rhizosphere colonization. *Frontiers in Microbiology* 7, 1341. https://doi.org/10.3389/fmicb.2016.01341.

Lopes, M. J. dos S., Dias-Filho, M. B. and Gurgel, E. S. C. 2021. Successful plant growth-promoting microbes: Inoculation methods and abiotic factors. *Frontiers in Sustainable Food Systems* 5, 606454. https://doi.org/10.3389/fsufs.2021.606454.

Macarisin, D., Santín, M., Bauchan, G. and Fayer, R. 2010. Infectivity of Cryptosporidium parvum oocysts after storage of experimentally contaminated apples. *Journal of Food Protection* 73(10), 1824–1829. https://doi.org/10.4315/0362-028X-73.10.1824.

Majowicz, S. E., Musto, J., Scallan, E., Angulo, F. J., Kirk, M., O'Brien, S. J., Jones, T. F., Fazil, A., Hoekstra, R. M. and International Collaboration on Enteric Disease 'Burden of

Illness' Studies. 2010. The global burden of nontyphoidal *Salmonella* gastroenteritis. *Clinical Infectious Diseases* 50(6), 882–889. https://doi.org/10.1086/650733.

Marine, S. C., Pagadala, S., Wang, F., Pahl, D. M., Melendez, M. V., Kline, W. L., Oni, R. A., Walsh, C. S., Everts, K. L., Buchanan, R. L. and Micallef, S. A. 2015. The growing season, but not the farming system, is a food safety risk determinant for leafy greens in the mid-Atlantic region of the United States. *Applied and Environmental Microbiology* 81(7), 2395–2407. https://doi.org/10.1128/AEM.00051-15.

Massalha, H., Korenblum, E., Tholl, D. and Aharoni, A. 2017. Small molecules below-ground: The role of specialized metabolites in the rhizosphere. *Plant Journal: For Cell and Molecular Biology* 90(4), 788–807. https://doi.org/10.1111/tpj.13543.

Mathew, E. N., Muyyarikkandy, M. S., Kuttappan, D. and Amalaradjou, M. A. 2018. Attachment of *Salmonella enterica* on mangoes and survival under conditions simulating commercial mango packing house and importer facility. *Frontiers in Microbiology* 9, 1519. https://doi.org/10.3389/fmicb.2018.01519.

McCormic, Z. D., Patel, K., Higa, J., Bancroft, J., Donovan, D., Edwards, L., Cheng, J., Adcock, B., Bond, C., Pereira, E., Doyle, M., Wise, M. E., Gieraltowski, L. and Salmonella Newport Investigation Team. 2022. Bi-national outbreak of *Salmonella* Newport infections linked to onions: The United States experience. *Epidemiology and Infection* 150, e199. https://doi.org/10.1017/S0950268822001571.

McEgan, R., Mootian, G., Goodridge, L. D., Schaffner, D. W. and Danyluk, M. D. 2013. Predicting *Salmonella* populations from biological, chemical, and physical indicators in Florida Surface Waters. *Applied and Environmental Microbiology* 79(13), 4094–4105. https://doi.org/10.1128/AEM.00777-13.

Melotto, M., Brandl, M. T., Jacob, C., Jay-Russell, M. T., Micallef, S. A., Warburton, M. L. and Van Deynze, A. 2020. Breeding crops for enhanced food safety. *Frontiers in Plant Science* 11, 428. https://doi.org/10.3389/fpls.2020.00428.

Melotto, M., Underwood, W., Koczan, J., Nomura, K. and He, S. Y. 2006. Plant stomata function in innate immunity against bacterial invasion. *Cell* 126(5), 969–980. https://doi.org/10.1016/j.cell.2006.06.054.

Meng, F., Altier, C. and Martin, G. B. 2013. *Salmonella* colonization activates the plant immune system and benefits from association with plant pathogenic bacteria. *Environmental Microbiology* 15(9), 2418–2430. https://doi.org/10.1111/1462-2920.12113.

Micallef, S. A., Han, S. and Martinez, L. 2022. Tomato cultivar Nyagous fruit surface metabolite changes during ripening affect *Salmonella* Newport. *Journal of Food Protection* 85(11), 1604–1613. https://doi.org/10.4315/JFP-22-160.

Micallef, S. A., Rosenberg Goldstein, R. E., George, A., Kleinfelter, L., Boyer, M. S., McLaughlin, C. R., Estrin, A., Ewing, L., Jean-Gilles Beaubrun, J., Hanes, D. E., Kothary, M. H., Tall, B. D., Razeq, J. H., Joseph, S. W. and Sapkota, A. R. 2012. Occurrence and antibiotic resistance of multiple *Salmonella* serotypes recovered from water, sediment and soil on mid-Atlantic tomato farms. *Environmental Research* 114, 31–39. https://doi.org/10.1016/j.envres.2012.02.005.

Miyahira, R. F. and Antunes, A. E. C. 2021. Bacteriological safety of sprouts: A brief review. *International Journal of Food Microbiology* 352, 109266. https://doi.org/10.1016/j.ijfoodmicro.2021.109266.

Monier, J. M. and Lindow, S. E. 2004. Frequency, size, and localization of bacterial aggregates on bean leaf surfaces. *Applied and Environmental Microbiology* 70(1), 346–355. https://doi.org/10.1128/AEM.70.1.346-355.2004.

Morris, C. E., Monier, J. and Jacques, M. 1997. Methods for observing microbial biofilms directly on leaf surfaces and recovering them for isolation of culturable microorganisms. *Applied and Environmental Microbiology* 63(4), 1570–1576. https://doi.org/10.1128/aem.63.4.1570-1576.1997.

Munck, N., Smith, J., Bates, J., Glass, K., Hald, T. and Kirk, M. D. 2020. Source attribution of *Salmonella* in macadamia nuts to animal and environmental reservoirs in Queensland, Australia. *Foodborne Pathogens and Disease* 17(5), 357–364. https://doi.org/10.1089/fpd.2019.2706.

Narasimhan, K., Basheer, C., Bajic, V. B. and Swarup, S. 2003. Enhancement of plant-microbe interactions using a rhizosphere metabolomics-driven approach and its application in the removal of polychlorinated biphenyls. *Plant Physiology* 132(1), 146–153. https://doi.org/10.1104/pp.102.016295.

Nasirian, H. 2019. Contamination of cockroaches (Insecta: Blattaria) by medically important bacteriae: A systematic review and meta-analysis. *Journal of Medical Entomology* 56(6), 1534–1554. https://doi.org/10.1093/jme/tjz095.

Naz, S., Siddiqi, R., Ahmad, S., Rasool, S. A. and Sayeed, S. A. 2007. Antibacterial activity directed isolation of compounds from *Punica granatum*. *Journal of Food Science* 72(9), M341–M345. https://doi.org/10.1111/j.1750-3841.2007.00533.x.

Neumann, C., Fraiture, M., Akum, F. N., Virlogeux-Payant, I., Chen, Y., Pateyron, S., Colcombet, J., Kogel, K.-H., Hirt, H., Brunner, F. and Schikora, A. 2014. The *Salmonella* effector protein SpvC, a phosphothreonine lyase is functional in plant cells. *Frontiers in Microbiology* 5, 548. https://doi.org/10.3389/fmicb.2014.00548.

Nohynek, L. J., Alakomi, H. L., Kähkönen, M. P., Heinonen, M., Helander, I. M., Oksman-Caldentey, K. M. and Puupponen-Pimiä, R. H. 2006. Berry phenolics: Antimicrobial properties and mechanisms of action against severe human pathogens. *Nutrition and Cancer* 54(1), 18–32. https://doi.org/10.1207/s15327914nc5401_4.

Nygård, K., Lassen, J., Vold, L., Andersson, Y., Fisher, I., Löfdahl, S., Threlfall, J., Luzzi, I., Peters, T., Hampton, M., Torpdahl, M., Kapperud, G. and Aavitsland, P. 2008. Outbreak of *Salmonella* thompson infections linked to imported Rucola lettuce. *Foodborne Pathogens and Disease* 5(2), 165–173. https://doi.org/10.1089/fpd.2007.0053.

Ottesen, A. R., González Peña, A., White, J. R., Pettengill, J. B., Li, C., Allard, S., Rideout, S., Allard, M., Hill, T., Evans, P., Strain, E., Musser, S., Knight, R. and Brown, E. 2013. Baseline survey of the anatomical microbial ecology of an important food plant: *Solanum lycopersicum* (tomato). *BMC Microbiology* 13, 114. https://doi.org/10.1186/1471-2180-13-114.

Park, C. H., Baskar, T. B., Park, S. Y., Kim, S. J., Valan Arasu, M., Al-Dhabi, N. A., Kim, J. K. and Park, S. U. 2016. Metabolic profiling and antioxidant assay of metabolites from three radish cultivars (*Raphanus sativus*). *Molecules* 21(2), 157. https://doi.org/10.3390/molecules21020157.

Patel, J. and Sharma, M. 2010. Differences in attachment of Salmonella enterica serovars to cabbage and lettuce leaves. *International Journal of Food Microbiology* 139(1–2), 41–47. https://doi.org/10.1016/j.ijfoodmicro.2010.02.005.

Plessis, E. M. D. du , Govender, S., Pillay, B. and Korsten, L. 2017. Exploratory study into the microbiological quality of spinach and cabbage purchased from street vendors and retailers in Johannesburg, South Africa. *Journal of Food Protection* 80(10), 1726–1733. https://doi.org/10.4315/0362-028X.JFP-16-540.

Pornsukarom, S. and Thakur, S. 2016. Assessing the impact of manure application in commercial swine farms on the transmission of antimicrobial resistant *Salmonella* in the environment. *PLOS ONE* 11(10), e0164621. https://doi.org/10.1371/journal.pone.0164621.

Potnis, N., Soto-Arias, J. P., Cowles, K. N., van Bruggen, A. H. C., Jones, J. B. and Barak, J. D. 2014. *Xanthomonas perforans* colonization influences *Salmonella enterica* in the tomato phyllosphere. *Applied and Environmental Microbiology* 80(10), 3173–3180. https://doi.org/10.1128/AEM.00345-14.

Poza-Carrion, C., Suslow, T. and Lindow, S. 2013. Resident bacteria on leaves enhance survival of immigrant cells of *Salmonella enterica*. *Phytopathology* 103(4), 341–351. https://doi.org/10.1094/PHYTO-09-12-0221-FI.

Quiroz-Santiago, C., Rodas-Suárez, O. R., Carlos, R. V., Fernández, F. J., Quiñones-Ramírez, E. I. and Vázquez-Salinas, C. 2009. Prevalence of *Salmonella* in vegetables from Mexico. *Journal of Food Protection* 72(6), 1279–1282. https://doi.org/10.4315/0362-028X-72.6.1279.

Reed, E., Ferreira, C. M., Bell, R., Brown, E. W. and Zheng, J. 2018. Plant-microbe and abiotic factors influencing *Salmonella* survival and growth on alfalfa sprouts and Swiss chard microgreens. *Applied and Environmental Microbiology* 84(9), e02814-17. https://doi.org/10.1128/AEM.02814-17.

Rezende, A. C. B., Crucello, J., Moreira, R. C., Silva, B. S. and Sant'Ana, A. S. 2016. Incidence and growth of *Salmonella enterica* on the peel and pulp of avocado (*Persea americana*) and custard apple (*Annona squamosa*). *International Journal of Food Microbiology* 235, 10–16. https://doi.org/10.1016/j.ijfoodmicro.2016.06.034.

Roy, D., Panchal, S., Rosa, B. A. and Melotto, M. 2013. *Escherichia coli* O157:H7 induces stronger plant immunity than *Salmonella enterica* typhimurium SL1344. *Phytopathology* 103(4), 326–332. https://doi.org/10.1094/PHYTO-09-12-0230-FI.

Sant'Ana, A. S., Landgraf, M., Destro, M. T. and Franco, B. D. G. M. 2011. Prevalence and counts of *Salmonella* spp. in minimally processed vegetables in São Paulo, Brazil. *Food Microbiology* 28(6), 1235–1237. https://doi.org/10.1016/j.fm.2011.04.002.

Scher, K., Romling, U. and Yaron, S. 2005. Effect of heat, acidification, and chlorination on *Salmonella enterica* serovar Typhimurium Cells in a biofilm formed at the air-liquid interface. *Applied and Environmental Microbiology* 71(3), 1163–1168. https://doi.org/10.1128/AEM.71.3.1163-1168.2005.

Schikora, A., Carreri, A., Charpentier, E. and Hirt, H. 2008. The dark side of the salad: Salmonella typhimurium overcomes the innate immune response of Arabidopsis thaliana and shows an endopathogenic lifestyle. *PLoS ONE* 3(5), e2279. https://doi.org/10.1371/journal.pone.0002279.

Schikora, A., Virlogeux-Payant, I., Bueso, E., Garcia, A. V., Nilau, T., Charrier, A., Pelletier, S., Menanteau, P., Baccarini, M., Velge, P. and Hirt, H. 2011. Conservation of *Salmonella* infection mechanisms in plants and animals. *PLoS ONE* 6(9), e24112. https://doi.org/10.1371/journal.pone.0024112.

Schlechter, R. O., Miebach, M. and Remus-Emsermann, M. N. P. 2019. Driving factors of epiphytic bacterial communities: A review. *Journal of Advanced Research* 19, 57–65. https://doi.org/10.1016/j.jare.2019.03.003.

Sharma, M., Handy, E. T., East, C. L., Kim, S., Jiang, C., Callahan, M. T., Allard, S. M., Micallef, S., Craighead, S., Anderson-Coughlin, B., Gartley, S., Vanore, A., Kniel, K. E., Haymaker, J., Duncan, R., Foust, D., White, C., Taabodi, M., Hashem, F., Parveen, S., May, E., Bui, A., Craddock, H., Kulkarni, P., Murray, R. T. and Sapkota, A. R. 2020. Prevalence of Salmonella and Listeria monocytogenes in non-traditional irrigation waters in the mid-Atlantic United States is affected by water type, season, and recovery method. *PLoS ONE* 15(3), e0229365. https://doi.org/10.1371/journal.pone.0229365.

Sheth, A. N., Hoekstra, M., Patel, N., Ewald, G., Lord, C., Clarke, C., Villamil, E., Niksich, K., Bopp, C., Nguyen, T. A., Zink, D. and Lynch, M. 2011. A national outbreak of *Salmonella* serotype Tennessee infections from contaminated peanut butter: A new food vehicle for salmonellosis in the United States. *Clinical Infectious Diseases* 53(4), 356–362. https://doi.org/10.1093/cid/cir407.

Shirron, N. and Yaron, S. 2011. Active suppression of early immune response in tobacco by the human pathogen *Salmonella* typhimurium. *PLoS ONE* 6(4), e18855. https://doi.org/10.1371/journal.pone.0018855.

Sikin, A. M., Zoellner, C. and Rizvi, S. S. H. 2013. Current intervention strategies for the microbial safety of sprouts. *Journal of Food Protection* 76(12), 2099–2123. https://doi.org/10.4315/0362-028X.JFP-12-437.

Singh, B. R., Singh, P., Agrawal, S., Teotia, U., Verma, A., Sharma, S., Chandra, M., Babu, N. and Kant Agarwal, R. 2007. Prevalence of multidrug resistant *Salmonella* in coriander, mint, carrot, and radish in Bareilly and Kanpur, Northern India. *Foodborne Pathogens and Disease* 4(2), 233–240. https://doi.org/10.1089/fpd.2006.0082.

Soto, M., Chavez, G., Baez, M., Martinez, C. and Chaidez, C. 2007. Internalization of *Salmonella typhimurium* into mango pulp and prevention of fruit pulp contamination by chlorine and copper ions. *International Journal of Environmental Health Research* 17(6), 453–459. https://doi.org/10.1080/09603120701695593.

Strawn, L. K., Gröhn, Y. T., Warchocki, S., Worobo, R. W., Bihn, E. A. and Wiedmann, M. 2013. Risk factors associated with *Salmonella* and *Listeria monocytogenes* contamination of produce fields. *Applied and Environmental Microbiology* 79(24), 7618–7627. https://doi.org/10.1128/AEM.02831-13.

Sun, Z., Zhang, X., Wu, H., Wang, H., Bian, H., Zhu, Y., Xu, W., Liu, F., Wang, D. and Fu, L. 2020. Antibacterial activity and action mode of chlorogenic acid against *Salmonella enteritidis*, a foodborne pathogen in chilled fresh chicken. *World Journal of Microbiology and Biotechnology* 36(2), 24. https://doi.org/10.1007/s11274-020-2799-2.

Tang, S., Orsi, R. H., Luo, H., Ge, C., Zhang, G., Baker, R. C., Stevenson, A. and Wiedmann, M. 2019. Assessment and comparison of molecular subtyping and characterization methods for *Salmonella*. *Frontiers in Microbiology* 10, 1591. https://doi.org/10.3389/fmicb.2019.01591.

Tindall, B. J., Grimont, P. A. D., Garrity, G. M. and Euzéby, J. P. 2005. Nomenclature and taxonomy of the genus *Salmonella*. *International Journal of Systematic and Evolutionary Microbiology* 55(1), 521–524. https://doi.org/10.1099/ijs.0.63580-0.

Topalcengiz, Z., Jeamsripong, S., Spanninger, P. M., Persad, A. K., Wang, F., Buchanan, R. L., Lejeune, J. E. F. F., Kniel, K. E., Jay-Russell, M. T. and Danyluk, M. D. 2020. Survival of *Salmonella* in various wild animal feces that may contaminate produce. *Journal of Food Protection* 83, 651–660. https://doi.org/10.4315/0362-028X.JFP-19-302.

Uesugi, A. R., Danyluk, M. D. and Harris, L. J. 2006. Survival of *Salmonella* enteritidis phage Type 30 on inoculated almonds stored at −20, 4, 23, and 35°C. *Journal of Food Protection* 69(8), 1851–1857. https://doi.org/10.4315/0362-028X-69.8.1851.

Ukuku, D. O. 2004. Effect of hydrogen peroxide treatment on microbial quality and appearance of whole and fresh-cut melons contaminated with *Salmonella* spp. *International Journal of Food Microbiology* 95(2), 137–146. https://doi.org/10.1016/j.ijfoodmicro.2004.01.021.

Ukuku, D. O. and Sapers, G. M. 2001. Effect of sanitizer treatments on *Salmonella* Stanley attached to the surface of cantaloupe and cell transfer to fresh-cut tissues during cutting practices. *Journal of Food Protection* 64(9), 1286–1291. https://doi.org/10.4315/0362-028X-64.9.1286.

Van Beeck, W., Verschueren, C., Wuyts, S., van den Broek, M. F. L., Uyttendaele, M. and Lebeer, S. 2020. Robustness of fermented carrot juice against *Listeria monocytogenes*, *Salmonella* Typhimurium and *Escherichia coli* O157:H7. *International Journal of Food Microbiology* 335, 108854. https://doi.org/10.1016/j.ijfoodmicro.2020.108854.

Van der Linden, I., Eriksson, M., Uyttendaele, M. and Devlieghere, F. 2016. Is there a relation between the microscopic leaf morphology and the association of *Salmonella* and *Escherichia coli* O157:H7 with iceberg lettuce leaves? *Journal of Food Protection* 79(10), 1784–1788. https://doi.org/10.4315/0362-028X.JFP-15-590.

Van Pelt, A. E., Quiñones, B., Lofgren, H. L., Bartz, F. E., Newman, K. L. and Leon, J. S. 2018. Low prevalence of human pathogens on fresh produce on farms and in packing facilities: A systematic review. *Frontiers in Public Health* 6, 40. https://doi.org/10.3389/fpubh.2018.00040.

Vančura, V. and Hanzlíková, A. 1972. Root exudates of plants: IV. Differences in chemical composition of seed and seedlings exudates. *Plant and Soil* 36(1–3), 271–282. https://doi.org/10.1007/BF01373482.

Vandamm, J. P., Li, D., Harris, L. J., Schaffner, D. W. and Danyluk, M. D. 2013. Fate of *Escherichia coli* O157:H7, *Listeria monocytogenes*, and *Salmonella* on fresh-cut celery. *Food Microbiology* 34(1), 151–157. https://doi.org/10.1016/j.fm.2012.11.016.

Voelker, R. 2021. Contaminated water is suspected in massive *Salmonella* outbreak. *JAMA* 325(23), 2336. https://doi.org/10.1001/jama.2021.9227.

Wales, A. D., Carrique-Mas, J. J., Rankin, M., Bell, B., Thind, B. B. and Davies, R. H. 2010. Review of the carriage of zoonotic bacteria by arthropods, with special reference to *Salmonella* in mites, flies and litter beetles. *Zoonoses and Public Health* 57(5), 299–314. https://doi.org/10.1111/j.1863-2378.2008.01222.x.

Walsh, K. A., Bennett, S. D., Mahovic, M. and Gould, L. H. 2014. Outbreaks associated with cantaloupe, watermelon, and honeydew in the United States, 1973–2011. *Foodborne Pathogens and Disease* 11(12), 945–952. https://doi.org/10.1089/fpd.2014.1812.

Walters, S. P., Gonzï¿½lez-Escalona, N., Son, I., Melka, D. C., Sassoubre, L. M. and Boehm, A. B. 2013. *Salmonella enterica* diversity in central californian coastal waterways. *Applied and Environment Microbiology* 79, 4199–4209. https://doi.org/10.1128/AEM.00930-13.

Wang, H., Feng, H., Liang, W., Luo, Y. and Malyarchuk, V. 2009. Effect of surface roughness on retention and removal of *Escherichia coli* O157:H7 on surfaces of selected fruits. *Journal of Food Science* 74(1), E8–E15. https://doi.org/10.1111/j.1750-3841.2008.00998.x.

White, A. P., Gibson, D. L., Kim, W., Kay, W. W. and Surette, M. G. 2006. Thin aggregative fimbriae and cellulose enhance long-term survival and persistence of *Salmonella*. *Journal of Bacteriology* 188(9), 3219–3227. https://doi.org/10.1128/JB.188.9.3219-3227.2006.

Winthrop, K. L., Palumbo, M. S., Farrar, J. A., Mohle-Boetani, J. C., Abbott, S., Beatty, M. E., Inami, G. and Werner, S. B. 2003. Alfalfa sprouts and *Salmonella Kottbus* infection: A multistate outbreak following inadequate seed disinfection with heat and chlorine.

Journal of Food Protection 66(1), 13–17. https://doi.org/10.4315/0362-028x-66.1.13.

Yang, L., Liu, Q., Wang, Y. and Liu, L. 2017. Identification and characterization of a glossy mutant in Welsh onion (*Allium fistulosum* L.). *Scientia Horticulturae* 225, 122–127. https://doi.org/10.1016/j.scienta.2017.05.014.

Yang, L., Zhang, C., Su, Z., Zhao, L., Wu, J., Sun, X., Zhang, X. and Hu, X. 2022. Inactivation of *Salmonella typhimurium* SL1344 by chlorogenic acid and the impairment of cellular integrity. *Frontiers in Microbiology* 13, 887950. https://doi.org/10.3389/fmicb.2022.887950.

Yuan, J., Zhang, N., Huang, Q., Raza, W., Li, R., Vivanco, J. M. and Shen, Q. 2015. Organic acids from root exudates of banana help root colonization of PGPR strain *Bacillus amyloliquefaciens* NJN-6. *Scientific Reports* 5, 13438. https://doi.org/10.1038/srep13438.

Zhang, D. and Hamauzu, Y. 2004. Phenolic compounds and their antioxidant properties in different tissues of carrots (*Daucus carota* L.). *Journal of Food, Agriculture and Environment* 2, 95–100.

Zhang, G., Hu, L., Melka, D., Wang, H., Laasri, A., Brown, E. W., Strain, E., Allard, M., Bunning, V. K., Musser, S. M., Johnson, R., Santillana Farakos, S., Scott, V. N., Pouillot, R., Doren, J. M. and Hammack, T. S. 2017. Prevalence of *Salmonella* in cashews, hazelnuts, macadamia nuts, pecans, pine nuts, and walnuts in the United States. *Journal of Food Protection* 80(3), 459–466. https://doi.org/10.4315/0362-028X.JFP-16-396.

Zhang, N., Wang, D., Liu, Y., Li, S., Shen, Q. and Zhang, R. 2014. Effects of different plant root exudates and their organic acid components on chemotaxis, biofilm formation and colonization by beneficial rhizosphere-associated bacterial strains. *Plant and Soil* 374(1–2), 689–700. https://doi.org/10.1007/s11104-013-1915-6.

Zheng, J., Allard, S., Reynolds, S., Millner, P., Arce, G., Blodgett, R. J. and Brown, E. W. 2013. Colonization and internalization of *Salmonella enterica* in tomato plants. *Applied and Environmental Microbiology* 79(8), 2494–2502. https://doi.org/10.1128/AEM.03704-12.

Zhu, L., Olsen, C., McHugh, T., Friedman, M., Jaroni, D. and Ravishankar, S. 2014. Apple, carrot, and hibiscus edible films containing the plant antimicrobials carvacrol and cinnamaldehyde inactivate *Salmonella* Newport on organic leafy greens in sealed plastic bags. *Journal of Food Science* 79(1), M61–M66. https://doi.org/10.1111/1750-3841.12318.

Chapter 2

Advances in understanding and presenting contamination of fresh produce by *Listeria monocytogenes*

Xinyi Zhou and Wei Zhang, Illinois Institute of Technology, USA

1 Introduction

Fresh vegetables and fruits are considered to be nutritionally dense foods, providing an excellent source of vitamins, minerals, and fiber. Diets rich in fresh produce can reduce the risk of many chronic diseases, including obesity, Type 2 diabetes, hypertension, and cardiovascular diseases. WHO recommends at least 400 g of fruits and vegetables per day excluding starchy roots for adults to maintain a healthy diet (WHO, 2020a). In the United States (U.S.), two to four cups of vegetables per day and one-and-a-half to two-and-a-half cups of fruits per day for adults are suggested by the 2021–2025 Dietary Guidelines of Americans based on a 2000 calories per day diet (USDA, 2020). Australian

http://dx.doi.org/10.19103/AS.2023.0121.02

Dietary Guidance recommends five to six servings of vegetables and two servings of fruits per day for adults (Australian Bureau of Statistics, 2022).

However, only 10% of people in the U.S. met these fresh produce daily intake requirements and the number is even lower in Australia where only 6.1% of adults meet the intake recommendations for both fruits and vegetables in 2020 (Australian Bureau of Statistics, 2022; Lee-Kwan et al., 2019; USDA, 2020). With increasing awareness of the importance of a healthy diet, consumer demand for fresh produce has increased rapidly over the last few decades. The International Fresh Produce Association (IFPA) estimated that the fresh produce market in the U.S. in 2021 was worth US$71.6 billion, a 14.4% increase compared to total fresh produce sales in 2019 (Roerink, 2021).

With the substantial growth in the fresh produce market, the industry is now faced with new challenges. One of the most notable challenges is to ensure the safety of fresh produce. To preserve optimum nutritional value, texture, and freshness, minimal processing techniques are often used in the post-harvest processing of fresh produce. Because minimal processing does not have a single 'kill' step commonly implemented for other types of processed foods such as pasteurized milk, fresh produce recalls, and outbreaks due to the contamination from various foodborne bacterial pathogens, including *Salmonella enterica*, *Escherichia coli*, and *Listeria monocytogenes*, pose a particular threat to public safety.

The reported incidence of *L. monocytogenes* contamination in fresh produce is relatively low compared to other enteric pathogens. However, the outcome of infection can be devastating due to the high fatality rate among infected individuals. In the past two decades, *L. monocytogenes* has caused over 472 hospitalizations and 93 deaths linked to contaminated fresh produce in the U.S. (CDC, 2022b). There are numerous possible transmission routes for *L. monocytogenes* to contaminate fresh produce and cause foodborne illness. For example, at the pre-harvest stage, *L. monocytogenes* can be introduced to and become established on the surface of fresh produce by contaminated irrigation water, soil, and animal feces (Beuchat, 1996). Contamination can also occur during both harvest and post-harvest stages from food handlers and contaminated equipment. Post-harvest handling processes, in particular, such as storage, transportation, washing, and packaging, provide multiple potential contamination points (Beuchat, 1996). A common practice in the industry to control *L. monocytogenes* and to minimize cross-contamination e.g. is through washing with cold water supplemented with a chemical sanitizer (Zhou et al., 2009).

Under the direction of the Food Safety Modernization Act of 2011 (FSMA), the U.S. Food and Drug Administration (FDA) promulgated the Produce Safety Rule (PSR, 21 CFR 112, Standards for the Growing, Harvesting, Packing, and Holding of Produce for Human Consumption). These standards were designed for use by the fresh (un-processed) produce industry to minimize the risk of serious adverse health consequences or death from the consumption of contaminated produce. Similarly, the European Union (EU) passed the Food Law and the hygiene rules (Regulation (EC) No 852/2004 of the European Parliament and the Council of Ministers of April 29, 2004) on the hygiene of foodstuffs to improve fresh produce safety.

However, the low efficiency of existing control measures has led to ongoing research efforts to develop novel technologies to improve the safety of fresh produce. However, many of these new techniques still need to be validated before commercial application. This chapter reviews recent research on foodborne outbreaks and control and prevention strategies for *L. monocytogenes* contamination of fresh produce, covering topics such as:

- pathogenicity, virulence, and listeriosis;
- outbreaks associated with fresh produce;
- sources of contamination;
- case study;
- intervention strategies in the fresh produce processing industry; and
- *L. monocytogenes* gene regulation and stress response.

The chapter concludes by looking at new research and trends.

2 Pathogenicity and virulence

The genus *Listeria* was first discovered by Murray et al. (1926) in rabbits and pigs and named after Lord Lister, a pioneer in antiseptic surgery (Pirie, 1940). The genus consists of approximately 26 recognized species as of 2021, with newer species continually being identified and added to this list. The genus is categorized into two groups (Carlin et al., 2021; Orsi and Wiedmann, 2016; Quereda et al., 2020):

1. *Listeria* sensu strictu, which includes *L. monocytogenes, L. innocua, L. welshimeri, L. seeligeri, L. ivanovii, L. marthii, L. farberi, L. immobilis,* and *L. cossartiae*; and
2. *Listeria* sensu lato, which includes *L. grayi, L. rocourtiae, L. fleischmannii, L. weihenstephanensis, L. floridensis, L. aquatica, L. cornellensis, L. riparia, L. grandensis, L. booriae, L. newyorkensis, L. costaricensis, L. goaensis, L. thailandensis, L. valentina, L. portnoyi,* and *L. rustica*.

Two species, *L. monocytogenes* and *L. ivanovii* are recognized as pathogens to humans and other mammals (Orsi et al., 2011):

- *L. ivanovii* causes sporadic infections, mainly in ruminants.
- *L. monocytogenes* is responsible for most infections associated with humans and animals.

Thirteen serotypes of *L. monocytogenes* can be distinguished by using somatic (O) and flagellar (H) antigens, of which 1/2a, 1/2b, 1/2c, and 4b contribute primarily to the listeriosis associated with humans (Ragon et al., 2008). A few years after the discovery of *L. monocytogenes*, Nyfeldt reported the first case of listeriosis in humans in 1929. However, *L. monocytogenes* was not recognized as a foodborne pathogen until a reported listeriosis outbreak in Canada in 1981 linked to the consumption of contaminated food (Gasanov et al., 2005).

L. monocytogenes is a gram-positive, rod-shaped, non-spore-forming, and facultative anaerobic bacterium. The size of the *L. monocytogenes* cell varies from 1 μm to 2 μm in length and from 0.4 μm to 0.5 μm in diameter (Wang and Oris, 2013). It is motile when the temperature is between 10°C and 25°C due to flagella (Vázquez-Boland et al., 2001). The temperature range for *L. monocytogenes* proliferation is between −2°C and 45°C, with optimal growth at 30–37°C. The bacterium is also able to survive and grow in a wide pH range from 4.0 to 9.5 and can tolerate salt (NaCl) concentrations up to 10% (w/v) (Hernandez-Milian and Payeras-Cifre, 2014; Liu et al., 2005). *L. monocytogenes* is widespread in the environment and can be found in soil, water, plants, sewage, and the feces of many animals (Freitag et al., 2009). The biofilm-forming capability of *L. monocytogenes* also supports its persistence and survival on both biotic and abiotic surfaces (Mazaheri et al., 2021). The surface of fresh produce provides an ideal environment for this pathogen to establish a population and subsequently cause foodborne infections in humans.

Listeriosis is an infectious disease caused by the consumption of food contaminated with *L. monocytogenes*. It can be divided into two types based on severity (Allerberger and Wagner, 2010):

- Non-invasive gastrointestinal listeriosis; and
- Invasive listeriosis.

The incident rate of listeriosis is between the range of 0.1 and 10 cases per 1 000 000 persons per year, depending on the country (WHO, 2018). Healthy adults usually display symptoms of non-invasive listeriosis rather than invasive listeriosis. Non-invasive listeriosis causes mild gastrointestinal symptoms, including diarrhea, headache, muscle pain, fatigue, and fever, and is generally self-limiting (Allerberger and Wagner, 2010; WHO, 2018). In contrast, invasive listeriosis is more severe and the mortality rate is as high as 20–30%. Immunocompromised people, including pregnant women, people with AIDS or other immune deficiencies, diabetes, chronic liver and kidney disease, and those who undergo cancer treatments, are more likely to develop this type of listeriosis (WHO, 2018). The clinical manifestations of invasive listeriosis include sepsis, meningitis, and death if the disease is not treated properly. For pregnant women, listeriosis may lead to miscarriage or stillbirth. The incubation time for invasive listeriosis is generally between 7 and 14 days, but for pregnancy-related cases, the incubation time can be as long as 6 weeks WHO, 2018).

Newborns can also acquire listeriosis, namely neonatal listeriosis, from an infected mother through the placenta. There are two clinic forms of neonatal listeriosis (Allerberger and Wagner, 2010):

- Early-onset listeriosis within 7 days of birth; and
- Late-onset listeriosis from one to several weeks after birth.

Blood and placenta cultures can be used to diagnose perinatal listeriosis. For diagnosis of non-pregnancy-related listeriosis, detection of *L. monocytogenes* in cerebrospinal fluid, feces, or vomitus can be conducted (WHO, 2018). Antibiotics, such as ampicillin and gentamicin, can be used to treat listeriosis (Fernández Guerrero et al., 2012).

The stages of infection by *L. monocytogenes* have been well studied and are depicted in Fig. 1. *L. monocytogenes* is able to cross the intestinal barrier, the blood-brain barrier, and the maternofetal barrier. After ingestion, *L. monocytogenes* can migrate to the liver and spleen from the intestine via the lymphatic system and blood, where it starts replication. The blood then delivers *L. monocytogenes* to the brain and placenta by passing the blood-brain barrier and maternofetal barrier, respectively. Successful invasion into the brain and placenta also depends on the unique capability of *L. monocytogenes* to evade phagocytic cells (Cossart and Toledo-Arana, 2008).

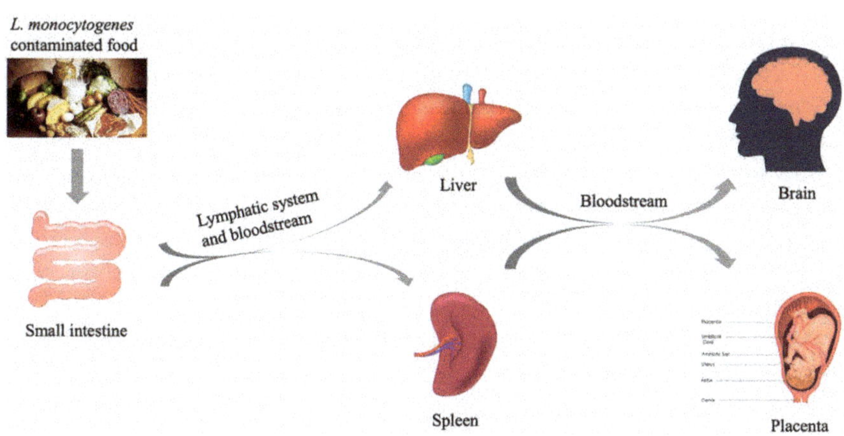

Figure 1 Successive steps of human listeriosis (Cossart and Toledo-Arana, 2008).

3 *Listeria monocytogenes* outbreaks linked to fresh produce

According to Centers for Disease Control and Prevention (CDC), it is estimated that *L. monocytogenes* causes approximately 1600 infections and 260 deaths annually in the U.S. due to consumption of contaminated food (CDC, 2021b). Among outbreaks with confirmed serotypes, serotypes 1/2a, 1/2b, and 4b account for 19%, 11%, and 28% of total infections, respectively (CDC, 2022a). In the 20 years from 2000 to 2020, the incidence of foodborne *L. monocytogenes* infection fluctuated between 0.21 and 0.33 per 100 000 persons in the U.S. (CDC, 2022a). In the EU, the incidence rate of listeriosis is higher than that in the U.S., fluctuating between 0.42 and 0.48 from 2015 to 2020 (EFSA and ECDC, 2021a,b). Although the average incidence of *L. monocytogenes* infection is much lower than those of *Salmonella* and *E. coli* O157:H7, hospitalization and death rates are much higher, reaching 93% and 17% in the U.S., respectively. The high hospitalization rate and fatality are also observed in other countries, such as the EU and China (EFSA and ECDC, 2016, 2017, 2018, 2019, 2021a, 2021b; Fan et al., 2019). Infections are not evenly distributed throughout age groups. People younger than 5 years or older than 65 years are more susceptible than people in other age groups (CDC, 2022a). The Interagency Food Safety Analytics Collaboration (IFSAC) designed a statistical modeling method to estimate the contribution of different types of food to pathogen outbreaks (Batz et al., 2021).

Based on the most recent report in 2019, over 90% of the *L. monocytogenes* outbreaks were attributed to dairy products and fresh produce in the U.S. Fruits, vegetable row crops, sprouts, and other produce accounted for 25.5%, 15.8%, 3.7%, and 14.8% of cases, respectively (CDC, 2021a). In the EU, fish and fish products, meat products, and dairy products contributed to most of the *L. monocytogenes* outbreaks (EFSA and ECDC, 2021b). From 2008 to 2018, there were a total of 15 reported listeriosis outbreaks linked to fresh produce in the U.S. and reported to the NORS database; these outbreaks are listed in Table 1 (CDC, 2022b). Three hundred and twenty illnesses, 297 hospitalizations, and 62 deaths were identified due to these outbreaks. Among the 15 outbreaks, leafy greens, sprouts, and fungi were involved in four, three, and two of the outbreaks, respectively, whereas the six remaining outbreaks were all associated with fruits. Table 2 includes selected outbreaks linked to fresh produce from other countries.

For example, a 2010 listeriosis outbreak was linked to contaminated diced celery in a chicken salad in Texas, leading to 10 cases of illness. The age range of infected persons ranged from 56 to 93, with a median age of 80 years. All of the cases were hospitalized, and half resulted in deaths. Pulsed-field gel electrophoresis (PFGE) analysis identified a common *L. monocytogenes*

Table 1 *Listeria monocytogenes* outbreaks associated with fruits and vegetables in the U.S. between 2008 and 2018 reported in the NORS database

Year	Location	Illnesses	Hospitalizations	Deaths	Serotype	Food vehicle
2008	Multistate	20	16	0	1/2 Garner and Kathariou (2016)	Sprouts
2010	Texas	10	10	5	1/2a Gaul et al. (2013)	Celery
2011	Multistate	147	143	33	1/2a and 1/2b Laksanalamai et al. (2012a)	Cantaloupe, whole
2014	Multistate	2	2	1	1/2b and 4b Chen et al. (2016a)	Peaches and nectarines
2014	Multistate	5	3	2	4b Garner and Kathariou (2016)	Mung bean sprouts
2014	Multistate	35	34	7	4b Angelo et al. (2017)	Apples (prepacked caramel apples)
2014	Virginia	2	2	0	ND	Sprouts
2014	Multistate	9	9	3	4b Self et al. (2019)	Leafy greens
2015	Multistate	19	19	1	ND	Lettuce
2016	Multistate	36	31	4	ND	Enoki mushroom
2017	Multistate	5	4	1	ND	Enoki mushrooms
2016	Multistate	10	9	1	ND	Avocado
2017	Multistate	3	3	0		Caramel apple
2017	Multistate	10	7	3	ND	Leafy greens
2018	Multistate	7	5	1	ND	Stone fruits

ND, not determined. Source: CDC (2022b).

Table 2 *Listeria monocytogenes* outbreaks associated with fruits and vegetables in other countries

Year	Location	Cases	Deaths	Food vehicle	Reference
1981	Canada	41	17	Vegetable mix for coleslaw	Zhu et al. (2017)
1997	Italy	2930	0	Corn	Aureli et al. (2000)
1998	Australia	6	5	Commercially prepared fruit salad	Food Directorate (2010)
2018	Australia	22	7	Rockmelon (Cantaloupe)	Das (2019)
2018	Austria, Denmark, Finland, Sweden, and the United Kingdom	53	10	Frozen corn	McLauchlin et al. (2021)
2017–2020	Canada, Australia, and French	16	0	Enoki mushroom	WHO (2020b)
2013	Switzerland	32	Unknown	Ready-to-eat salad	Stephan et al. (2015)

serotype 1/2a strain from patients, diced celery, and the environment of the facility where the food was prepared where poor sanitation and inadequate handling of the celery were observed. The outbreak led to the production facility shutting down and recalling their products (Goulet et al., 2013).

From November 2016 to June 2020, a listeriosis outbreak associated with enoki mushrooms resulted in a total of 36 illnesses and 31 hospitalizations reported from 17 states. Four deaths were recorded in California (two), Hawaii (one), and New Jersey (one). Two fetal losses were also reported out of six affected pregnant women. Of the interviewed people, 55% reported consumption of mushrooms before the illness. Subsequently, whole genome sequencing (WGS) identified the outbreak strain from enoki mushrooms produced by Green Co. LTD in the Republic of Korea. Three distribution companies conducted recalls of enoki mushrooms supplied by this producer (FDA, 2020).

4 Sources of contamination in the fresh produce supply chain

L. monocytogenes is a saprophytic environmental organism that is common in soil, vegetation, plant residues, wastewater, and sewage (Buchanan et al., 2017). Since produce is often grown in contact with (or in close proximity to) the ground, contamination with *L. monocytogenes* or other soil-dwelling foodborne pathogens is possible during pre-harvest stage. Routes for *L. monocytogenes* contamination of fresh produce also exist during the harvest and post-harvest handling stages. This section will provide an overview of common routes of *Listeria* contamination throughout the farm-to-fork continuum (Fig. 2).

4.1 Pre-harvest

Fresh produce can readily be contaminated with foodborne pathogens during growing, by contact with soil or soil amendments, irrigation water, humans, animals or insects, tools and equipment, or other environmental factors such as wind-blown contaminated dust particles. Farms that grow, harvest, and offer for sale non-processed produce must be in compliance with the Produce Safety Rule. FDA also publishes guidance for the industry to assist in complying with the rule and to provide additional information on how to reduce the risk of microbial contamination of fruits and vegetables (FDA, 1998).

Soil, including soils used for growing produce, is a known environmental niche for *L. monocytogenes*; the pathogen can be easily isolated from soil and water in dairy, fruit, and vegetable farms (Linke et al., 2014; Strawn et al., 2013a; Vivant et al., 2013). In one study, *L. monocytogenes* was isolated from

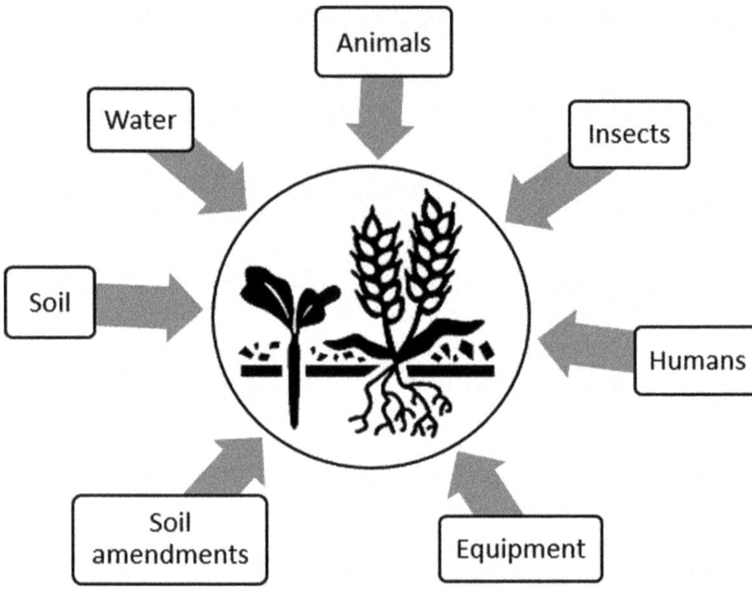

Figure 2 Potential pre-harvest routes of *Listeria monocytogenes* contamination of fresh produce.

9% of soil samples (n = 178) collected from fruit and vegetable farms in New York, USA (Strawn et al., 2013a). Although soil alone may be a reservoir for *L. monocytogenes* at very low levels, soil can become further contaminated by feces of ruminants and other wildlife animals, which are known vehicles of *L. monocytogenes* (Rodriguez et al., 2021; Strawn et al., 2013b). The pathogen is also found in sewage, recycled sludge, compost, or biological soil amendments of animal origin. Biological soil amendments of animal origin, including manure, are often added to crop soil to improve water retention and chemical or physical condition, provide additional nutrients, and ultimately improve yield. If untreated, biological soil amendments are used, contamination of soil and crops with *L. monocytogenes* is likely to occur.

L. monocytogenes has been documented to survive for months or even years in soil (Locatelli et al., 2013; Vivant et al., 2013). Once present in the soil, *L. monocytogenes* can contaminate crops by coming into contact with the harvestable portion of the plant, sometimes due to water splash (Monaghan and Hutchison, 2012). Other foodborne pathogens, such as *Salmonella* or *E. coli*, have also been documented to contaminate crops like spinach and lettuce through the root structure (Karmakar et al., 2018; Solomon et al., 2002; Wright et al., 2013). Although research on the internalization of *L. monocytogenes* in crops via roots is scarce, a recent study determined that the pathogen was capable of colonizing *Arabidopsis thaliana* through root internalization (Schoenborn et al., 2021), indicating its potential to also infiltrate fresh produce crops in this manner.

Crop irrigation water is an important source of *L. monocytogenes* contamination of fresh produce. *L. monocytogenes* has been detected in various types of agricultural water sources (Cooley et al., 2014; Falardeau et al., 2017; Weller et al., 2015). Irrigation water can be sourced from rivers, lakes, reservoirs, collected rainwater, groundwater, and reuse of wastewater. If wastewater is not properly treated, there is the potential for contamination with foodborne pathogens, which may cross-contaminate fresh produce prior to human consumption. One study reported fecal coliform levels to be as high as 4 log CFU/mL in irrigation water samples on Texas melon farms (Materon et al., 2007). Another study collected irrigation and non-irrigation water samples for detecting *L. monocytogenes*, finding that the pathogen was present in 9% of irrigation water samples and in 39% of non-irrigation water samples, respectively (Strawn et al., 2013b). Another study collected six different types of irrigation water samples over multiple seasons in the Mid-Atlantic U.S. and identified *L. monocytogenes* in 31% of samples (n = 170) (Sharma et al., 2020). The type of irrigation has been shown to impact the microbiological quality of fresh produce (Suslow et al., 2003).

Humans, animals, and insects are additional sources of contamination of fresh produce by *L. monocytogenes*. Farm workers can disseminate foodborne pathogens if good hygiene is not practiced or if they work when ill. It is estimated that 20% of all foodborne bacterial outbreaks are due to food handler transmission (Greig et al., 2007). Farm workers with appropriate hygiene training and skills are key to the safe production of fresh produce, including access to restrooms and handwashing facilities.

Farm animals such as cows, pigs, sheep, and chickens are also natural reservoirs of foodborne pathogens and can shed *L. monocytogenes* in their feces, leading to cross-contamination of fresh produce if adequate exclusion or mitigation measures are not in place (Lyautey et al., 2007; Rodriguez et al., 2021; Strawn et al., 2013b). Birds, either local or migratory, may be vehicles for foodborne pathogens as they use farms as sources of food and shelter (Cernicchiaro et al., 2012). Reptiles, amphibians, rodents, helminths, and insects such as flies and ticks can also serve as potential vehicles for fresh produce contamination (Beuchat, 2006; Kulesh et al., 2019). Wind-driven dust might be a possible route of pathogen transmission on mixed-use farms (Pang et al., 2017) or when animal operations are adjacent or close to farms producing fresh produce.

4.2 Harvest

Contamination of fresh produce with *L. monocytogenes* can occur during harvest via contaminated equipment, containers, water and ice, transport vehicles, and farm personnel. As at the pre-harvest stage, workers at the harvest stage should adhere to good hygienic practices and should not work when sick. During harvest, it is often necessary for fresh produce to undergo manipulation such as removal of outer leaves, coring, and cutting, either by hand or via harvesting equipment. Harvesting equipment, including mechanical cutters and corers, are potential sources of *L. monocytogenes* contamination as they can harbor foodborne pathogens and disseminate these organisms to fresh produce (Taormina et al., 2009; Yang et al., 2012). The act of cutting produce also exposes inner plant tissues and nutrients, leading to increased water activity, and can enhance bacterial survival and proliferation. *L. monocytogenes* is known to also attach preferentially to cut plant tissues. Foodborne pathogens may infiltrate produce via damaged or bruised surfaces. If produce is harvested from the ground (i.e. carrots, potatoes) or close to the ground (i.e. lettuce and cantaloupe), there is the potential for some soil retention, which may lead to contamination with *L. monocytogenes* or other soil-resident foodborne pathogens.

Containers, ice, and transport vehicles also play a role in the potential contamination of fresh produce. Harvested produce is often placed into containers in bulk and may be cooled using water or ice. These containers may be contaminated with *L. monocytogenes* from produce residues or soil, especially if produce roots are also harvested. Regular sanitization of containers and transport equipment is necessary to prevent contamination or cross-contamination. Since foodborne pathogens, including *L. monocytogenes*, can readily proliferate on cut produce at ambient temperatures (Salazar et al., 2017), processed or cut produce must be cooled immediately after harvest to minimize bacterial growth and increase shelf-life. Water, ice, or cooled transport vehicles have previously been identified as sources of *L. monocytogenes* during contamination events (Suslow et al., 2003). Water quality used for ice is essential as using poor quality water may contribute to *L. monocytogenes* contamination. Melting water may also assist in cross-contamination. Since it is not uncommon for fresh produce to be transported in reusable containers or in trucks that previously held meat or fish products, proper sanitization is critical to prevent contamination (Suslow et al., 2003).

4.3 Post-harvest

Sources for fresh produce contamination with *L. monocytogenes* during the post-harvest stage may include processing and storage. Fresh produce may be minimally processed, which can include washing, shredding or cutting, drying, and packaging. Fresh produce can become contaminated from equipment used in these processes or from cross-contamination with other produce, the processing facility, or from food contact surfaces. Studies have frequently found *Listeria* spp. in drains, entryways, storage shelves, dump tanks and flumes, brush and drying rollers, and other equipment in fresh and fresh-cut produce handling and processing facilities (Dunn et al., 2022; John et al., 2020; Leong et al., 2014; Ruiz-Llacsahuanga et al., 2021; Simonetti et al., 2021; Sullivan and Wiedmann, 2020).

Listeria sp, *L. monocytogenes*, is often found in processing environments. In one study that evaluated the prevalence of *L. monocytogenes* in seven produce handling and processing facilities in the Pacific Northwest U.S., five of the facilities were found to harbor the pathogen with prevalence values ranging from 2% to 26% (John et al., 2020). Another study found a 2.3% prevalence of *L. monocytogenes* on conveyer belts and cooler surfaces in multiple cabbage-packing sheds (Prazak et al., 2002).

In FDA's document Current Good Manufacturing Practice, Hazard Analysis, and Risk-Based Preventive Controls for Human Food, facilities are required to test for environmental pathogens during hazard analysis whenever a ready-to-eat food is exposed to the environment prior to packaging, the packaged food does not receive a subsequent decontamination treatment or an alternative control measure (such as a formulation lethal to the pathogen) that would significantly minimize the pathogen (21 CFR 117.130(c)(ii)). When contamination by environmental pathogens, such as *L. monocytogenes*, requires preventive control, facilities are further required to establish environmental monitoring procedures for the pathogen or an appropriate surrogate (21 CFR 117.165(a)(3)).

Identifying *L. monocytogenes* harborage sites is of utmost importance as *L. monocytogenes* can evade cleaning and sanitizers via biofilm formation and persist for long periods of time. Harborage sites can develop through poorly-designed or old equipment. Persistent strains of *L. monocytogenes* present in these niches can lead to contamination of produce during washing, cutting, or packaging. One study determined that *L. monocytogenes* could be transferred from equipment to celery during mechanical dicing (Kaminski et al., 2014).

L. monocytogenes contamination of produce can often be traced back to the environment where handling, processing, packaging, or holding and storage occurs. Various outbreaks associated with *L. monocytogenes* contamination of fresh produce have been traced back to either the processing or packinghouse environment. For example, a listeriosis outbreak in 2014 associated with stone fruits, including peaches and nectarines, was linked to contamination of the packinghouse environment (Chen et al., 2016b; Jackson et al., 2015). Stone fruits were considered unlikely vehicles for *L. monocytogenes* transmission and had not previously been implicated in human listeriosis cases prior to 2014. In the epidemiological investigation, *L. monocytogenes* was detected on the surfaces of different types of stone fruits in the packinghouse as well as in the packinghouse environment. Other notable listeriosis outbreaks where the *L. monocytogenes* contamination was traced back to the product processing environment have been associated with cantaloupe and apples (CDC, 2011, 2015a). A high prevalence of *L. monocytogenes* (33%, n = 39) was reported in a packinghouse linked to a cantaloupe listeriosis outbreak (FDA, 2011).

In addition to processing and packing environments, *L. monocytogenes* contamination of fresh produce can occur at retail, restaurant, and consumer stages in the supply chain, mainly through cross-contamination (Table 3). Fresh produce is often cut or shredded by the end user and consumed in its raw state (i.e. cut leafy greens in salads) without further treatment. *L. monocytogenes* can cross-contaminate produce during cutting from sources such as hands, knives, and food contact surfaces such as cutting boards. For example, *L. monocytogenes* has been documented to readily transfer to fresh-cut cantaloupe from gloves (Qi et al., 2020) and between contaminated lettuce and kitchen knives (Zilelidou et al., 2015). *L. monocytogenes* can also adhere to cutting board surfaces, leading to cross-contamination during fresh produce cutting (Beltrame et al., 2015).

Table 3 *Listeria monocytogenes* post-harvest contamination sources associated with fresh produce listeriosis outbreaks

Country	Food vehicle	Source of contamination	Year	Reference
U.S.	Packaged salads	Processing environment	2021	CDC (2021c); FDA (2021)
U.S.	Packaged salads	Processing environment	2021	CDC (2021d)
U.S.	Packaged salads	Processing environment	2016	CDC (2016)
U.S.	Caramel apples	Processing environment	2014	CDC (2015a)
U.S.	Mung bean sprouts	Processing environment	2014	CDC (2015b)
U.S.	Stone fruits	Packinghouse environment	2014	Chen et al. (2016b); Jackson et al. (2015)
U.S.	Cantaloupe	Processing environment	2011	CDC (2011)
U.S.	Celery	Processing environment	2010	Gaul et al. (2013)
Australia	Cantaloupe	Packinghouse environment	2018	Das (2019)
Switzerland	Ready-to-eat salads	Processing environment	2013	Stephan et al. (2015)

5 Case study: *Listeria monocytogenes* in cantaloupes

From August to October in 2011, a listeriosis outbreak associated with whole cantaloupes involved 147 cases and 28 U.S. states, with the most cases in Colorado (40), Texas (18), New Mexico (15), Oklahoma (12), and Kansas (11). This was the first listeriosis outbreak linked to cantaloupe, involving the third-highest number of cases among recorded listeriosis outbreaks in the U.S. (CDC, 2011). According to the CDC, the outbreak resulted in a 96% hospitalization rate and a 22% death rate. In addition, a fetal loss of a pregnant woman was reported. For all the hospitalized cases, ages ranged from <1 to 96 years, with a median age of 78 years; 86% of ill people were older than 60 years of age (CDC, 2012).

PFGE was used to determine the DNA fingerprint of the isolated *L. monocytogenes* strain. Initially, three subtypes of *L. monocytogenes* associated with this outbreak were identified from patients in Colorado. Later, a fourth subtype was isolated from patients in another state and was confirmed through PulseNet. These four subtypes encompassed 146 total cases. Finally, a fifth subtype with different PFGE patterns was isolated from a cantaloupe sample and matched the fingerprint of the isolate from a patient who was excluded from the outbreak in the PulseNet database. Consequently, the case number for this outbreak went up from 146 to 147 in total (CDC, 2012). A total of five subtypes of *L. monocytogenes* were involved in this outbreak, which included two serotypes, 1/2a and 1/2b (CDC, 2011; Laksanalamai et al., 2012b).

During the interview process, 134 ill persons reported consumption of whole cantaloupe within the month of illness, and some pointed out that the type of cantaloupes they ate were Rocky Ford cantaloupes. The FDA and the Colorado Department of Public Health and Environment (CDPHE) isolated the *L. monocytogenes* outbreak strain from cantaloupe samples collected from grocery stores, patient residences, and a packing house. The outbreak strain was also detected from the environment, equipment, and food contact surfaces in the packing house. As a result, whole cantaloupes from the packinghouse were confirmed as the source of this outbreak. The root cause analysis investigation by FDA and CDPHE suggested that a truck next to the processing line which traveled between the firm and a nearby cattle operation introduced *L. monocytogenes* to the packing house. Additionally, poor sanitation and design of the handling equipment facilitated the transmission and growth of *L. monocytogenes* on the cantaloupes (McCollum et al., 2013).

Since *L. monocytogenes* present on the cantaloupe rind can be transmitted to the flesh during cutting, providing the pathogen with a suitable environment to survive and proliferate, FDA suggests that consumers wash all produce thoroughly under running water before preparing cantaloupes for eating (FDA, 2022b). However, washing at home may not be sufficient to remove all pathogenic cells on the rind. Rigorous good agricultural practices (GAPs) during pre-harvest cultivation and good manufacturing practices (GMPs) during post-harvest processing should be implemented by the industry to serve as the first line of defense to prevent pathogen contamination. More details about the industry intervention strategies are discussed later in this chapter.

Since this large cantaloupe outbreak, many researchers have investigated the attachment, survival, and growth of *L. monocytogenes* on whole cantaloupes as well as fresh-cut cantaloupes (Danyluk et al., 2014; Martinez et al., 2016; Nyarko et al., 2016; Salazar et al., 2017; Ukuku et al., 2012). Novel technologies and sanitizers, such as steam vapor, octenidine dihydrochloride, chlorine dioxide gas, and blue light-emitting diodes, have been developed and evaluated to inactivate *L. monocytogenes* on the rind (Forney et al., 2018; Josewin et al., 2018; Kaur et al., 2015; Mahmoud, 2012; Ukuku et al., 2016; Upadhyay et al., 2016).

6 Strategies to control *Listeria monocytogenes* in the fresh produce supply chain

As noted, to preserve nutrients, texture, flavor, and appearance to the largest extent, fresh produce often undergoes minimal processing. After harvesting at the farm, fresh produce is often precooled to prevent the growth of bacteria, followed by transport to the processing facility for sorting, washing, and packing. Generally, harvested produce is subjected to various processing steps such as inspection, sorting, washing, drying, etc. Among these procedures, inspection and sorting improve product quality by removing damaged and sub-standard produce. The washing step can be accomplished by submerging produce in one or multiple water tanks filled with chilled water supplemented with or without sanitizers, followed by water spraying. Washing with chilled water supplemented with or without sanitizers may be the only step to reduce pathogen loads on fresh produce (Gil et al., 2015; Pietrysiak et al., 2019), although all such processes would require validation to determine actual pathogen reduction.

Gil et al. (2015) have described a washing system for fresh leafy greens, consisting of three washing water tanks. The first tank is designed to contain water solely for removing soil, dust, and debris from fresh produce. The shear force generated by water can also remove some pathogens attached to the surface of the fresh produce. However, if pathogens are present on the surface of the produce, their concentration in the wash water can increase rapidly, which introduces the potential for cross-contamination of products when wash water is recirculating (Banach et al., 2015). Refreshing of water and the addition of antimicrobial agents is needed to control cross-contamination (Gil et al., 2009).

After an initial wash, fresh produce is transported to the second tank filled with water and antimicrobial agent to reduce the pathogen load on fresh produce. The most commonly used sanitizers will be discussed in the following section. Finally, the water in the third wash tank rinses off the residual sanitizer and cools the products. A more efficient and water-saving washing unit which contains two showers and one water tank has also been developed (Gil et al., 2009). The application of sanitizers to wash fresh produce is critical, as water alone is only able to reduce *L. monocytogenes* by 0–3 log (Hoelzer et al., 2014). The addition of sanitizers can also reduce the risk of cross-contamination.

Two types of antimicrobial agents are commonly used in the industry:

- Chlorine-containing compounds; and
- Peracetic acid (PAA).

Chlorine is the most common sanitizer to control the microbial load in wash water because it is inexpensive and easy to obtain. The FDA approves the use of chlorinated water to wash fresh produce (CFR - Code of Federal Regulations Title 21, n.d.). Commonly used concentrations of chlorine range from 50 ppm to 200 ppm, with exposure times from 1 min to 2 min. Chlorinated water can be prepared by the addition of chlorine gas, sodium hypochlorite solution, and calcium hypochlorite powder into water (Gehringer and Kaletunç, 2017).

The oxidizing and antimicrobial activity of chlorinated water is based on free chlorine concentration, which also depends on the pH of the chlorinated water (Suslow, 1997); the pH of the solution determines the existing form of chlorine once dissolved in water (Fig. 3). When the pH ranges between 6 and 7.5, the chlorine exists mainly in the form of hypochlorous acid (HOCl) which has the best potential to inactivate pathogens. HOCl can react with most cellular macromolecules, especially cystine, methionine, glutathione, and primary and secondary amines to form irreversible chemical bonds which can lead to protein dysfunction. The bacterial cell membrane is damaged by the loss of permeability, and enzymes and DNA are destroyed, which eventually causes lethal effects (Gray et al., 2013; Pietrysiak et al., 2019). However, when the

$$NaOCl + H_2O \rightarrow OCl^- + OH^- + H^+ \qquad \text{Equation 1}$$

$$OCl^- + H^+ + OH^- \rightleftharpoons HOCl + OH^- \qquad \text{Equation 2}$$

$$HOCl + H^+ + Cl^- \rightarrow Cl_2 + H_2O \qquad \text{Equation 3}$$

Figure 3 Forms of chlorine in the water. Equation 1. After sodium hypochlorite (NaOCl) solution is dissolved in water, it will release a hypochlorite ion (OCl^-). Equation 2. If pH is between 6 and 7.5, OCl^- ions react with hydrogen ions (H^+) in water to form hypochlorous acid (HOCl). However, if the pH is greater than 7.5, HOCl can react with hydroxide ions (OH^-) to release OCl^-. Equation 3. When the environment becomes acidic, introduced H^+ ions react with HOCl releasing chlorine gas (Cl_2).

solution becomes acidic, toxic chlorine gas can be produced and the chlorinated water becomes corrosive to processing equipment. Conversely, when the pH goes above 7.5, chlorine mainly exists in the form of OCl, which can significantly reduce its effectiveness (Pietrysiak et al., 2019). The effectiveness of chlorine is also affected by temperature and organic compounds. Chlorine can react with organic matter, such as fresh produce and soil in the washing unit, causing a rapid loss of oxidizing power due to degradation (Pietrysiak et al., 2019; Suslow, 1997). As a result, the pH and concentration of chlorine should be monitored frequently during processing for the best antimicrobial effectiveness.

Health concerns regarding the byproducts of chlorine have been raised. An incomplete reaction of chlorine and organic compounds can form byproducts such as chloroform and trihalomethanes, which are potential carcinogens (Gadelha et al., 2019; Suslow, 1997). HOCl can also react with primary and secondary amines to produce chloramine which is harmful to humans (Gray et al., 2013). Due to these undesirable byproducts, countries in Europe, including the Netherlands, Sweden, Germany, and Belgium, have prohibited the use of chlorinated water to wash fresh produce (Pérez-Gregorio et al., 2011; Rico et al., 2007).

Peracetic acid (PAA) is another commonly used sanitizer in the fresh produce industry. It is produced by the reaction of hydrogen peroxide and acetic acid. It exists at an equilibrium state which includes hydrogen peroxide, acetic acid, water, and peracetic acid with a strong vinegar odor (Banach et al., 2015). It shows excellent effectiveness when the pH is between 3 and 7.5, and the temperature is between 0°C and 40°C (USDA, 2016). The ability of PAA to inactivate pathogens is mainly attributed to generated reactive oxygen species which can oxidize sulfhydryl and disulfide bonds to increase cell membrane permeability. DNA, enzymatic and transport systems, proteins, and enzymes are all damaged or denatured (Banach et al., 2015, 2020; Vandekinderen et al., 2009). The inclusion of hydrogen peroxide in commercially available PAA partially contributes to its antimicrobial properties (Banach et al., 2015). PAA is superior to chlorine-based sanitizers because it does not generate harmful byproducts and its effectiveness is less affected by organic matter such as soil, dirt, and debris (Haute et al., 2015; Huang et al., 2015). However, compared to chlorine-based sanitizer, PAA is more expensive. Unpleasant odors on treated fresh produce and instability of PAA at high concentrations are also limitations of use in the fresh produce industry (Rodgers et al., 2004).

7 *Listeria monocytogenes* gene regulation and survival mechanisms

Understanding the genetics of *L. monocytogenes* can elucidate the mechanisms by which this pathogen survives and persists in produce-associated environments and its attachment, survival, and proliferation in produce tissues. From the first sequenced *L. monocytogenes* strain, EGD-e, it was recognized that a high percentage of the genome (11.6%) coded for proteins involved in transport, of which approximately 25% are responsible for carbohydrate transport and the phosphoenolpyruvate-dependent phosphotransferase (PTS) systems (Glaser et al., 2001). This suggests the ability to utilize various carbohydrate sources from different environments, conferring a fitness advantage. Among the fully sequenced *L. monocytogenes* genomes currently available, there is a high degree of synteny, suggesting a stable evolutionary core genome (den Bakker et al., 2013). The pan-genome of *L. monocytogenes* comprises approximately 4400 genes, of which 2230 to 2465 are core putative protein-coding genes (den Bakker et al., 2013; Deng et al., 2010). Within the pan-genome of *L. monocytogenes*, genes encoding regulatory proteins, including transcription factors, are well conserved.

Two of the most widely studied *L. monocytogenes* transcription factors are the positive regulatory factor A (PrfA) and sigma factor B (SigB). These transcription factors are known to regulate virulence and survival mechanisms. PrfA regulates at least 145 other *L. monocytogenes* genes which can be divided into three groups, as determined via transcriptomic or proteomic assays (Milohanic et al., 2003; Port and Freitag, 2007). These genes are major virulence determinants and play a role in infection and the switch from soil saprophyte to intracellular survival. While PrfA mainly regulates virulence genes, it also regulates genes involved in the survival of *L. monocytogenes* in adverse environments, including exposure to oxidative stress (Mains et al., 2021), and promotes biofilm formation (Zhou et al., 2011). Lm0753, a transcription factor similar to PrfA, has also been shown to play a role in biofilm formation in *L. monocytogenes*, in addition to soil survival and surface attachment to fresh produce, including lettuce and cantaloupe rinds (Salazar et al., 2013).

SigB is a major transcriptional regulator of stress response genes in *L. monocytogenes* and is activated under adverse environmental conditions, including oxidizing conditions, nutrient scarcity, and pH and osmotic changes (Ferreira et al., 2001, 2003). SigB is known to play a role in the adaptation and survival of *L. monocytogenes* in the soil environment and to plant root systems (Gorski et al., 2011; Marinho et al., 2020). When artificially inoculated into the soil, an *L. monocytogenes sigB* gene knockout mutant strain was reduced in population by approximately 1–2 log CFU compared to the wild-type strain over 4 weeks. In addition, SigB plays an important role in *L. monocytogenes* biofilm formation and its resistance to disinfectants and detergents, promoting the pathogen's survival in the food processing environment (Ryan et al., 2008; Smith et al., 2018; van der Veen and Abee, 2010).

In addition to the core *L. monocytogenes* genome, plasmids are also common and confer their own types of resistance, further protecting the pathogen from the environment. Genes on three *L. monocytogenes* plasmids (in strains of ST5, ST8, and ST121) are known to play a role in the survival of the pathogen in food processing environments (Naditz et al., 2019). The presence of these three plasmids in *L. monocytogenes* contributes to its tolerance and adaptation to high temperatures, salt, acid, and oxidative stresses. The three plasmids confer increased tolerance to hydrogen peroxide, lactic acid, and high salt concentrations of up to 15%. The plasmid in an ST5 strain also contributes to *L. monocytogenes* heat tolerance at 55°C and benzalkonium chloride (BC) stress survival, which is a common quaternary ammonium compound used for sanitation in industry.

L. monocytogenes has adapted to survival in the soil and vegetative environment and has genes that play important roles in the attachment to and colonization of fresh produce. For example, the gene encoding the *Listeria* cellulose binding protein (Lcp) has been reported to play a role in the adherence of *L. monocytogenes* to fresh produce, including spinach, cantaloupe rinds, and lettuce (Bae et al., 2013). Other genes, including *flaA*, which encodes the flagellin, and *motAB*, which encodes part of the flagellar motor in *L. monocytogenes*, are known to play a role in surface attachment and colonization of alfalfa and radish sprouts (Gorski et al., 2009). Overall, the genetics of *L. monocytogenes* suggests it is well suited to survival and persistence in both the food processing environment and on the surface of fresh produce.

8 *Listeria monocytogenes* adaptation to stress

8.1 Cold shock

Cold storage is a common practice for short-term fresh produce transportation and preservation. At temperatures below the optimal temperature (30–37°C), *L. monocytogenes* suffers from reduced membrane fluidity, misfolding proteins, stalled gene expression, and reduced enzymatic activity, all of which prevent the bacterium from surviving under cold shock. Under cold stress, the gene expression of *L. monocytogenes* is altered to overcome these conditions and adapt to the environment (Soni et al., 2011; Wiedmann, 2016; Wiktorczyk-Kapischke et al., 2021). Shortly after cold stress, exposure genes that encode cold shock proteins (e.g. *cspA*, *cspB*, and *cspD*) are induced. They can revive the secondary structure of RNA which in turn promotes the resumption of gene expression. Mutations of *cspA* and *cspD* significantly compromise the cold stress response of *L. monocytogenes* (Schmid et al., 2009). After prolonged cold stress exposure, cold acclimation proteins are also induced to facilitate the survival of the pathogen. Both cold shock proteins and cold acclimation proteins play critical roles in the cold stress response mechanisms of *L.*

monocytogenes (Wiktorczyk-Kapischke et al., 2021). It has been demonstrated that the transcriptional level of the genes encoding RNA helicases was higher at 4°C than that at 37°C (Chan et al., 2007). The RNA helicases share similar functions to cold shock proteins in that they facilitate translation activities. Other mechanisms involved in the cold stress response of *L. monocytogenes* include modification of lipid membrane fluidity, upregulation of genes that encode ABC transporters, and oligopeptide permease-mediated peptide transport (Borezee et al., 2000; Hingston et al., 2017; Wiedmann, 2016).

8.2 Chlorine

Sodium hypochlorite is a commonly used chlorine-based sanitizer used in the fresh produce industry to disinfect food-contact surfaces and in washing solutions to prevent cross-contamination. However, *L. monocytogenes* can develop stress adaption to chlorine-based sanitizers at sublethal concentrations. Highmore et al. (2018) found that chlorine-based sanitizers can induce a viable-but-nonculturable (VBNC) state in *L. monocytogenes*. The authors found that a 2 min treatment of 50 ppm chlorinated water completely changed the state of *L. monocytogenes* to VBNC in spinach. Despite being non-culturable, the pathogenicity of VBNC *L. monocytogenes* remained. This was observed in the reduced life span of *Caenorhabditis elegans* after ingestion of VBNC *L. monocytogenes*. Overall, the mechanisms that *L. monocytogenes* adopt to adapt to chlorine stress are not fully understood and further research is needed.

8.3 Peracetic acid

As noted earlier, peracetic acid (PAA) is an alternative to chlorine-based sanitizer in the fresh produce industry. Even though PAA is a good choice to reduce pathogen loads on fresh produce, PAA was also reported to induce high sub-lethal injury to *L. monocytogenes* cells (Siderakou et al., 2021). Injured cells could potentially cause safety concerns as they can regain their pathogenicity. During adaption to PAA stress, *L. monocytogenes* is capable of developing altered virulence and stress resistance (Siderakou et al., 2022). Siderakou et al. (2022) investigated the stress response of *L. monocytogenes* to PAA in terms of attachment and virulence, as well as transcriptomic assessment. The injured state of *L. monocytogenes* did not significantly affect its attachment to stainless steel surfaces. Injured *L. monocytogenes* is still capable of invading the host cell, but the virulence is strain dependent. At the transcriptomic level, to adapt to PAA stress, high expression levels of virulence genes (*inlA* and *plcA*), flagellum motility genes, and attachment-related genes (*motB* and *flaA*) were observed.

8.4 Lactic acid

Lactic acid is an organic acid often used in the fresh produce industry. To investigate the acid tolerance response of *L. monocytogenes*, RNA-sequencing of two strains, 6179 and R479a, exposed to 1% (v/v) lactic acid for 30 min was performed to evaluate differentially expressed (DE) genes (Cortes et al., 2020). The two strains showed similar stress responses to lactic acid exposure with 1742 shared DE genes. The virulence genes, including *inlA*, *inlB*, *prfA*, and *hly* were significantly upregulated in both strains. It was also demonstrated that the upregulated genes encoding glutamate decarboxylase, antiporters, and arginine deiminase also participated in the stress response to lactic acid.

8.5 High-pressure processing

High-pressure processing (HPP) is a novel non-thermal technology in the food industry to reduce pathogens such as *L. monocytogenes*. Genomic microarray analysis was used to elucidate the mechanisms of the stress response of *L. monocytogenes* treated with HPP at 400 and 600 MPa for 5 min (Bowman et al., n.d.). The results suggested that genes related to DNA repair, protein secretion, flagella synthesis, cell division, and cell wall component synthesis were upregulated. On the other hand, HPP suppressed the expression of genes associated with catabolism and virulence, especially the SigB and PrfA regulons. However, another study on the transcriptome response of HPP showed the opposite regulation for some genes (Duru et al., 2021). For example, genes in the SigB regulon were induced by HPP and cell-division-related genes were downregulated. The reason for these contradictory results is unclear and future studies could further investigate the differential gene regulation of *L. monocytogenes* during HPP treatment.

9 Biocontrol of *Listeria* monocytogenes

Novel technologies to assist in mitigating the risk of fresh produce contamination and resulting listeriosis outbreaks are constantly under development. Recent research has focused on *Listeria* biocontrol agents, such as bacteriophages and bacteriocin-producing cultures for application to crops, food packaging, or ready-to-eat foods including fresh produce. Bacteriophages (phages) are viruses that infect bacterial cells. Over 500 different *Listeria* phages have been discovered and isolated from various sources including plants, sewage, and food processing environments (Stone et al., 2020). Most *Listeria* phages are temperate, meaning they have both lytic and lysogenic cycles. In the lysogenic cycle, the phages integrate into the *L. monocytogenes* genome; in the lytic cycle, they replicate inside the host cell and kill the host. Lytic phages of *L. monocytogenes* can be used as natural antimicrobials against the pathogen in foods or in the food-processing environment. Two *Listeria* phage products were introduced in 2006 and are currently commercially available: Listex P100 and ListShield (a cocktail of six lytic phages). Both products were granted Generally Recognized As Safe (GRAS) status by the FDA and USDA in 2007 for use in foods (Hagens and Loessner, 2014). Since 2007, other phage products have also become commercially available (Polaska and Sokolowska, 2019).

Many studies have examined the efficacy of these commercial products and other phages against *L. monocytogenes* in both fresh produce and the associated processing environment. One study examined phages P100 and A511 for biocontrol of *L. monocytogenes* on sliced cabbage and lettuce leaves and determined that both were effective at reducing pathogen populations on the two product types by approximately 2-log CFU/g after 1 day of storage at 6°C (Guenther et al., 2009). Another study assessed the biocontrol of certain strains of *L. monocytogenes* on spinach using P61 and determined that the phage was effective at reducing the pathogen by approximately 2-log CFU at 8°C and 12°C and 3-log CFU at 25°C over 5 days (Stone et al., 2020). Since *L. monocytogenes* biofilms are of concern in the produce processing industry, studies have also examined phage biocontrol of *L. monocytogenes* biofilms on food contact surface types, such as stainless steel and polypropylene. For example, P100 was effective at reducing *L. monocytogenes* by 3.5–5.4-log CFU/cm^2 on stainless steel (Soni and Nannapaneni, 2010).

In addition to phages, biocontrol of *L. monocytogenes* can also be conducted with bacteriocin-producing cultures, like lactic acid bacteria (LAB). Bacteriocins are synthesized polypeptides that have bactericidal or bacteriostatic activities. LAB, including *Lactococcus* and *Lactobacillus*, are profuse producers of bacteriocins targeting *L. monocytogenes*. LAB have garnered attention for their use in the dairy industry to reduce or eliminate levels of *L. monocytogenes* in milk, yogurt, and cheese products (Falardeau et al., 2021; Silva et al., 2018). In addition to cheese, LAB has also been examined for bacteriocin production in the biocontrol of *L. monocytogenes* on fresh produce. Most notably, LAB efficacy against *L. monocytogenes* was reported in a study examining both pre-harvest and post-harvest cantaloupe (Yin et al., 2020). For the pre-harvest portion, *L. innocua*, a surrogate for *L. monocytogenes*, was spray inoculated onto field cantaloupes, followed by spray treatment with LAB; *L. monocytogenes* was reduced by 2.1-log CFU after 5 days. For post-harvest experiments, *L. monocytogenes* was reduced on cantaloupe by more than 2-log CFU after 7 days at 25°C.

10 Novel processing technologies to control *Listeria monocytogenes*

10.1 Power ultrasound

Ultrasound is a non-thermal, eco-friendly, and cost-effective technology (Kentish and Ashokkumar, 2011). Ultrasound can be classified into two groups based on its frequency: high-frequency ultrasound and power ultrasound. High-frequency ultrasound is an ultrasound with a frequency range between 2 and 20 MHz, which has been widely used in medical imaging (Zhou et al., 2009). Power ultrasound uses lower frequencies between 20 and 100 kHz but has higher energy. It has been successfully applied to the medical field to clean medical equipment (FDA, 2022a; Kovach, 2019).

Power ultrasound is a form of wave and, when it passes through a medium, expansion and compression of the medium will generate mechanical energy. Due to the expansion and compression, gases dissolved in the liquid medium quickly participate in bubble formation, growth, and collapse. This whole process is known as cavitation which is the main mechanism of the inactivation of bacteria because of the extremely high temperature, high pressure, and high energy generated during cavitation (Gallo et al., 2018; Knorr et al., 2004). Moreover, free radicals are also produced during this treatment. As a result, power ultrasound can inactivate *L. monocytogenes* by destroying membrane structures and DNA (Birmpa et al., 2013; Chen et al., 2019; São José et al., 2014). The shear force generated by ultrasound also helps the removal of pathogenic bacteria tightly attached to the product surface (Huang et al., 2018; São José et al., 2014; Zhang et al., 2020). Taken together, power ultrasound technology can contribute to the reduction of *L. monocytogenes* in fresh produce.

The potential of power ultrasound to reduce pathogen load on fresh produce has been documented by several studies on a laboratory scale. It was demonstrated that sonication only can cause an approximate 1-log reduction in *L. monocytogenes* (Alexandre et al., 2013; Elizaquível et al., 2012; Huang et al., 2006; Seymour et al., 2002; Zhang et al., 2020; Zhou et al., 2009). Therefore, to achieve higher pathogen reduction, power ultrasound is usually used in combination with other technologies or sanitizers (Ding et al., 2015; Özcan and Demirel, 2016; Rafeeq and Ovissipour, 2021; Sun et al., 2022). Table 4 summarizes the efficiency of ultrasound treatment to reduce *L. monocytogenes* on different fresh produce. Despite many projects on ultrasound as a novel technology to control pathogens in fresh produce, large-scale experiments are still lacking to demonstrate the use of this technology and its possible integration into commercialized washing lines in the fresh produce industry (Bilek and Turantaş, 2013).

Table 4 Recent studies using power ultrasound to reduce *Listeria monocytogenes* on fresh produce

Produce type	Treatment conditions	Log reductions (log CFU/g)	References
Lettuce	40 kHz ultrasound + 2% lactic/citric/malic acid, 5 min	2.9	Sagong et al. (2011)
Mixture of lettuce, parsley, and dill	20 kHz ultrasound + 2% cinnamon essential oil, 5 min	0.89	Özcan and Demirel (2016)
Lettuce Strawberry	37 kHz ultrasound, 45 min	5.72 6.12	Birmpa et al. (2013)
Spinach	Ultrasound + nanobubbles, 20 min Ultrasound alone, 20 min	>2 0.5	Rafeeq and Ovissipour (2021)

10.2 Chlorine dioxide gas

Chlorine dioxide gas is an alternative disinfectant to traditional chlorine-based sanitizers, such as sodium hypochlorite. Unlike traditional chlorine-based sanitizers, pH is not a significant variable that affects the antimicrobial efficiency of chlorine dioxide gas. Chlorine dioxide gas also shows a better potential to inactivate pathogens, including spore-forming bacteria, and higher oxidative power compared to chlorinated water (Bhagat et al., 2010; Reina et al., 1995). Bhagat et al. (2010) studied the bactericidal power of chlorine dioxide gas on the tomato surface. A 12-minute chlorine dioxide gas treatment at 0.1, 0.3, and 0.5 mg/L reduced *L. monocytogenes* on tomato surfaces by over 5-log CFU. The treatment also delayed the growth of natural microflora on tomatoes, delaying decay and increasing shelf-life. The appearance and nutritional value of tomatoes also did not change significantly in the 28 days after treatment. Gaseous chlorine dioxide can also significantly reduce *L. monocytogenes* by 3.60-, 5.88-, 1.53-, and 5.5-log CFU on fresh-cut cabbage, carrot, lettuce, and whole apples, respectively (Du et al., 2002; Sy et al., 2005). Han et al. (2004) reported that continuous gaseous chlorine dioxide was more effective than batch treatment to reduce the *L. monocytogenes* load on strawberry surfaces. These studies suggest that chlorine dioxide gas is an effective agent to control *L. monocytogenes* in fresh produce.

10.3 Other novel technologies and disinfecting agents

Despite the fact that traditional thermal processing can achieve a 5-log reduction of pathogens, it also can destroy the appearance of food and its nutritional value, making such techniques unsuitable for treating fresh produce. Novel technologies used to reduce *L. monocytogenes* in fresh produce are exclusively non-thermal processing methods, including cold plasma, ionizing irradiation, ultraviolet light, and HPP (Fan and Wang, 2022). The effectiveness of these technologies to inactivate *L. monocytogenes* has been demonstrated by various studies (Adhikari et al., 2015; Bari et al., 2005; Critzer et al., 2007; Kim et al., 2006, 2013; Min et al., 2016; Mintier and Foley, 2006; Montgomery and Banerjee, 2015; Todoriki et al., 2009; Wu et al., 2012; Zhang et al., 2013; Ziuzina et al., 2015). However, all of these studies were conducted on a laboratory scale. There are some limitations to these novel technologies in terms of commercialization. In the case of ionizing irradiation, safety concerns about irradiated food, legal approval, and customer acceptance are among the biggest challenges. HPP is also not suitable for porous fresh produce, and it will damage the tissues. In addition, the high cost of this technology is also one limitation to prevent commercialization (Fan and Wang, 2020). For ultraviolet light, because of its limited permeability, uniformity of the ultraviolet light on fresh produce surface needs to be improved before commercialization (Esua et al., 2020).

Because of the corrosion to food-processing equipment and the potentially carcinogenic by-products of traditional chlorine-based sanitizers, and, more importantly, the ineffectiveness of chlorinated water to control pathogens on fresh produce, novel disinfectant agents have been studied as alternatives. Kang and Song (2019) investigated the antibacterial activity of noni fruit extract against *L. monocytogenes*; they observed 1.47-3.38-log CFU reductions on romaine lettuce, spinach, and kale with 0.5% noni fruit extract. Essential oils, such as lemon and cinnamon, have been suggested to have the ability to reduce *L. monocytogenes* by 2-4 log CFU on fresh produce surfaces (Özcan and Demirel, 2016). Organic acids, such as lactic, malic, and citric acid, also attracted attention because of their antimicrobial and GRAS characteristics (Akbas and Ölmez, 2007; Park et al., 2011; Raybaudi-Massilia et al., 2009). The effectiveness of organic acids as antimicrobial agents to treat fresh produce has been determined by several studies; however, their impact on sensory properties remains an issue (Akbas and Ölmez, 2007; Park et al., 2011; Raybaudi-Massilia et al., 2009).

11 Where to look for further information

- Pre- and postharvest preventive measures and intervention strategies to control microbial food safety hazards of fresh leafy vegetables, by Gil et al. (2015).
- Sources and survival of *L. monocytogenes* on fresh, leafy produce, by Smith et al. (2018).
- The contribution of transcriptomic and proteomic analysis in elucidating stress adaptation responses of *L. monocytogenes*, by Soni et al. (2011).

12 Acknowledgements

We would particularly like to acknowledge Dr. Joelle K. Salazar, U.S. Food and Drug Administration, Bedford Park, IL, USA, for her valuable suggestions in organizing this chapter as well as her constructive comments which greatly improved the manuscript.

13 References

Adhikari, A., Syamaladevi, R. M., Killinger, K. and Sablani, S. S. (2015). Ultraviolet-C light inactivation of Escherichia coli O157:H7 and Listeria monocytogenes on organic fruit surfaces. *International Journal of Food Microbiology*. Elsevier 210: 136–142. DOI: 10.1016/J.IJFOODMICRO.2015.06.018.

Akbas, M. Y. and Ölmez, H. (2007). Inactivation of Escherichia coli and Listeria monocytogenes on iceberg lettuce by dip wash treatments with organic acids. *Letters in Applied Microbiology*. John Wiley & Sons, Ltd 44(6): 619–624. DOI: 10.1111/j.1472-765X.2007.02127.x.

Alexandre, E. M. C., Brandão, T. R. S. and Silva, C. L. M. (2013). Impact of non-thermal technologies and sanitizer solutions on microbial load reduction and quality factor retention of frozen red bell peppers. *Innovative Food Science and Emerging Technologies*. Elsevier 17: 99–105. DOI: 10.1016/J.IFSET.2012.11.009.

Allerberger, F. and Wagner, M. (2010). Listeriosis: A resurgent foodborne infection. *Clinical Microbiology and Infection*. Blackwell Publishing Ltd 16(1): 16–23. DOI: 10.1111/j.1469-0691.2009.03109.x.

Andrade, N. J. de, Ramos, A. M., Ramos, A. M., Vanetti, M. C. D., Stringheta, P. C. and Chaves, J. B. P. (2014). Decontamination by ultrasound application in fresh fruits and vegetables. *Food Control*. Elsevier Ltd 45: 36–50. DOI: 10.1016/j.foodcont.2014.04.015.

Angelo, K. M., Conrad, A. R., Saupe, A., Dragoo, H., West, N., Sorenson, A., Barnes, A., Doyle, M., Beal, J., Jackson, K. A., Stroika, S., Tarr, C., Kucerova, Z., Lance, S., Gould, L. H., Wise, M. and Jackson, B. R. (2017). Multistate outbreak of Listeria monocytogenes infections linked to whole apples used in commercially produced, prepackaged caramel apples: United States, 2014–2015. *Epidemiology and Infection* 145(5): 848–856. DOI: 10.1017/S0950268816003083.

Aureli, P., Fiorucci, G. C., Caroli, D., Marchiaro, G., Novara, O., Leone, L. and Salmaso, S. (2000). An outbreak of febrile gastroenteritis associated with corn contaminated by Listeria monocytogenes. *The New England Journal of Medicine* 342(17): 1236–1241. DOI: 10.1056/NEJM200004273421702.

Australian Bureau of Statistics. (2022). Dietary behaviour. Available at: https://www.abs.gov.au/statistics/health/health-conditions-and-risks/dietary-behaviour/latest-release (accessed 27 June 2022).

Bae, D., Seo, K. S., Zhang, T. and Wang, C. (2013). Characterization of a potential Listeria monocytogenes virulence factor associated with attachment to fresh produce. *Applied and Environmental Microbiology* 79(22): 6855–6861. DOI: 10.1128/AEM.01006-13.

Banach, J. L., Sampers, I., van Haute, S. and van der Fels-Klerx, H. J. (2015). Effect of disinfectants on preventing the cross-contamination of pathogens in fresh produce washing water. *International Journal of Environmental Research and Public Health*. Multidisciplinary Digital Publishing Institute 12(8): 8658–8677. DOI: 10.3390/ijerph120808658.

Banach, J. L., van Bokhorst-van de Veen, H., van Overbeek, L. S., van der Zouwen, P. S., Zwietering, M. H. and van der Fels-Klerx, H. J. (2020). Effectiveness of a peracetic acid solution on Escherichia coli reduction during fresh-cut lettuce processing at the laboratory and industrial scales. *International Journal of Food Microbiology*. Elsevier 321: 108537. DOI: 10.1016/J.IJFOODMICRO.2020.108537.

Bari, M. L., Nakauma, M., Todoriki, S., Juneja, V. K., Isshiki, K. and Kawamoto, S. (2005). Effectiveness of irradiation treatments in inactivating Listeria monocytogenes on fresh vegetables at refrigeration temperature. *Journal of Food Protection*. Allen Press 68(2): 318–323. DOI: 10.4315/0362-028X-68.2.318.

Batz, M. B., Richardson, L. C., Bazaco, M. C., Parker, C. C., Chirtel, S. J., Cole, D., Golden, N. J., Griffin, P. M., Gu, W., Schmitt, S. K., Wolpert, B. J., Kufel, J. S. Z. and Hoekstra, R. M. (2021). Recency-weighted statistical modeling approach to attribute illnesses caused by 4 pathogens to food sources using outbreak data, United States. *Emerging Infectious Diseases*. Centers for Disease Control and Prevention (CDC) 27(1): 214–222. DOI: 10.3201/EID2701.203832.

Beltrame, C. A., Martelo, E. B., Mesquita, R. A., Barbosa, J., Steffens, C., Toniazzo, G., Valduga, E. and Cansian, R. L. (2015). Adhesion of Listeria monocytogenes to cutting board surfaces and removal by different sanitizers. *Journal für Verbraucherschutz und Lebensmittelsicherheit* 10(1): 41–47.

Beuchat, L. R. (1996). Pathogenic microorganisms associated with fresh produce. *Journal of Food Protection* 59(2): 204–216. DOI: 10.4315/0362-028X-59.2.204.

Beuchat, L. R. (2006). Vectors and conditions for preharvest contamination of fruits and vegetables with pathogens capable of causing enteric diseases. *British Food Journal*. Emerald Group Publishing Limited 108(1): 38–53. DOI: 10.1108/00070700610637625.

Bhagat, A., Mahmoud, B. S. M. and Linton, R. H. (2010). Inactivation of Salmonella enterica and Listeria monocytogenes inoculated on hydroponic tomatoes using chlorine dioxide gas. *Foodborne Pathogens and Disease* 7(6): 677–685. DOI: 10.1089/FPD.2009.0466. Available at: https://home.liebertpub.com/fpd.

Bilek, S. E. and Turantaş, F. (2013). Decontamination efficiency of high power ultrasound in the fruit and vegetable industry, a review. *International Journal of Food Microbiology*. Elsevier 166(1): 155–162. DOI: 10.1016/J.IJFOODMICRO.2013.06.028.

Birmpa, A., Sfika, V. and Vantarakis, A. (2013). Ultraviolet light and ultrasound as non-thermal treatments for the inactivation of microorganisms in fresh ready-to-eat foods. *International Journal of Food Microbiology* 167(1): 96–102. DOI: 10.1016/j.ijfoodmicro.2013.06.005.

Borezee, E., Pellegrini, E. and Berche, P. (2000). OppA of Listeria monocytogenes, an oligopeptide-binding protein required for bacterial growth at low temperature and involved in intracellular survival. *Infection and Immunity* 68(12): 7069–7077. DOI: 10.1128/IAI.68.12.7069-7077.2000.

Bowman, J. P. (n.d.). Differential gene expression of Listeria monocytogenes during high hydrostatic pressure processing, Bittencourt CR and Correspondence TR. DOI: 10.1099/mic.0.2007/010314-0.

Buchanan, R. L., Gorris, L. G. M., Hayman, M. M., Jackson, T. C. and Whiting, R. C. (2017). A review of Listeria monocytogenes: An update on outbreaks, virulence, dose-response, ecology, and risk assessments. *Food Control*. Elsevier Ltd 75: 1–13. DOI: 10.1016/j.foodcont.2016.12.016.

Carlin, C. R., Liao, J., Weller, D. L., Guo, X., Orsi, R. and Wiedmann, M. (2021). *Listeria cossartiae* sp. nov., *Listeria immobilis* sp. nov., *Listeria portnoyi* sp. nov. and *Listeria rustica* sp. nov., isolated from agricultural water and natural environments. *International Journal of Systematic and Evolutionary Microbiology* 71(5). Microbiology Society. DOI: 10.1099/IJSEM.0.004795.

CDC. (2012). Multistate outbreak of Listeriosis linked to whole cantaloupes from Jensen farms. Available at: https://www.cdc.gov/listeria/outbreaks/cantaloupes-jensen-farms/index.html (accessed 17 February 2022).

CDC. (2015a). Multistate outbreak of listeriosis linked to commercially produced, prepackaged caramel apples made from Bidart bros. Apples. Available at: https://www.cdc.gov/listeria/outbreaks/caramel-apples-12-14/index.html (accessed 19 February 2022).

CDC. (2015b). Wholesome Soy Products, Inc. sprouts and investigation of human listeriosis cases (final update). Available at: https://www.cdc.gov/listeria/outbreaks/bean-sprouts-11-14/index.html.

CDC. (2016). Multistate outbreak of listeriosis linked to packaged salads produced at Springfield, Ohio Dole processing facility (final update). Available at: https://www.cdc.gov/listeria/outbreaks/bagged-salads-01-16/index.html.

CDC. (2021a). Foodborne illness source attribution estimates for Salmonella, Escherichia coli O157, Listeria monocytogenes, and campylobacter using multi-year outbreak surveillance data. Available at: https://www.cdc.gov/foodsafety/ifsac/annual-reports.html (accessed 17 February 2022).

CDC. (2021b). Listeria (listeriosis). Available at: https://www.cdc.gov/listeria/index.html (accessed 17 February 2022).

CDC. (2021c). Listeria outbreak linked to packaged salads produced by Dole. Available at: https://www.cdc.gov/listeria/outbreaks/packaged-salad-mix-12-21/index.html.

CDC. (2021d). Listeria outbreak linked to packaged salads Produced by Fresh Express. Available at: https://www.cdc.gov/listeria/outbreaks/packaged-salad-12-21-b/index.html (accessed 19 February 2022).

CDC. (2022a). FoodNet fast. Available at: https://wwwn.cdc.gov/FoodNetFast/ (accessed 17 February 2022).

CDC. (2022b). National outbreak reporting system (NORS). Available at: https://wwwn.cdc.gov/norsdashboard/ (accessed 19 January 2022).

Centers for Disease Control and Prevention. (2011). Multistate outbreak of listeriosis associated with Jensen farms cantaloupe – United States, August–September 2011. *MMWR: Morbidity and Mortality Weekly Report* 60(39): 1357–1358. Available at: https://www.cdc.gov/mmwr/preview/mmwrhtml/mm6039a5.htm (accessed 17 February 2022).

Cernicchiaro, N., Pearl, D. L., McEwen, S. A., Harpster, L., Homan, H. J., Linz, G. M. and Lejeune, J. T. (2012). Association of wild bird density and farm management factors with the prevalence of *E. coli* O157 in dairy herds in Ohio (2007–2009). *Zoonoses and Public Health* 59(5): 320–329. DOI: 10.1111/j.1863-2378.2012.01457.x.

CFR. (n.d.). Code of federal regulations title 21. Available at: https://www.accessdata.fda.gov/scripts/cdrh/cfdocs/cfcfr/CFRSearch.cfm?fr=173.315 (accessed 17 February 2022).

Chan, Y. C., Raengpradub, S., Boor, K. J. and Wiedmann, M. (2007). Microarray-based characterization of the Listeria monocytogenes cold regulon in log- and stationary-phase cells. *Applied and Environmental Microbiology*. American Society for Microbiology 73(20): 6484–6498. DOI: 10.1128/AEM.00897-07.

Chen, F. (2019). Application of ultrasound technology in processing of ready-to-eat fresh food: A review. *Ultrasonics Sonochemistry*. Elsevier 63: 104953. DOI: 10.1016/j.ultsonch.2019.104953.

Chen, Y., Burall, L. S., Luo, Y., Timme, R., Melka, D., Muruvanda, T., Payne, J., Wang, C., Kastanis, G., Maounounen-Laasri, A., De Jesus, A. J., Curry, P. E., Stones, R., K'Aluoch, O., Liu, E., Salter, M., Hammack, T. S., Evans, P. S., Parish, M., Allard, M. W., Datta, A., Strain, E. A. and Brown, E. W. (2016a). Listeria monocytogenes in stone fruits linked to a multistate outbreak: Enumeration of cells and whole-genome sequencing. *Applied and Environmental Microbiology* 82(24): 7030–7040. DOI: 10.1128/AEM.01486-16.

Chen, Y., Burall, L. S., Luo, Y., Timme, R., Melka, D., Muruvanda, T., Payne, J., Wang, C., Kastanis, G., Maounounen-Laasri, A., De Jesus, A. J., Curry, P. E., Stones, R., K'Aluoch, O., Liu, E., Salter, M., Hammack, T. S., Evans, P. S., Parish, M., Allard, M. W., Datta, A., Strain, E. A. and Brown, E. W. (2016b). Listeria monocytogenes in stone fruits linked to a multistate outbreak: Enumeration of cells and whole-genome sequencing. *Applied and Environmental Microbiology*. American Society for Microbiology 82(24): 7030–7040. DOI: 10.1128/AEM.01486-16.

Cooley, M. B., Quinones, B., Oryang, D., Mandrell, R. E. and Gorski, L. (2014). Prevalence of Shiga toxin producing *Escherichia coli*, *Salmonella enterica*, and Listeria monocytogenes at public access watershed sites in a California Central Coast agricultural region. *Frontiers in Cellular and Infection Microbiology* 4(30): 30.

Cortes, B. W., Naditz, A. L., Anast, J. M. and Schmitz-Esser, S. (2020). Transcriptome sequencing of Listeria monocytogenes reveals major gene expression changes in response to lactic acid stress exposure but a less pronounced response to oxidative stress. *Frontiers in Microbiology* 10: 3110. DOI: 10.3389/FMICB.2019.03110/BIBTEX.

Cossart, P. and Toledo-Arana, A. (2008). Listeria monocytogenes, a unique model in infection biology: An overview. *Microbes and Infection* 10(9): 1041–1050. DOI: 10.1016/j.micinf.2008.07.043.

Critzer, F. J., Kelly-Wintenberg, K., South, S. L. and Golden, D. A. (2007). Atmospheric plasma inactivation of foodborne pathogens on fresh produce surfaces. *Journal of Food Protection* 70(10): 2290–2296. Available at: http://meridian.allenpress.com/jfp/article-pdf/70/10/2290/1677173/0362-028x-70_10_2290.pdf (accessed 15 February 2022).

Danyluk, M. D., Friedrich, L. M. and Schaffner, D. W. (2014). Modeling the growth of Listeria monocytogenes on cut cantaloupe, honeydew and watermelon. *Food Microbiology* 38: 52–55. DOI: 10.1016/J.FM.2013.08.001.

Das, A. (2019). Listeriosis in Australia - January to July 2018. *Global Biosecurity*. University of New South Wales 1(1): 150. DOI: 10.31646/GBIO.9.

den Bakker, H. C., Desjardins, C. A., Griggs, A. D., Peters, J. E., Zeng, Q., Young, S. K., Kodira, C. D., Yandava, C., Hepburn, T. A., Haas, B. J., Birren, B. W. and Wiedmann, M. (2013). Evolutionary dynamics of the accessory genome of Listeria monocytogenes. *PLoS ONE* 8(6): e67511. DOI: 10.1371/journal.pone.0067511.

Deng, X., Phillippy, A. M., Li, Z., Salzberg, S. L. and Zhang, W. (2010). Probing the pan-genome of Listeria monocytogenes: New insights into intraspecific niche expansion and genomic diversification. *BMC Genomics* 11(1): 500. DOI: 10.1186/1471-2164-11-500.

Ding, T., Ge, Z., Shi, J., Xu, Y., Jones, C. L. and Liu, D. (2015). Impact of slightly acidic electrolyzed water (SAEW) and ultrasound on microbial loads and quality of fresh fruits. *LWT - Food Science and Technology*. Academic Press 60(2): 1195–1199. DOI: 10.1016/J.LWT.2014.09.012.

Du, J., Han, Y. and Linton, R. H. (2002). Inactivation by chlorine dioxide gas (ClO_2) of Listeria monocytogenes spotted onto different apple surfaces. *Food Microbiology*. Academic Press 19(5): 481–490. DOI: 10.1006/FMIC.2002.0501.

Dunn, L. L., Friedrich, L. M., Strawn, L. K. and Danyluk, M. D. (2022). Prevalence of Listeria monocytogenes and indicator microorganisms in Florida cantaloupe packinghouses, 2013–2014. *Food Microbiology* 104: 103970.

Duru, I. C., Bucur, F. I., Andreevskaya, M., Nikparvar, B., Ylinen, A., Grigore-Gurgu, L., Rode, T. M., Crauwels, P., Laine, P., Paulin, L. and Løvdal, T. (2021). High-pressure processing-induced transcriptome response during recovery of Listeria monocytogenes. *BMC Genomics*. BioMed Central Ltd 22(1): 1–20. DOI: 10.1186/S12864-021-07407-6/TABLES/1.

EFSA and ECDC. (2016). The European Union summary report on trends and sources of zoonoses, zoonotic agents and food-borne outbreaks in 2015. *EFSA Journal* 14(12). DOI: 10.2903/j.efsa.2016.4634.

EFSA and ECDC. (2018). The European Union summary report on trends and sources of zoonoses, zoonotic agents and food-borne outbreaks in 2017. *EFSA Journal: European Food Safety Authority*. Wiley-Blackwell Publishing Ltd 16(12): e05500. DOI: 10.2903/j.efsa.2018.5500.

EFSA and ECDC. (2019). The European Union one health 2018 zoonoses report. *EFSA Journal: European Food Safety Authority*. Wiley-Blackwell Publishing Ltd 17(12): e05926. DOI: 10.2903/j.efsa.2019.5926.

EFSA and ECDC. (2021a). The European Union one health 2019 zoonoses report. *EFSA Journal: European Food Safety Authority*. Blackwell Publishing Ltd 19(2): e06406. DOI: 10.2903/j.efsa.2021.6406.

EFSA and ECDC. (2021b). The European Union one health 2020 zoonoses report. *EFSA Journal: European Food Safety Authority*. John Wiley and Sons Inc 19(12): e06971. DOI: 10.2903/j.efsa.2021.6971.

Elizaquível, P., Sánchez, G., Selma, M. V. and Aznar, R. (2012). Application of propidium monoazide-qPCR to evaluate the ultrasonic inactivation of Escherichia coli O157:H7 in fresh-cut vegetable wash water. *Food Microbiology*. Elsevier Ltd 30(1): 316–320. DOI: 10.1016/j.fm.2011.10.008.

Esua, O. J., Chin, N. L., Yusof, Y. A. and Sukor, R. (2020). A review on individual and combination technologies of UV-C radiation and ultrasound in postharvest handling of fruits and vegetables. *Processes*. Multidisciplinary Digital Publishing Institute 8(11): 1433. DOI: 10.3390/PR8111433.

European Food Safety Authority and European Centre for Disease Prevention and Control (2017). The European Union summary report on trends and sources of zoonoses, zoonotic agents and food-borne outbreaks in 2016. *EFSA Journal: European Food Safety Authority* 15(12): e05077. DOI: 10.2903/j.efsa.2017.5077.

Falardeau, J., Johnson, R. P., Pagotto, F. and Wang, S. (2017). Occurrence, characterization, and potential predictors of verotoxigenic Escherichia coli, Listeria monocytogenes, and Salmonella in surface water used for produce irrigation in the Lower Mainland of British Columbia, Canada. *PLoS ONE* 12(9): e0185437.

Falardeau, J., Trmcic, A. and Wang, S. (2021). The occurrence, growth, and biocontrol of Listeria monocytogenes in fresh and surface-ripened soft and semisoft cheeses. *Comprehensive Reviews in Food Science and Food Safety* 20(4): 4019–4048. DOI: 10.1111/1541-4337.12768.

Fan, X. and Wang, W. (2022). Quality of fresh and fresh-cut produce impacted by nonthermal physical technologies intended to enhance microbial safety. *Critical Reviews in Food Science and Nutrition*. Taylor & Francis 62(2): 362–382. DOI: 10.1080/10408398.2020.1816892.

Fan, Z., Xie, J., Li, Y. and Wang, H. (2019). Listeriosis in mainland China: A systematic review. *International Journal of Infectious Diseases*. Elsevier 81: 17–24. DOI: 10.1016/J.IJID.2019.01.007.

FDA. (1998). Guidance for industry guide to minimize microbial food safety hazards for fresh fruits and vegetables. Available at: https://www.fda.gov/regulatory-information/search-fda-guidance-documents/guidance-industry-guide-minimize-microbial-food-safety-hazards-fresh-fruits-and-vegetables (accessed 18 January 2022).

FDA. (2011). Environmental assessment: Factors potentially contributing to the contamination of fresh whole cantaloupe implicated in a multi-state outbreak of listeriosis. Available at: http://wayback.archive-it.org/7993/20171114022308/https:/www.fda.gov/Food/RecallsOutbreaksEmergencies/Outbreaks/ucm276247.htm.

FDA. (2020). Outbreak investigation of Listeria monocytogenes: Enoki mushrooms. Available at: https://www.fda.gov/food/outbreaks-foodborne-illness/outbreak-investigation-listeria-monocytogenes-enoki-mushrooms-march-2020 (accessed 17 February 2022).

FDA. (2021). Dole fresh vegetables announces voluntary recall for salads processed at its Bessemer City, NC and Yuma, AZ facilities due to possible health risk from Listeria monocytogenes. Available at: https://www.fda.gov/safety/recalls-market-withdrawals-safety-alerts/dole-fresh-vegetables-announces-voluntary-recall-salads-processed-its-bessemer-city-nc-and-yuma-az.

FDA. (2022a). CFR 880.6150 – Ultrasonic cleaner for medical instruments (Vol. 21). Available at: https://www.ecfr.gov/current/title-21/chapter-I/subchapter-H/part-880/subpart-G/section-880.6150 (accessed 2 April 2022).

FDA. (2022b). Selecting and serving produce safely. Available at: https://www.fda.gov/food/buy-store-serve-safe-food/selecting-and-serving-produce-safely (accessed 17 February 2022).

Fernández Guerrero, M. L., Torres, R., Mancebo, B., González-López, J. J., Górgolas, M., Jusdado, J. J. and Roblas, R. F. (2012). Antimicrobial treatment of invasive non-perinatal human listeriosis and the impact of the underlying disease on prognosis. *Clinical Microbiology and Infection*. Elsevier 18(7): 690–695. DOI: 10.1111/J.1469-0691.2011.03616.X.

Ferreira, A., O'Byrne, C. P. and Boor, K. J. (2001). Role of sigma(B) in heat, ethanol, acid, and oxidative stress resistance and during carbon starvation in Listeria monocytogenes. *Applied and Environment Microbiology* 67(10): 4454–4457. DOI: 10.1128/AEM.67.10.4454-4457.2001.

Ferreira, A., Sue, D., O'Byrne, C. P. and Boor, K. J. (2003). Role of Listeria monocytogenes sigma(B) in survival of lethal acidic conditions and in the acquired acid tolerance response. *Applied and Environment Microbiology* 69(5): 2692–2698. DOI: 10.1128/AEM.69.5.2692-2698.2003.

Food Directorate. (2010). Health products and food branch Health Canada. *Archived: Policy on Listeria Monocytogenes in Ready-to-Eat Foods*. Available at: https://www.canada.ca/en/health-canada/services/food-nutrition/public-involvement-partnerships/listeria-monocytogenes-ready-eat-foods-2010/document.html (accessed 27 June 2022).

Forney, C. F., Fan, L., Bezanson, G. S., Ells, T. C., LeBlanc, D. I. and Fillmore, S. (2018). Impact of listeria inoculation and aerated steam sanitization on volatile emissions of whole fresh cantaloupes. *Journal of Food Science* 83(4): 1017–1024. DOI: 10.1111/1750-3841.14092.

Freitag, N. E., Port, G. C. and Miner, M. D. (2009). Listeria monocytogenes - From saprophyte to intracellular pathogen. *Nature Reviews. Microbiology* 7(9): 623–628. DOI: 10.1038/NRMICRO2171.

Gadelha, J. R., Allende, A., López-Gálvez, F., Allende, A., López-Gálvez, F., Fernández, P., Gil, M. I. and Egea, J. A. (2019). Chemical risks associated with ready-to-eat vegetables: Quantitative analysis to estimate formation and/or accumulation of disinfection byproducts during washing. *EFSA Journal. European Food Safety Authority*. John Wiley & Sons 17 (Suppl. 2): e170913. DOI: 10.2903/J.EFSA.2019.E170913.

Gallo, M., Ferrara, L. and Naviglio, D. (2018). Application of ultrasound in food science and technology: A perspective. *Foods*. Multidisciplinary Digital Publishing Institute (MDPI) 7(10): 164. DOI: 10.3390/FOODS7100164.

Garner, D. and Kathariou, S. (2016). Fresh produce-associated listeriosis outbreaks, sources of concern, teachable moments, and insights. *Journal of Food Protection*. Allen Press 79(2): 337–344. DOI: 10.4315/0362-028X.JFP-15-387.

Gasanov, U., Hughes, D. and Hansbro, P. M. (2005). Methods for the isolation and identification of Listeria spp. and Listeria monocytogenes: A review. *FEMS Microbiology Reviews*. Oxford Academic 29(5): 851–875. DOI: 10.1016/J.FEMSRE.2004.12.002.

Gaul, L. K., Farag, N. H., Shim, T., Kingsley, M. A., Silk, B. J. and Hyytia-Trees, E. (2013). Hospital-acquired listeriosis outbreak caused by contaminated diced celery–Texas, 2010. *Clinical Infectious Diseases*. Oxford Academic 56(1): 20–26. DOI: 10.1093/CID/CIS817.

Gehringer, R. A. and Kaletunç, G. (2017). Chlorinated water sanitation of leafy green vegetables for fresh produce processors. Available at: https://ohioline.osu.edu/factsheet/aex-262 (accessed 17 February 2022).

Gil, M. I., Selma, M., López-Gálvez, F. and Allende, A. (2009). Fresh-cut product sanitation and wash water disinfection: Problems and solutions. *International Journal of Food Microbiology*. Elsevier 134(1–2): 37–45. DOI: 10.1016/J.IJFOODMICRO.2009.05.021.

Gil, M. I., Selma, M. V., Suslow, T., Jacxsens, L., Uyttendaele, M. and Allende, A. (2015). Pre- and postharvest preventive measures and intervention strategies to control microbial food safety hazards of fresh leafy vegetables. *Critical Reviews in Food Science and Nutrition*. Taylor and Francis Inc. 55(4): 453–468. DOI: 10.1080/10408398.2012.657808.

Glaser, P., Frangeul, L., Buchrieser, C., Rusniok, C., Amend, A., Baquero, F., Berche, P., Bloecker, H., Brandt, P., Chakraborty, T., Charbit, A., Chetouani, F., Couvé, E., de Daruvar, A., Dehoux, P., Domann, E., Domínguez-Bernal, G., Duchaud, E., Durant, L., Dussurget, O., Entian, K. D., Fsihi, H., García-del Portillo, F., Garrido, P., Gautier, L.,

Goebel, W., Gómez-López, N., Hain, T., Hauf, J., Jackson, D., Jones, L. M., Kaerst, U., Kreft, J., Kuhn, M., Kunst, F., Kurapkat, G., Madueno, E., Maitournam, A., Vicente, J. M., Ng, E., Nedjari, H., Nordsiek, G., Novella, S., de Pablos, B., Pérez-Diaz, J. C., Purcell, R., Remmel, B., Rose, M., Schlueter, T., Simoes, N., Tierrez, A., Vázquez-Boland, J. A., Voss, H., Wehland, J. and Cossart, P. (2001). Comparative genomics of Listeria species. *Science* 294(5543): 849–852. DOI: 10.1126/science.1063447.

Gorski, L., Duhe, J. M. and Flaherty, D. (2009). The use of flagella and motility for plant colonization and fitness by different strains of the foodborne pathogen Listeria monocytogenes. *PLoS ONE* 4(4): e5142.

Gorski, L., Duhe, J. M. and Flaherty, D. (2011). The Sigma B operon is a determinant of fitness for a Listeria monocytogenes serotype 4b strain in soil. *Foodborne Pathogens and Disease* 8(6): 699–704.

Goulet, V., King, L. A., Vaillant, V. and de Valk, H. (2013). What is the incubation period for listeriosis? *BMC Infectious Diseases*. BioMed Central 13(1): 1–7. DOI: 10.1186/1471-2334-13-11/TABLES/3.

Gray, M. J., Wholey, W. Y. and Jakob, U. (2013). Bacterial responses to reactive chlorine species. *Annual Review of Microbiology* 67: 141–160. DOI: 10.1146/ANNUREV-MICRO-102912-142520.

Greig, J. D., Todd, E. C., Bartleson, C. A. and Michaels, B. S. (2007). Outbreaks where food workers have been implicated in the spread of foodborne disease. Part 1. Description of the problem, methods, and agents involved. *Journal of Food Protection* 70(7): 1752–1761.

Guenther, S., Huwyler, D., Richard, S. and Loessner, M. J. (2009). Virulent bacteriophage for efficient biocontrol of Listeria monocytogenes in ready-to-eat foods. *Applied and Environment Microbiology* 75(1): 93–100. DOI: 10.1128/AEM.01711-08.

Hagens, S. and Loessner, M. J. (2014). Phages of Listeria offer novel tools for diagnostics and biocontrol. *Frontiers in Microbiology* 5: 159. DOI: 10.3389/fmicb.2014.00159.

Han, Y., Selby, T. L., Schultze, K. K., Nelson, P. E. and Linton, R. H. (2004). Decontamination of strawberries using batch and continuous chlorine dioxide gas treatments. *Journal of Food Protection* 67(11): 2450–2455. Available at: http://meridian.allenpress.com/jfp/article-pdf/67/11/2450/1674392/0362-028x-67_11_2450.pdf (accessed 15 February 2022).

Haute, S., Sampers, I., Jacxsens, L. and Uyttendaele, M. (2015). Selection criteria for water disinfection techniques in agricultural practices. *Critical Reviews in Food Science and Nutrition* 55(11): 1529–1551. DOI: 10.1080/10408398.2012.705360.

Hernandez-Milian, A. and Payeras-Cifre, A. (2014). What is new in listeriosis? *BioMed Research International*. Hindawi Publishing Limited 2014: 358051. DOI: 10.1155/2014/358051.

Highmore, C. J., Warner, J. C., Rothwell, S. D., Wilks, S. A. and Keevil, C. W. (2018). Viable-but-nonculturable Listeria monocytogenes and Salmonella enterica serovar Thompson induced by chlorine stress remain infectious. *mBio* 9(2). DOI: 10.1128/mBio.00540-18.

Hingston, P., Chen, J., Allen, K., Truelstrup Hansen, L. and Wang, S. (2017). Strand specific RNA-sequencing and membrane lipid profiling reveals growth phase-dependent cold stress response mechanisms in Listeria monocytogenes. *PLoS ONE*. Public Library of Science 12(6): e0180123. DOI: 10.1371/JOURNAL.PONE.0180123.

Hoelzer, K., Pouillot, R., van Doren, J. M. and Dennis, S. (2014). Reduction of Listeria monocytogenes contamination on produce e A quantitative analysis of common liquid fresh produce wash compounds. *Food Control* 46: 430–440. DOI: 10.1016/j.foodcont.2014.06.011.

Huang, J., Luo, Y. and Nou, X. (2015). Growth of *Salmonella enterica* and listeria monocytogenes on fresh-cut cantaloupe under different temperature abuse scenarios. *Journal of Food Protection* 78(6): 1125–1131. DOI: 10.4315/0362-028X.JFP-14-468.

Huang, K., Wrenn, S., Tikekar, R. and Nitin, N. (2018). Efficacy of decontamination and a reduced risk of cross-contamination during ultrasound-assisted washing of fresh produce. *Journal of Food Engineering*. Elsevier 224: 95–104. DOI: 10.1016/J.JFOODENG.2017.11.043.

Huang, T.-S., Xu, C., Walker, K., West, P., Zhang, S. and Weese, J. (2006). Decontamination efficacy of combined chlorine dioxide with ultrasonication on apples and lettuce. *Journal of Food Science*. John Wiley & Sons 71(4): M134–M139. DOI: 10.1111/J.1750-3841.2006.00015.X.

Jackson, B. R., Salter, M., Tarr, C., Conrad, A., Harvey, E., Steinbock, L., Saupe, A., Sorenson, A., Katz, L., Stroika, S., Jackson, K. A., Carleton, H., Kucerova, Z., Melka, D., Strain, E., Parish, M., Mody, R. K. and Centers for Disease Control and Prevention (CDC). (2015). Notes from the field: Listeriosis associated with stone fruit–United States, 2014. *MMWR: Morbidity and Mortality Weekly Report* 64(10): 282–283. Available at: https://www.ncbi.nlm.nih.gov/pubmed/25789745.

John, J., Joy, W. C. and Jovana, K. (2020). Prevalence of Listeria spp. in produce handling and processing facilities in the Pacific Northwest. *Food Microbiology* 90: 103468. DOI: 10.1016/j.fm.2020.103468.

Josewin, S. W., Kim, M. J. and Yuk, H. G. (2018). Inactivation of Listeria monocytogenes and *Salmonella* spp. on cantaloupe rinds by blue light emitting diodes (LEDs). *Food Microbiology* 76: 219–225. DOI: 10.1016/J.FM.2018.05.012.

Kaminski, C. N., Davidson, G. R. and Ryser, E. T. (2014). Listeria monocytogenes transfer during mechanical dicing of celery and growth during subsequent storage. *Journal of Food Protection* 77(5): 765–771. DOI: 10.4315/0362-028X.JFP-13-382.

Kang, J. H. and Song, K. B. (2019) Antibacterial activity of the noni fruit extract against Listeria monocytogenes and its applicability as a natural sanitizer for the washing of fresh-cut produce. *Food Microbiology*. Academic Press 84: 103260. DOI: 10.1016/J.FM.2019.103260.

Karmakar, K., Nath, U., Nataraja, K. N. and Chakravortty, D. (2018). Root mediated uptake of Salmonella is different from phyto-pathogen and associated with the colonization of edible organs. *BMC Plant Biology* 18(1): 344.

Kaur, S., Smith, D. J. and Morgan, M. T. (2015). Chloroxyanion residue quantification in cantaloupes treated with chlorine dioxide gas. *Journal of Food Protection* 78(9): 1708–1718. DOI: 10.4315/0362-028X.JFP-14-576.

Kentish, S. and Ashokkumar, M. (2011). The physical and chemical effects of ultrasound. In: Feng, H., Barbosa-Canovas, G. and Weiss, J. (Eds). *Ultrasound Technologies for Food and Bioprocessing*. Springer, pp. 1–12. DOI: 10.1007/978-1-4419-7472-3_1.

Kim, J., Lee, J., Kim, J., Seo, J., Han, S., Chung, H. and Byun, M. (2006). Effect of gamma irradiation on Listeria ivanovii inoculated to iceberg lettuce stored at cold temperature. *Food Control*. Elsevier 17(5): 397–401. DOI: 10.1016/J.FOODCONT.2005.01.008.

Kim, Y. H., Jeong, S. G., Back, K. H., Park, K. H., Chung, M. S. and Kang, D. H. (2013). Effect of various conditions on inactivation of Escherichia coli O157:H7, Salmonella Typhimurium, and Listeria monocytogenes in fresh-cut lettuce using ultraviolet radiation. *International Journal of Food Microbiology*. Elsevier 166(3): 349–355. DOI: 10.1016/J.IJFOODMICRO.2013.08.010.

Knorr, D., Zenker, M., Heinz, V. and Lee, D. (2004). Applications and potential of ultrasonics in food processing. *Trends in Food Science and Technology*. Elsevier 15(5): 261–266. DOI: 10.1016/J.TIFS.2003.12.001.

Kovach, S. M. (2019). Research: Ensuring cavitation in a medical device ultrasonic cleaner. *Biomedical Instrumentation and Technology*. Allen Press 53(4): 280–285. DOI: 10.2345/0899-8205-53.4.280.

Kulesh, R., Shinde, S., Khan, W. A., Chaudhari, S. P., Patil, A. R., Kurkure, N. V., Paliwal, N., Likhite, A. V., Zade, N. N. and Barbuddhe, S. B. (2019). The occurrence of Listeria monocytogenes in goats, farm environment and invertebrates. *Biological Rhythm Research*. Taylor & Francis: 1–10. DOI: 10.1080/09291016.2019.1660836.

Laksanalamai, P., Joseph, L. A., Silk, B. J., Burall, L. S., L Tarr, C., Gerner-Smidt, P. and Datta, A. R. (2012a). Genomic characterization of Listeria monocytogenes Strains involved in a multistate listeriosis outbreak associated with cantaloupe in US. *PLoS ONE*. Public Library of Science 7(7): e42448. DOI: 10.1371/JOURNAL.PONE.0042448.

Laksanalamai, P., Joseph, L. A., Silk, B. J., Burall, L. S., Tarr, C., Gerner-Smidt, P. and Datta, A. R. (2012b). Genomic characterization of Listeria monocytogenes Strains involved in a multistate listeriosis outbreak associated with cantaloupe in US. *PLoS ONE*. Public Library of Science 7(7): e42448. DOI: 10.1371/JOURNAL.PONE.0042448.

Lee-Kwan, S. H., Moore, L., Blanck, H. M., Harris, D. M. and Galuska, D. (2019). Disparities in state-specific adult fruit and vegetable consumption – United States, 2015. *MMWR: Morbidity and Mortality Weekly Report*. Centers for Disease Control 66(45): 1241-1247. DOI: 10.15585/MMWR.MM6645A1.

Leong, D., Alvarez-Ordonez, A. and Jordan, K. (2014). Monitoring occurrence and persistence of Listeria monocytogenes in foods and food processing environments in the Republic of Ireland. *Frontiers in Microbiology* 5: 436. DOI: 10.3389/fmicb.2014.00436.

Linke, K., Ruckerl, I., Brugger, K., Karpiskova, R., Walland, J., Muri-Klinger, S., Tichy, A., Wagner, M. and Stessl, B. (2014). Reservoirs of listeria species in three environmental ecosystems. *Applied and Environment Microbiology* 80(18): 5583–5592. DOI: 10.1128/AEM.01018-14.

Liu, D., Lawrence, M. L., Ainsworth, A. J. and Austin, F. W. (2005). Comparative assessment of acid, alkali and salt tolerance in Listeria monocytogenes virulent and avirulent strains. *FEMS Microbiology Letters*. Oxford Academic 243(2): 373-378. DOI: 10.1016/J.FEMSLE.2004.12.025.

Locatelli, A., Spor, A., Jolivet, C., Piveteau, P. and Hartmann, A. (2013). Biotic and abiotic soil properties influence survival of Listeria monocytogenes in soil. *PLoS ONE* 8(10): e75969. DOI: 10.1371/journal.pone.0075969.

Lyautey, E., Hartman, A., Pagotto, F., Tyler, K., Lapen, D. R., Wilkes, G., Piveteau, P., Rieu, A., Robertson, W. J., Medeiros, D. T., Edge, T. A., Gannon, V. and Topp, E. (2007). Characteristics and frequency of detection of fecal Listeria monocytogenes shed by livestock, wildlife, and humans. *Canadian Journal of Microbiology* 53(10): 1158–1167.

Mahmoud, B. S. M. (2012). Effects of X-ray treatments on pathogenic bacteria, inherent microflora, color, and firmness on whole cantaloupe. *International Journal of Food Microbiology* 156(3): 296–300. DOI: 10.1016/J.IJFOODMICRO.2012.04.001.

Mains, D. R., Eallonardo, S. J. and Freitag, N. E. (2021). Identification of Listeria monocytogenes Genes contributing to oxidative stress resistance under conditions relevant to host infection. *Infection and Immunity* 89(4): e00700-20. DOI: 10.1128/IAI.00700-20.

Marinho, C. M., Garmyn, D., Ga, L., Brunhede, M. Z., O'Byrne, C. and Piveteau, P. (2020). Investigation of the roles of AgrA and sigmaB regulators in Listeria monocytogenes adaptation to roots and soil. *FEMS Microbiology Letters* 367(3): fnaa036. DOI: 10.1093/femsle/fnaa036.

Martinez, M. R., Osborne, J., Jayeola, V. O., Katic, V. and Kathariou, S. (2016). Capacity of Listeria monocytogenes strains from the 2011 cantaloupe outbreak to adhere, survive, and grow on cantaloupe. *Journal of Food Protection* 79(5): 757–763. DOI: 10.4315/0362-028X.JFP-15-498.

Materon, L. A., Martinez-Garcia, M. and McDonald, V. (2007). Identification of sources of microbial pathogens on cantaloupe rinds from pre-harvest to post-harvest operations. *World Journal of Microbiology and Biotechnology* 23(9): 1281–1287.

Mazaheri, T., Cervantes-Huamán, B. R. H., Bermúdez-Capdevila, M., Ripolles-Avila, C. and Rodríguez-Jerez, J. J. (2021). Listeria monocytogenes biofilms in the food industry: Is the current hygiene program sufficient to combat the persistence of the pathogen? *Microorganisms*. MDPI AG 9(1): 181. DOI: 10.3390/microorganisms9010181.

McCollum, J. T., Cronquist, A. B., Silk, B. J., et al. (2013). Multistate outbreak of listeriosis associated with cantaloupe. *New England Journal of Medicine* 369(10): 944–953. DOI: 10.1056/NEJMOA1215837/SUPPL_FILE/NEJMOA1215837_DISCLOSURES.PDF.

McLauchlin, J., Aird, H., Amar, C., Barker, C., Dallman, T., Lai, S., Painset, A. and Willis, C. (2021). An outbreak of human listeriosis associated with frozen sweet corn consumption: Investigations in the UK. *International Journal of Food Microbiology*. Elsevier 338: 108994. DOI: 10.1016/J.IJFOODMICRO.2020.108994.

Milohanic, E., Glaser, P., Coppee, J. Y., Frangeul, L., Vega, Y., Vázquez-Boland, J. A., Kunst, F., Cossart, P. and Buchrieser, C. (2003). Transcriptome analysis of Listeria monocytogenes identifies three groups of genes differently regulated by PrfA. *Molecular Microbiology* 47(6): 1613–1625. DOI: 10.1046/j.1365-2958.2003.03413.x.

Min, S. C., Roh, S. H., Niemira, B. A., Sites, J. E., Boyd, G. and Lacombe, A. (2016). Dielectric barrier discharge atmospheric cold plasma inhibits *Escherichia coli* O157:H7, Salmonella, Listeria monocytogenes, and Tulane virus in Romaine lettuce. *International Journal of Food Microbiology*. Elsevier 237: 114–120. DOI: 10.1016/J.IJFOODMICRO.2016.08.025.

Mintier, A. M. and Foley, D. M. (2006). Electron beam and gamma irradiation effectively reduce Listeria monocytogenes populations on chopped romaine lettuce. *Journal of Food Protection*. Allen Press 69(3): 570–574. DOI: 10.4315/0362-028X-69.3.570.

Monaghan, J. M. and Hutchison, M. L. (2012). Distribution and decline of human pathogenic bacteria in soil after application in irrigation water and the potential for soil-splash-mediated dispersal onto fresh produce. *Journal of Applied Microbiology* 112(5): 1007–1019. DOI: 10.1111/j.1365-2672.2012.05269.x.

Montgomery, N. L. and Banerjee, P. (2015). Inactivation of *Escherichia coli* O157:H7 and Listeria monocytogenes in biofilms by pulsed ultraviolet light. *BMC Research Notes*. BioMed Central Ltd. 8(1): 1–12. DOI: 10.1186/S13104-015-1206-9/FIGURES/5.

Murray, E. G. D., Webb, R. A. and Swann, M. B. R. (1926). A disease of rabbits characterised by a large mononuclear leucocytosis, caused by a hitherto undescribed bacillus

Bacterium monocytogenes (n.sp.). *The Journal of Pathology and Bacteriology*. John Wiley & Sons, Ltd 29(4): 407–439. DOI: 10.1002/PATH.1700290409.

Naditz, A. L., Dzieciol, M., Wagner, M. and Schmitz-Esser, S. (2019). Plasmids contribute to food processing environment-associated stress survival in three Listeria monocytogenes ST121, ST8, and ST5 strains. *International Journal of Food Microbiology* 299: 39–46. DOI: 10.1016/j.ijfoodmicro.2019.03.016.

Nyarko, E., Kniel, K. E., Millner, P. D., Luo, Y., Handy, E. T., Reynnells, R., East, C. and Sharma, M. (2016). Survival and growth of Listeria monocytogenes on whole cantaloupes is dependent on site of contamination and storage temperature. *International Journal of Food Microbiology* 234: 65–70. DOI: 10.1016/J.IJFOODMICRO.2016.06.030.

Orsi, R. H., den Bakker, H. C. and Wiedmann, M. (2011) Listeria monocytogenes lineages: Genomics, evolution, ecology, and phenotypic characteristics. *International Journal of Medical Microbiology* 301(2): 79–96. DOI: 10.1016/j.ijmm.2010.05.002.

Orsi, R. H. and Wiedmann, M. (2016). Characteristics and distribution of Listeria spp., including Listeria species newly described since 2009. *Applied Microbiology and Biotechnology*. Springer Verlag 100(12): 5273–5287. DOI: 10.1007/s00253-016-7552-2.

Özcan, G. and Demirel, N. N. (2016). Combined effect of ultrasound and essential oils to reduce Listeria monocytogenes on fresh produce. *Food Science and Technology International*. SAGE Publications Inc. 22(4): 353–362. DOI: 10.1177/1082013215604478.

Pang, H., McEgan, R., Mishra, A., Micallef, S. A. and Pradhan, A. K. (2017). Identifying and modeling meteorological risk factors associated with pre-harvest contamination of Listeria species in a mixed produce and dairy farm. *Food Research International* 102: 355–363. DOI: 10.1016/j.foodres.2017.09.029.

Park, S. H., Choi, M. R., Park, J. W., Park, K. H., Chung, M. S., Ryu, S. and Kang, D. H. (2011). Use of organic acids to inactivate *Escherichia coli* O157:H7, salmonella typhimurium, and Listeria monocytogenes on organic fresh apples and lettuce. *Journal of Food Science*. John Wiley & Sons 76(6): M293–M298. DOI: 10.1111/J.1750-3841.2011.02205.X.

Pérez-Gregorio, M. R., González-Barreiro, C., Rial-Otero, R. and Simal-Gándara, J. (2011). Comparison of sanitizing technologies on the quality appearance and antioxidant levels in onion slices. *Food Control*. Elsevier 22(12): 2052–2058. DOI: 10.1016/J.FOODCONT.2011.05.028.

Pietrysiak, E., Smith, S. and Ganjyal, G. M. (2019). Food safety interventions to control Listeria monocytogenes in the fresh apple packing industry: A review. *Comprehensive Reviews in Food Science and Food Safety*. Blackwell Publishing Inc. 18(6): 1705–1726. DOI: 10.1111/1541-4337.12496.

Pirie, J. H. (1940). The genus listerella PIRIE. *Science (New York, N.Y.)* 91(2364): 383. DOI: 10.1126/SCIENCE.91.2364.383.

Polaska, M. and Sokolowska, B. (2019). Bacteriophages-a new hope or a huge problem in the food industry. *AIMS Microbiology* 5(4): 324–346. DOI: 10.3934/microbiol.2019.4.324.

Port, G. C. and Freitag, N. E. (2007). Identification of novel Listeria monocytogenes secreted virulence factors following mutational activation of the central virulence regulator, PrfA. *Infection and Immunity*. American Society for Microbiology 75(12): 5886–5897. DOI: 10.1128/IAI.00845-07.

Prazak, A. M., Murano, E. A., Mercado, I. and Acuff, G. R. (2002). Prevalence of Listeria monocytogenes during production and postharvest processing of cabbage. *Journal of Food Protection* 65(11): 1728–1734. DOI: 10.4315/0362-028x-65.11.1728.

Qi, Y., He, Y., Beuchat, L. R., Zhang, W. and Deng, X. (2020). Glove-mediated transfer of Listeria monocytogenes on fresh-cut cantaloupe. *Food Microbiology* 88: 103396. DOI: 10.1016/j.fm.2019.103396.

Quereda, J. J., Leclercq, A., Moura, A., Vales, G., Gómez-Martín, Á., García-Muñoz, Á., Thouvenot, P., Tessaud-Rita, N., Bracq-Dieye, H. and Lecuit, M. (2020). Listeria valentina sp. Nov., isolated from a water trough and the faeces of healthy sheep. *International Journal of Systematic and Evolutionary Microbiology* 70(11): 5868–5879. DOI: 10.1099/IJSEM.0.004494/CITE/REFWORKS.

Rafeeq, S. and Ovissipour, R. (2021). The effect ultrasound and surfactants on nanobubbles efficacy against *Listeria innocua* and *Escherichia coli* O157:H7, in cell suspension and on fresh produce surfaces. *Foods*. Multidisciplinary Digital Publishing Institute 10(9): 2154. DOI: 10.3390/FOODS10092154.

Ragon, M., Wirth, T., Hollandt, F., Lavenir, R., Lecuit, M., Le Monnier, A. and Brisse, S. (2008). A new perspective on Listeria monocytogenes evolution. *PLoS Pathogens*. Public Library of Science 4(9): e1000146. DOI: 10.1371/JOURNAL.PPAT.1000146.

Raybaudi-Massilia, R. M., Mosqueda-Melgar, J., Sobrino-López, A., Soliva-Fortuny, R. and Martín-Belloso, O. (2009). Use of malic acid and other quality stabilizing compounds to assure the safety of fresh-cut "FUJI" apples by inactivation of *Listeria monocytogenes, Salmonella enteritidis* and *Escherichia coli* O157:H7. *Journal of Food Safety*. John Wiley & Sons, Ltd 29(2): 236–252. DOI: 10.1111/J.1745-4565.2009.00153.X.

Reina, L. D., Fleming, H. P. and Humphries, E. G. (1995). Microbiological control of cucumber hydrocooling water with chlorine dioxide. *Journal of Food Protection* 58(5): 541–546. DOI: 10.4315/0362-028X-58.5.541.

Rico, D., Martín-Diana, A. B., Barat, J. M. and Barry-Ryan, C. (2007). Extending and measuring the quality of fresh-cut fruit and vegetables: A review. *Trends in Food Science and Technology*. Elsevier 18(7): 373–386. DOI: 10.1016/J.TIFS.2007.03.011.

Rodgers, S. L., Cash, J. N., Siddiq, M. and Ryser, E. T. (2004). A comparison of different chemical sanitizers for inactivating *Escherichia coli* O157:H7 and *Listeria monocytogenes* in solution and on apples, lettuce, strawberries, and cantaloupe. *Journal of Food Protection* 67(4): 721–731. DOI: 10.4315/0362-028X-67.4.721.

Rodriguez, C., Taminiau, B., Garcia-Fuentes, E., Daube, G. and Korsak, N. (2021). Listeria monocytogenes dissemination in farming and primary production: Sources, shedding and control measures. *Food Control* 120: 107540.

Roerink, A.-M. (2021). Surpasses 2020. In: *Dollars But Volume Was Not Able to Hold the Line*. Available at: https://www.freshproduce.com/siteassets/files/reports/retail/iri/ifpa-producesales-week42.pdf (accessed 16 February 2022).

Ruiz-Llacsahuanga, B., Hamilton, A., Zaches, R., Hanrahan, I. and Critzer, F. (2021). Prevalence of Listeria species on food contact surfaces in Washington State Apple packinghouses. *Applied and Environment Microbiology* 87(9): e02932-20. DOI: 10.1128/AEM.02932-20.

Ryan, E. M., Gahan, C. G. and Hill, C. (2008). A significant role for Sigma B in the detergent stress response of Listeria monocytogenes. *Letters in Applied Microbiology* 46(2): 148–154. DOI: 10.1111/j.1472-765X.2007.02280.x.

Sagong, H. G., Lee, S. Y., Chang, P. S., Heu, S., Ryu, S., Choi, Y. J. and Kang, D. H. (2011). Combined effect of ultrasound and organic acids to reduce *Escherichia coli*

O157:H7, *Salmonella Typhimurium*, and *Listeria monocytogenes* on organic fresh lettuce. *International Journal of Food Microbiology*. Elsevier 145(1): 287–292. DOI: 10.1016/J.IJFOODMICRO.2011.01.010.

Salazar, J. K., Sahu, S. N., Hildebrandt, I. M., Zhang, L., Qi, Y., Liggans, G., Datta, A. R. and Tortorello, M. L. (2017). Growth kinetics of Listeria monocytogenes in cut produce. *Journal of Food Protection* 80(8): 1328–1336. DOI: 10.4315/0362-028X.JFP-16-516.

Salazar, J. K., Wu, Z., David McMullen, P. D., Luo, Q., Freitag, N. E., Tortorello, M. L., Hu, S. and Zhang, W. (2013). Prfa-like transcription factor gene Lmo0753 contributes to L-rhamnose utilization in Listeria monocytogenes Strains associated with human food-borne infections. *Applied and Environmental Microbiology* 79(18): 5584–5592. DOI: 10.1128/AEM.01812-13.

Schmid, B., Klumpp, J., Raimann, E., Loessner, M. J., Stephan, R. and Tasara, T. (2009). Role of cold shock proteins in growth of Listeria monocytogenes under cold and osmotic stress conditions. *Applied and Environmental Microbiology* 75(6): 1621–1627. DOI: 10.1128/AEM.02154-08.

Schoenborn, A. A., Clapper, H., Eckshtain-Levi, N. and Shank, E. A. (2021). Rhizobacteria impact colonization of Listeria monocytogenes on Arabidopsis thaliana Roots. *Applied and Environment Microbiology* 87(23): e0141121. DOI: 10.1128/AEM.01411-21.

Self, J. L., Conrad, A., Stroika, S., Jackson, A., Whitlock, L., Jackson, K. A., Beal, J., Wellman, A., Fatica, M. K., Bidol, S., Huth, P. P., Hamel, M., Franklin, K., Tschetter, L., Kopko, C., Kirsch, P., Wise, M. E. and Basler, C. (2019). Multistate outbreak of listeriosis associated with packaged leafy green salads, United States and Canada, 2015–2016. *Emerging Infectious Diseases*. Centers for Disease Control and Prevention (CDC) 25(8): 1461–1468. DOI: 10.3201/EID2508.180761.

Seymour, I. J., Burfoot, D., Smith, R. L., Cox, L. A. and Lockwood, A. (2002). Ultrasound decontamination of minimally processed fruits and vegetables. *International Journal of Food Science and Technology* 37(5): 547–557. DOI: 10.1046/j.1365-2621.2002.00613.x.

Sharma, M., Handy, E. T., East, C. L., Kim, S., Jiang, C., Callahan, M. T., Allard, S. M., Micallef, S., Craighead, S., Anderson-Coughlin, B., Gartley, S., Vanore, A., Kniel, K. E., Haymaker, J., Duncan, R., Foust, D., White, C., Taabodi, M., Hashem, F., Parveen, S., May, E., Bui, A., Craddock, H., Kulkarni, P., Murray, R. T. and Sapkota, A. R. (2020). Prevalence of Salmonella and Listeria monocytogenes in non-traditional irrigation waters in the Mid-Atlantic United States is affected by water type, season, and recovery method. *PLoS ONE* 15(3): e0229365.

Siderakou, D., Zilelidou, E., Poimenidou, S., Tsipra, I., Ouranou, E., Papadimitriou, K. and Skandamis, P. (2021). Assessing the survival and sublethal injury kinetics of Listeria monocytogenes under different food processing-related stresses. *International Journal of Food Microbiology*. Elsevier 346: 109159. DOI: 10.1016/J.IJFOODMICRO.2021.109159.

Siderakou, D., Zilelidou, E., Poimenidou, S., Paramithiotis, S., Mavrogonatou, E., Zoumpopoulou, G., Tsipra, I., Kletsas, D., Tsakalidou, E. and Skandamis, P. N. (2022). *In vitro* virulence potential, surface attachment, and transcriptional response of sublethally injured listeria monocytogenes following exposure to peracetic acid. *Applied and Environmental Microbiology* 88(2): e0158221. Available at: https://journals.asm.org/journal/aem.

Silva, C. C. G., Silva, S. P. M. and Ribeiro, S. C. (2018). Application of bacteriocins and protective cultures in dairy food preservation. *Frontiers in Microbiology* 9: 594. DOI: 10.3389/fmicb.2018.00594.

Simonetti, T., Peter, K., Chen, Y., Jin, Q., Zhang, G., LaBorde, L. F. and Macarisin, D. (2021). Prevalence and distribution of Listeria monocytogenes in three commercial tree fruit packinghouses. *Frontiers in Microbiology* 12: 652708. DOI: 10.3389/fmicb.2021.652708.

Smith, A., Moorhouse, E., Monaghan, J., Taylor, C. and Singleton, I. (2018). Sources and survival of Listeria monocytogeneson fresh, leafy produce. *Journal of Applied Microbiology* 125(4): 930–942.

Solomon, E. B., Yaron, S. and Matthews, K. R. (2002). Transmission of Escherichia coli O157:H7 from contaminated manure and irrigation water to lettuce plant tissue and its subsequent internalization. *Applied and Environment Microbiology* 68(1): 397–400. Available at: https://www.ncbi.nlm.nih.gov/pubmed/11772650.

Soni, K. A. and Nannapaneni, R. (2010). Removal of Listeria monocytogenes biofilms with bacteriophage P100. *Journal of Food Protection* 73(8): 1519–1524. DOI: 10.4315/0362-028x-73.8.1519.

Soni, K. A., Nannapaneni, R. and Tasara, T. (2011). The contribution of transcriptomic and proteomic analysis in elucidating stress adaptation responses of Listeria monocytogenes. *Foodborne Pathogens and Disease* 8(8): 843–852. DOI: 10.1089/fpd.2010.0746.

Stephan, R., Althaus, D., Kiefer, S., Lehner, A., Hatz, C., Schmutz, C., Jost, M., Gerber, N., Baumgartner, A., Hächler, H. and Mäusezahl-Feuz, M. (2015). Foodborne transmission of Listeria monocytogenes via ready-to-eat salad: A nationwide outbreak in Switzerland, 2013–2014. *Food Control*. Elsevier 57: 14–17. DOI: 10.1016/J.FOODCONT.2015.03.034.

Stone, E., Lhomet, A., Neve, H., Grant, I. R., Campbell, K. and McAuliffe, O. (2020). Isolation and characterization of Listeria monocytogenes phage vB_LmoH_P61, a phage with biocontrol potential on different food matrices. *Frontiers in Sustainable Food Systems* 4: 521645.

Strawn, L. K., Fortes, E. D., Bihn, E. A., Nightingale, K. K., Gröhn, Y. T., Worobo, R. W., Wiedmann, M. and Bergholz, P. W. (2013a). Landscape and meteorological factors affecting prevalence of three food-borne pathogens in fruit and vegetable farms. *Applied and Environmental Microbiology* 79(2): 588–600. DOI: 10.1128/AEM.02491-12.

Strawn, L. K., Gröhn, Y. T., Warchocki, S., Worobo, R. W., Bihn, E. A. and Wiedmann, M. (2013b). Risk factors associated with Salmonella and Listeria monocytogenes contamination of produce fields. *Applied and Environmental Microbiology* 79(24): 7618–7627. DOI: 10.1128/AEM.02831-13.

Sullivan, G. and Wiedmann, M. (2020). Detection and prevalence of listeria in U.S. produce packinghouses and fresh-cut facilities. *Journal of Food Protection* 83(10): 1656–1666. DOI: 10.4315/JFP-20-094.

Sun, Y., Wu, Z., Zhang, Y. and Wang, J. (2022). Use of aqueous ozone rinsing to improve the disinfection efficacy and shorten the processing time of ultrasound-assisted washing of fresh produce. *Ultrasonics Sonochemistry*. Elsevier 83: 105931. DOI: 10.1016/J.ULTSONCH.2022.105931.

Suslow, T. (1997). Postharvest chlorination: Basic properties & key points for effective distribution. *Postharvest Chlorination: Basic Properties & Key Points for Effective*

Distribution. University of California, Agriculture and Natural Resources. DOI: 10.3733/UCANR.8003.

Suslow, T., Oria, M. P., Beuchat, L. R., Garrett, E. H., Parish, M. E., Harris, L. J., Farber, J. N. and Busta, F. F. (2003). Chapter II: Production practices as risk factors in microbial food safety and fresh and fresh-cut produce. *CRFSFS* 2: 38–77.

Sy, K. V. v., Murray, M. B., Harrison, M. D. and Beuchat, L. R. (2005). Evaluation of gaseous chlorine dioxide as a sanitizer for killing Salmonella, Escherichia coli O157:H7, Listeria monocytogenes, and yeasts and molds on fresh and fresh-cut produce. *Journal of Food Protection* 68(6): 1176–1187. Available at: http://meridian.allenpress.com/jfp/article-pdf/68/6/1176/1676744/0362-028x-68_6_1176.pdf (accessed 15 February 2022).

Taormina, P. J., Beuchat, L. R., Erickson, M. C., Ma, L., Zhang, G. and Doyle, M. P. (2009). Transfer of Escherichia coli O157:H7 to iceberg lettuce via simulated field coring. *Journal of Food Protection* 72(3): 465–472. Available at: https://www.ncbi.nlm.nih.gov/pubmed/19343932.

Todoriki, S., Bari, L., Kitta, K., Ohba, M., Ito, Y., Tsujimoto, Y., Kanamori, N., Yano, E., Moriyama, T., Kawamura, Y. and Kawamoto, S. (2009). Effect of gamma-irradiation on the survival of Listeria monocytogenes and allergenicity of cherry tomatoes. *Radiation Physics and Chemistry*. Pergamon 78(7–8): 619–621. DOI: 10.1016/J.RADPHYSCHEM.2009.03.060.

Ukuku, D. O., Geveke, D. J., Chau, L. and Niemira, B. A. (2016). Microbial safety and overall quality of cantaloupe fresh-cut pieces prepared from whole fruit after wet steam treatment. *International Journal of Food Microbiology* 231: 86–92. DOI: 10.1016/J.IJFOODMICRO.2016.05.019.

Ukuku, D. O., Olanya, M., Geveke, D. J. and Sommers, C. H. (2012). Effect of native microflora, waiting period, and storage temperature on listeria monocytogenes serovars transferred from cantaloupe rind to fresh-cut pieces during preparation. *Journal of Food Protection* 75(11): 1912–1919. DOI: 10.4315/0362-028X.JFP-12-191.

Upadhyay, A., Chen, C. H., Yin, H., Upadhyaya, I., Fancher, S., Liu, Y., Nair, M. S., Jankelunas, L., Patel, J. R. and Venkitanarayanan, K. (2016). Inactivation of Listeria monocytogenes, Salmonella spp. and Escherichia coli O157:H7 on cantaloupes by octenidine dihydrochloride. *Food Microbiology* 58: 121–127. DOI: 10.1016/J.FM.2016.04.007.

USDA. (2016). Peracetic acid: Handling and processing. Available at: https://www.ams.usda.gov/sites/default/files/media/Peracetic%20Acid%20TR%203_3_2016%20Handling%20final.pdf (accessed 24 June 2022).

USDA. (2020). Dietary guidelines for Americans, 2020–2025. Available at: https://www.dietaryguidelines.gov/resources/2020-2025-dietary-guidelines-online-materials (accessed 19 January 2022).

van der Veen, S. and Abee, T. (2010). Importance of SigB for Listeria monocytogenes static and continuous-flow biofilm formation and disinfectant resistance. *Applied and Environment Microbiology* 76(23): 7854–7860. DOI: 10.1128/AEM.01519-10.

Vandekinderen, I., Devlieghere, F., de Meulenaer, B., Ragaert, P. and Van Camp, J. (2009). Optimization and evaluation of a decontamination step with peroxyacetic acid for fresh-cut produce. *Food Microbiology* 26(8): 882–888. DOI: 10.1016/J.FM.2009.06.004.

Vázquez-Boland, J. A., Kuhn, M., Berche, P., Chakraborty, T., Domínguez-Bernal, G., Goebel, W., González-Zorn, B., Wehland, J. and Kreft, J. (2001). Listeria pathogenesis and molecular virulence determinants. *Clinical Microbiology Reviews* 14(3): 584–640. DOI: 10.1128/CMR.14.3.584-640.2001.

Vivant, A. L., Garmyn, D. and Piveteau, P. (2013). Listeria monocytogenes, a down-to-earth pathogen. *Frontiers in Cellular and Infection Microbiology* 3: 87. DOI: 10.3389/fcimb.2013.00087.

Wang, S. and Oris, R. H. (2013). Listeria. In: *Foodborne Infections and Intoxications*. Elsevier Inc. Chapters. Google. Available at: https://books.google.com/books?hl=en&lr=&id=LVV1DAAAQBAJ&oi=fnd&pg=PP7&dq=Foodborne+Infections+and+Intoxications:+Chapter+11.+Listeria&ots=xq8t6ersH4&sig=Nd3hukS83lP1qMNWnpqZ-xBG4kk#v=onepage&q=Foodborne%20Infections%20and%20Intoxications%3A%20Chapter%2011.%20Listeria&f=false (accessed 17 February 2022).

Weller, D., Wiedmann, M. and Strawn, L. K. (2015). Irrigation is significantly associated with an increased prevalence of Listeria monocytogenes in produce production environments in New York state. *Journal of Food Protection* 78(6): 1132–1141.

WHO. (2020a). Healthy diet. Available at: https://www.who.int/news-room/fact-sheets/detail/healthy-diet (accessed 27 June 2022).

WHO. (2020b). INFOSAN quarterly summary. Available at: https://www.who.int/news/item/29-07-2020-infosan-quarterly-summary-2020-2 (accessed 28 June 2022).

Wiedmann, M. (2016). Physiology and genetics of Listeria monocytogenes survival and growth at cold temperatures view project 100K pathogen genome project view project. DOI: 10.1080/10408390701856272.

Wiktorczyk-Kapischke, N., Skowron, K., Grudlewska-Buda, K., Wałecka-Zacharska, E., Korkus, J. and Gospodarek-Komkowska, E. (2021). Adaptive response of Listeria monocytogenes to the stress factors in the food processing environment. *Frontiers in Microbiology* 12: 710085. DOI: 10.3389/FMICB.2021.710085.

World Health Organization. (2018). Listeriosis. Available at: https://www.who.int/news-room/fact-sheets/detail/listeriosis (accessed 16 February 2022).

Wright, K. M., Chapman, S., McGeachy, K., Humphris, S., Campbell, E., Toth, I. K. and Holden, N. J. (2013). The endophytic lifestyle of Escherichia coli O157:H7: Quantification and internal localization in roots. *Phytopathology* 103(4): 333–340. DOI: 10.1094/PHYTO-08-12-0209-FI.

Wu, Z. S., Zhang, M. and Wang, S. (2012). Effects of high pressure argon treatments on the quality of fresh-cut apples at cold storage. *Food Control*. Elsevier 23(1): 120–127. DOI: 10.1016/J.FOODCONT.2011.06.021.

Yang, Y., Luo, Y., Millner, P., Turner, E. and Feng, H. (2012). Assessment of Escherichia coli O157:H7 transference from soil to iceberg lettuce via a contaminated field coring harvesting knife. *International Journal of Food Microbiology* 153(3): 345–350. DOI: 10.1016/j.ijfoodmicro.2011.11.024.

Yin, H., Chen, C., Boomer, A., Byun, S., Venkitanarayanan, K., Macarisin, D. and Patel, J. (2020). Biocontrol of Listeria on cantaloupes with lactic acid bacteria. *Journal of Food Processing and Preservation* 44(6): e14465.

Zhang, H., Wang, S., Goon, K., Gilbert, A., Nguyen Huu, C., Walsh, M., Nitin, N., Wrenn, S. and Tikekar, R. V. (2020). Inactivation of foodborne pathogens based on synergistic effects of ultrasound and natural compounds during fresh produce washing. *Ultrasonics Sonochemistry*. Elsevier 64: 104983. DOI: 10.1016/j.ultsonch.2020.104983.

Zhang, M., Oh, J. K., Cisneros-Zevallos, L. and Akbulut, M. (2013). Bactericidal effects of nonthermal low-pressure oxygen plasma on S. typhimurium LT2 attached to fresh produce surfaces. *Journal of Food Engineering*. Elsevier Ltd 119(3): 425–432. DOI: 10.1016/j.jfoodeng.2013.05.045.

Zhou, B., Feng, H. and Luo, Y. (2009). Ultrasound enhanced sanitizer efficacy in reduction of Escherichia coli O157:H7 population on spinach leaves. *Journal of Food Science* 74(6): M308–M313. DOI: 10.1111/j.1750-3841.2009.01247.x.

Zhou, Q., Feng, F., Wang, L., Feng, X., Yin, X. and Luo, Q. (2011). Virulence regulator PrfA is essential for biofilm formation in Listeria monocytogenes but not in Listeria innocua. *Current Microbiology* 63(2): 186–192. DOI: 10.1007/S00284-011-9964-7.

Zhu, Q., Gooneratne, R. and Hussain, M. A. (2017). Listeria monocytogenes in fresh produce: Outbreaks, prevalence and contamination levels. *Foods*. Multidisciplinary Digital Publishing Institute (MDPI) 6(3): 1–11. DOI: 10.3390/FOODS6030021.

Zilelidou, E. A., Tsourou, V., Poimenidou, S., Loukou, A. and Skandamis, P. N. (2015). Modeling transfer of Escherichia coli O157:H7 and Listeria monocytogenes during preparation of fresh-cut salads: Impact of cutting and shredding practices. *Food Microbiology* 45(B): 254–265. DOI: 10.1016/j.fm.2014.06.019.

Ziuzina, D., Han, L., Cullen, P. J. and Bourke, P. (2015). Cold plasma inactivation of internalised bacteria and biofilms for Salmonella enterica serovar Typhimurium, Listeria monocytogenes and Escherichia coli. *International Journal of Food Microbiology*. Elsevier 210: 53–61. DOI: 10.1016/J.IJFOODMICRO.2015.05.019.

Chapter 3

Advances in understanding the contamination of fresh produce by pathogenic *Escherichia coli*

Karl R. Matthews, Rutgers University, USA

1 Introduction

The contamination of fresh produce is of concern in developing and developed countries. Produce is available in markets year-round through the import/export of fresh fruits and vegetables. Practices and regulations surrounding the production and handling of fresh fruits and vegetables vary by country; these differences may impact product quality and microbial safety. Regardless of this, no product, irrespective of whether it is lettuce, cucumber, mango, or tomato, is safe from potential microbial contamination at any step from the field to home.

Agricultural production practices and government and commodity group regulations designed to alleviate microbial contamination have been adopted globally. The global sourcing of fresh produce places additional pressure to ensure production and processing practices meet the regulatory standards of the country of import. Foodborne illness outbreaks associated with fresh produce are seemingly more common. Global awareness of the problem by public health agencies, increased consumption of fresh produce, and changes in production and distribution practices are the likely contributing factors.

http://dx.doi.org/10.19103/AS.2023.0121.03

The microorganisms linked to produce outbreaks vary considerably and include norovirus, *Cyclospora*, *Salmonella*, *Cryptosporidium*, *Listeria monocytogenes*, and Shiga toxin-producing *Escherichia coli* (STEC) (Carstens et al., 2019; Matthews, 2013). The commodities involved and the pathogen commodity pairings are ever shifting. Outbreaks associated with *Listeria monocytogenes* were typically linked to deli meats and soft cheeses, and outbreaks in the United States have been linked to leafy greens and packaged salad products (Self et al., 2016; Zhu et al., 2017; FDA, 2022b). In 2021, a *Listeria monocytogenes* outbreak in the United States was linked to iceberg lettuce (FDA, 2022b). Whole-genome sequencing analysis showed that an *Listeria monocytogenes* isolate recovered from harvesting equipment matched the strain causing illness in the outbreak. Foods may be contaminated with any of the pathotypes of human diarrheagenic *E. coli* (DEC). DEC contamination pre- and post-harvest may be associated with feces from infected humans or animals. Several decades ago, STEC foodborne illness outbreaks were predominantly linked to meat products, and contamination of leafy greens is now concerning. Large STEC outbreaks occurring from 2018 to 2020 linked to leafy greens accelerated efforts to ensure the microbiological safety of leafy greens grown in the United States (FDA, 2020b; CDC, 2020). Measures implemented to improve the microbial safety of leafy greens must continually be evaluated and re-evaluated to reduce the number of STEC outbreaks linked to leafy greens (e.g. lettuce and spinach) (FDA, 2022a).

Sources of pathogens include contaminated soil amendments, agricultural water, field workers, wildlife and domestic animals, contaminated harvesting equipment, processing plants, and retail handling (Matthews, 2013). The microbial safety of a product can be compromised through improper handling prior to and during processing. Fresh produce processing practices currently in use fail to eliminate all human pathogens that may contaminate a commodity. Since fresh produce are often consumed raw and minimally processed, the potential human health risks increase (Jung et al., 2014). Many processors have implemented robust testing programs to reduce the likelihood that contaminated products will reach the consumer (Olimpi et al., 2019). Microbiological testing of products prior to harvest is becoming more common and when coupled with 'test and hold' programs of harvested products prior to processing it provides additional barriers to foodborne illness outbreaks. Maintaining a cold chain from harvest to retail improves quality and importantly prevents or impedes the outgrowth of bacterial foodborne pathogens when present. In developing countries, ensuring continuity in the cold chain can be a challenge. Lack of refrigerated trucks and cold storage warehouses can result in temperature abuse of a product which is further exacerbated when product must be transported long distances particularly in tropical climates. Microbiological testing of fresh produce and processing facilities by industry and regulatory agencies can trigger product recall to prevent consumer exposure to product potentially contaminated with a human pathogen. Most countries have food recall systems in place as part of systems to prevent foodborne illness outbreaks.

Fresh produce contaminated with pathogenic *E. coli* continues to cause human illness globally. This chapter focuses on STEC since a majority of produce-related outbreaks are associated with this pathotype.

2 Pathogenicity and virulence

Human DEC is grouped based on clinical, microbiologic, and epidemiologic characteristics including mechanisms of pathogenicity, clinical syndromes, and distinct O:H serotypes (Matthews, 2017). Three major surface antigens allow serotyping: the O (somatic), H (flagellar), and K (capsule) antigens. DEC with combinations of specific virulence traits are categorized into specific groups. The DEC pathotypes include the enterotoxigenic *E. coli* (ETEC), enteroinvasive *E. coli* (EIEC), enteroaggregative *E. coli* (EAEC), enteropathogenic *E. coli* (EPEC), diffusely adherent *E. coli*, and STEC (Matthews, 2017; Jesser and Levy, 2020).

All DEC pathotypes described may be shed in the feces of infected humans or animals, creating the potential for spread to other humans, animals, and the environment. The DEC pathotypes EPEC, ETEC, EIEC, and STEC have been implicated in food- or waterborne illness. EPEC is often associated with outbreaks of watery diarrhea in nurseries in limited income countries. Humans are an important reservoir of the bacterium. ETEC is a major cause of infantile diarrhea in developing countries and most frequently responsible for traveler's diarrhea. Like EPEC, diarrhea caused by EIEC occurs worldwide and is common in children in developing countries. EIEC has clinical presentation, nonbloody diarrhea, and dysentery, similar to that caused by *Shigella* spp. STEC and EHEC (enterohemorrhagic *E. coli*) may incorrectly be used interchangeably. In short, all EHEC are STEC but not all STEC are EHEC (Carstens et al., 2019). STEC infections are usually self-limiting and associated with bloody diarrhea, fever, and vomiting, whereas EHEC strains carry additional virulence factors that can result in life-threatening hemolytic uremic syndrome (Matthews, 2017). Most produce-related outbreaks are linked to STEC contamination.

The most common STEC serotype linked to food-related outbreaks is *E. coli* O157:H7. Recognition of the O157 serotype as an important foodborne pathogen in the United States occurred in the 1980s when *E. coli* O157:H7 was identified as the cause of two outbreaks of bloody diarrhea (Carstens et al., 2019; Jesser and Levy, 2020). The United States in 1994 declared STEC O157 as an adulterant in beef; in 2012, six additional STEC serotypes (non-O157 STEC) were added: O26, O45, O103, O111, O121, and O145 (USDA, 2012). In Europe, the most common non-O157 serogroups associated with human illness include O26, O103, O91, O146, and O145. In 2011, an EAEC O104 strain that had acquired the Stx2 encoding bacteriophage caused a large outbreak in Europe (EFSA BIOHAZ Panel, 2020).

The production of Shiga toxins is characteristic of STEC infections. The *stx* toxin genes, which are encoded on bacteriophages, are integrated into the *E. coli* chromosome. There are two major types of Shiga toxin, Stx1 and Stx2, which are encoded by *stx*1 gene which has four subtypes and *stx*2 that has 12 subtypes (EFSA BIOHAZ Panel, 2020). Strains may carry both *stx*1 and *stx*2 (Gonzalez-Escalona and Kase, 2019). STEC strains require an additional set of genes beyond *stx*1 and *stx*2 to cause illness.

Genes that permit STEC to attach, colonize, and produce and secrete Shiga toxin protein are needed to cause illness. STECs that carry the locus of enterocyte effacement (LEE) pathogenicity island have the capacity to produce attaching and effacing (A/E) lesions on intestinal mucosa. Effectors present in the LEE island include EAE protein (intimin, coded by the *eae* gene), secreted effector proteins (Esp), a TIR receptor, and other T3SS. Other combinations of virulence genes have been reported in LEE-negative isolates (Gonzalez-Escalona and Kase, 2019). STECs can be divided into sub-pathotypes depending on adhesins, including Agg-STEC (which produces aggregative adherence fimbria, i.e. AAF/Hda adhesins) and Saa-STEC (which produces STEC auto-agglutinating adhesion, i.e. Saa) (Lee et al., 2017).

Food is the predominant route for transmission of STECs. Produce contaminated at the farm level remains a major concern. Researchers report that *E. coli* strains carrying Shiga toxins 1 and 2 (*stx*1 and *stx*2), intimin (*eae*), and hemolysin (*ehx*A) production genes were found in grass shoot, rhizosphere soil, and stable manure samples from a small-scale cattle farm located at the center of Netherlands (van Overbeek et al., 2020). A model was proposed in which human pathogens can circulate between plants and farm animals, using the plant as an alternative ecosystem. The model suggests that STECs were spread into pastures that were ingested by grazing cattle, multiplying in the gastrointestinal track with high numbers introduced onto the grassland when the cattle defecated. The STECs were then spread more broadly on the hooves of the grazing cattle. Therefore, grazed pasture should be considered as a potential carrier of STEC and possibly other pathogens. Moreover, STECs have been associated more recently with a number of new animal species and aquaculture industries (Kim et al., 2020). Lee et al. (2017) reported that more than half of the cattle farms (n = 110) in Japan were contaminated by STEC. Importantly, the isolates consisted of different subtypes based on pathogenicity islands. Sampling of swine and cattle in central China found a heterogeneous group of STECs composed of serotypes O100, O97, O91, O149, O26, O92, O102, O157, and O34 (Peng et al., 2019). Most of the STEC carried *stx*1, *stx*2, and *ehx*A simultaneously. The presence of virulence genes encoding intimin (*eae*) and enterohemorrhagic *E. coli* hemolysin (*hlyA*) and Shiga toxin gene (*stx*) subtype was determined in STEC isolated from raw pork in Canada. Only O157-STEC had virulence gene profiles containing both *eae* and *hly*A (Zhang et al., 2021). The investigators concluded that the occurrence of non-O157 STEC capable of causing severe human illness is rare in Canadian retail pork.

Strains of DEC can differ greatly in the ability to survive and persist in the environment and food. Strain characteristics will also influence pathogenicity and virulence and impact the severity of disease. Characterizing DEC associated with foodborne illness outbreaks and human clinical cases will enhance the development of strategies to improve clinical outcomes.

3 Produce contamination with pathogenic *Escherichia coli*

Produce may be contaminated with pathogenic *E. coli* pre- and post-harvest (Luna-Guevara et al., 2019). Since most produce is grown in open environments exposed to wind, dust, soil, wildlife, water, and agricultural inputs, the risk of pre-harvest contamination is great. The ability of STEC to survive adverse conditions and persist in the environment will impact the ability to contaminate crops in the field. Cattle and wildlife serve as a reservoir of STEC (Topalcengiz et al., 2020). There are many STEC serotypes that have been linked to produce-related outbreaks. The prevalence of serotypes involved in outbreaks often differs by country or geographic region. Serotype-specific differences in prevalence and survival outside of the gastrointestinal tract may be important to consider when evaluating potential exposure from wildlife, manure, or irrigation water. STEC strains representing seven STEC serotypes (O26, O45, O103, O111, O121, O145, and O157) were evaluated for differences in survival in microcosms simulating cattle feedlot runoff (Durso et al., 2021). STEC O26 and O45 strains remained culturable for the longest period (>25 months). The O157 strain remained culturable for 17 months, but STEC O111, O121, and O145 declined rapidly under the conditions evaluated.

Direct contact with animal feces or fecally contaminated water and soil in agricultural fields can contaminate crops. The survival of STEC in wild and domesticated animal feces may vary representing different potential risks. A study in the United States included the collection of cattle, feral pig, waterfowl, deer, and raccoon feces from sources in California, Delaware, Florida, and Ohio (Topalcengiz et al., 2020). Fecal samples were inoculated with an STEC cocktail and held at ambient temperature for up to 1 year. Microbiological testing revealed that STEC serotypes O103 and O104 were the predominant serotypes through the entire storage period in feces from all animals and all states. *E. coli* O104:H4 was evaluated for the ability to survive in diluvial sand or alluvial loam with and without manure addition at two different temperatures during a 12-week period (Detert and Schmidt, 2021). In addition, isogenic Δ*rpoS* and Δ*fliC* deletion mutants of the wildtype *E. coli* O104:H4 were included to investigate the general stress response and influence of flagella. The study demonstrated that under conditions evaluated, the *E. coli* O104:H4 strain was able to survive 12 weeks in soil environments and that RpoS was important for survival.

Fecal material or manure may be incorporated into soil through wild animal intrusion or direct application as a soil amendment. The soil type (loam, sandy-loam, and clay-loam), soil pH, temperature, and microbial diversity may influence STEC survival and persistence. The ability of STEC 026 to persist in various manure-amended soils held at 4°C and 20°C was evaluated (Fremaux et al., 2008). The research demonstrated that STEC O26 survived for 365 days in low temperatures, and the presence of clay and organic matter contributed to survival. The proper composting of manure prior to field application is essential to mitigate the contamination of soil with STEC. Researchers demonstrated that *E. coli* O157:H7 strains persisted twofold longer in autoclaved compared to natural soils (Baker et al., 2020). The authors concluded that microbial diversity of the natural soils positively correlated with the reduction of *E. coli* O157:H7. The association of *E. coli* O157:H7 with soil, organic fertilizers, irrigation water, harvest boxes, and workers' hands was determined on six farms in Brazil (Ceuppens et al., 2014). Results of the study suggest that controlling irrigation water quality and quality of composted organic fertilizers were key factors in improving the microbiological safety of lettuce.

Indeed, the soil may serve as a secondary habitat for *E. coli* (Xing et al., 2019). In an effort to separate the effects of microbial diversity from soil abiotic factors the microbial diversity of four different soils was manipulated (Xing et al., 2019). Overall, the results showed that *E. coli* O157:H7 declined more rapidly in soils with high microbial diversity and persisted longest in sterile soil.

Others have studied how abiotic (e.g. moisture, temperature, and pH) and biotic (e.g. indigenous microbial community) factors influence STEC survival in soil. Soils collected from fruit production areas in China were evaluated and models developed to understand STEC survival (Han et al., 2020). Results of the study suggest that soil texture and soil bacterial diversity might be targets in the control of STEC in the agricultural environment. The impact of anthropogenic influences in agricultural situations that lead to decreased soil diversity must be considered as factors that influence the survival of STEC.

On the farm flooding of fields, application of contaminated organic fertilizer, irrigation with water of inferior quality and large distances between the field and toilets, wildlife or cattle feces, runoff, high temperature, and irrigation water quality individually and collectively impact survival and persistence of STECs (Berry et al., 2019; van Overbeek et al., 2020). Depending on the conditions that STECs are exposed, the pathogen may enter a long-term survival (LTS) phase.

STEC in the LTS phase exhibit higher chlorine tolerance and less sublethal injury following exposure to chlorine treatment during lettuce processing (Bhullar et al., 2021). Therefore, pre-harvest conditions may influence the post-harvest survival of STEC.

The survival and persistence of STEC in the environment is far more complex than recognized previously. Subtle differences in soil composition, pH, and commensal microflora may collectively influence whether STEC survive and persist. Moreover, micro- and macro-soil environments may influence the STEC population and ultimately crop contamination.

4 Interaction of pathogenic *Escherichia coli* with produce

The interaction between plants and pathogenic *E. coli* on/in the leaf or other plant structures of edible crops may influence the persistence of the bacterium for days to weeks (Solomon et al., 2002; Solomon et al., 2003; Wang et al., 2021). The ability of *E. coli* to interact with a plant leaf surface or internalize is likely affected by plant age, plant immune response, plant variety, and specific bacterial structures including exocellular polysaccharide, cell surface charge, presence/absence of fimbriae, and hydrophobicity. Researchers are developing approaches to untangle plant and bacterial factors that influence plant-bacteria interaction.

Produce outbreaks linked to pathogenic *E. coli* contaminated lettuce continue to regularly occur globally (Machado-Moreira et al., 2019). Many different lettuce varieties have been linked to pathogenic *E. coli* outbreaks, with most involving Romaine lettuce. Eleven different lettuce genotypes including Salinas iceberg, Red loose leaf, Green Towers Romaine, Butterhead, and others were evaluated for bacterial attachment, internalization, and apoplastic persistence after inoculation with *E. coli* O157:H7 (Jacob and Melotto, 2019). A higher population of *E. coli* O157:H7 persisted on Serriola II variety at day 10 post-inoculation compared to other lettuce varieties evaluated. Perhaps not surprising, the internalization rate varied among varieties but was highest in Serriola II variety. Results suggest that lettuce genotype and associated variations in plant immune responses influence colonization.

The internalization of STEC in lettuce and other crops grown in soil or hydroponically presents a significant challenge to the efficacy of post-harvest measures designed to inactivate or remove human pathogens (Solomon et al., 2002; Solomon et al., 2003; Merget et al., 2019). Pioneering work by Solomon et al. (2002) demonstrated the ability of STEC to internalize within edible portions of lettuce through transmission from contaminated manure and irrigation water. Factors influencing internationalization including plant variety, STEC strain, growth matrix (soil, hydroponic), and other factors continue to be investigated. The internalization of STEC strains (*E. coli* O157:H7) occurred at higher levels in spinach roots than in lettuce roots (Merget et al., 2019). The impact of root injury on internalization of *E. coli* O157:H7 in hydroponically grown lettuce was investigated (Moriarty et al., 2019). The degree of root injury, achieved by making random cuts to the root system, did not affect the concentration of internalized *E. coli* O157:H7. Inoculation of the water in the hydroponic system to 5.3–6.0 log CFU/mL of *E. coli* O157:H7 resulted in *E. coli* O157:H7 internalized in lettuce at 2.4–4.0 log CFU/g. Wang et al. (2021) investigated the internalization of STEC in four plant species, basil (*Ocimum basilicum* L. cv. Genovese), cilantro (*Coriandrum sativum* L.), lettuce (*Lactuca sativa* cv. Cherokee), and kale (*Brassica oleracea* var. *sabellica*), grown in aquaponic and hydroponic systems. The roots of seedlings were damaged before transplanting and roots of mature plants damaged 3 weeks after transplanting. The aquaponic and hydroponic systems had been used previously in STEC studies, and the potential for cross-contamination was considered. In the hydroponic systems, STEC was not detected in the roots or leaves of all four plant species. In the aquaponic systems, STEC was found in the plant roots, leaves, and fish feces. STEC was found in the leaves of mature plants grown from seedlings that received root damage. These studies suggest that root damage on seedlings can result in the internalization of STEC in edible parts of leafy vegetables.

Bacterial structures that may influence adhesion or interaction with plant tissue include exocellular polysaccharide, cell surface charge, presence/absence of fimbriae, and hydrophobicity. The cell surface structures of STEC may also influence the induction of plant defense systems. *Arabidopsis thaliana* was used as a model plant system to study plant defense system response, and the population of an *E. coli* O157:H7 wildtype strain was significantly lower on wildtype *Arabidopsis* plants than that of an *E. coli* O157:H7 flagellum-deficient mutant (Seo and Matthews, 2012). These results suggest that the plant defense system recognized the *E. coli* O157:H7 flagellum. In the same study, a curli-deficient *E. coli* O157:H7 strain survived better on wildtype *Arabidopsis* plants than the curli-producing wildtype strain did. The authors concluded that the cell surface structures of *E. coli* O157:H7 have a significant impact on the induction of differential plant defense responses, thereby impacting the persistence or survival of the pathogen on plants.

A series of reports focused on the influence of surface polysaccharides of STEC on plant defense response and survival on *Arabidopsis thaliana* and lettuce (Jang and Matthews, 2018a; Jang and Matthews, 2018b). The researchers found that populations of three *E. coli* O104:H4 strains on *Arabidopsis* and lettuce were significantly greater than those of two *E. coli* O157:H7 strains at day 5 post-inoculation, suggesting *E. coli* O104:H4 may have better survival ability on the plants (Jang and Matthews, 2018a). The *E. coli* O104:H4 strains were shown to produce significantly higher amounts of capsular polysaccharide compared with the *E. coli* O157:H7 strains. These results suggest that different levels of capsular polysaccharide produced by the different STEC might influence bacterial survival or persistence in plants. The population of wildtype strain of *E. coli* O157:H7 was greater on lettuce compared with the colanic acid-deficient and lipopolysaccharide-truncated mutants (Jang and Matthews, 2018b). Lower production of capsular polysaccharide was associated with greater expression of pathogenesis-related gene (PR1) by *Arabidopsis* compared with the wildtype strain.

The ability of STEC fimbriae to recognize specific plant structure would add to the suite of factors that the STEC can use to facilitate attachment and surface colonization. Holmes et al. (2020) investigated the interaction of *E. coli* O157:H7 F9 fimbriae with plant cell wall hemicellulose. The F9 fimbriae specifically recognized the galactosylated side chains of xyloglucan. The researchers used *Arabidopsis thaliana* as a model plant system demonstrating that the F9 fimbriae influenced the expression of genes involved in the plant defense response. Collectively, research on STEC surface structures suggests a role in interaction with plant tissue and influence on the plant defense system.

The enigma surrounding the interaction of STEC with plant tissue is slowly changing. Characteristics of the organism and the plant must each be considered. Understanding factors that influence STEC plant interaction may lead to pre- and post-harvest strategies that improve the microbial safety of produce.

5 Produce outbreaks linked to pathogenic *Escherichia coli*

Produce-related foodborne outbreaks over the past several decades have increased globally (Bottichio et al., 2020; CDC, 2020; EFSA BIOHAZ Panel, 2020). The serotype O157 is most frequently associated with outbreaks, although more recently the proportion of non-O157 outbreaks has increased (Kim et al., 2020). Other common serogroups of STEC linked to produce outbreaks include O26, O103, O121, O145, and O146. Produce linked to STEC outbreaks are varied and include seed sprouts, celery, spinach, and other row crops (Gobin et al., 2018; Kintz et al., 2019; EFSA BIOHAZ Panel, 2020; Tack et al., 2021). In the United States and Canada, STEC outbreaks are predominantly linked to leafy greens, with only ground beef accounting for more outbreaks (Marshall et al., 2020). Romaine lettuce is often the leafy green involved in STEC outbreaks (Turner et al., 2019; Marshall et al., 2020; Irvin et al., 2021).

Domestic animals serve as a reservoir for O157 and non-O157 *E. coli* which spread to produce production areas through environmental and anthropogenic factors. Research suggests that STEC associated with domestic animal production operations in the vicinity of leafy greens growing operations may lead to in-field contamination of crops (Kim et al., 2020; Marshall et al., 2020). Irrigation water, dust, soil inputs, workers, equipment, and other factors directly or indirectly contaminate leafy greens and other commodities in the field. During an outbreak, investigation determining where or how *E. coli* was introduced to the contaminated product is often not possible; the product may no longer be available, the processing facility would have gone through cleaning/sanitizing procedures; or agronomic practices in the growing field would have been changed.

The types of commodities linked to STEC outbreaks are likely to expand based on the trend observed over the past 30 years. In 2021, an *E. coli* O103 outbreak was linked to minced celery. The outbreak which occurred in Canada was associated with celery imported from the United States (Smith et al., 2022). The minced celery was used in sandwiches prepared at two hospitals with the celery sourced from a single processor in California. An inspection of the processor revealed that *E. coli* testing was limited and that the chlorination wash step was confined to prior to mincing. The situation was likely exacerbated by temperature abuse of the sandwiches prior to consumption. Interestingly, coarsely chopped celery from the same batch and widely distributed was not linked to *E. coli* O103 outbreaks. Collectively, this may suggest risk associated with practices used to mince celery.

During the period of 2009 to 2018, 40 outbreaks linked to leafy greens occurred in Canada and the United States (Marshall et al., 2020). Of the 40 outbreaks, *E. coli* O157 was associated with the most outbreaks (n = 32) followed by STEC O145, O26, O111, and O126. More than 50% were associated with Romaine lettuce, with outbreaks occurring predominantly in the Fall (45%) and Spring (28%). Production of leafy greens in the United States occurs mainly in California and Arizona in the winter months (November–March) and California in the spring, summer, and fall months (April–October) (Marshall et al., 2020). The seasonality of outbreaks is beginning to be addressed as additional outbreaks linked to Romaine lettuce continue to occur in the United States (CDC, 2020; FDA, 2020a; FDA, 2020b).

The seasonal occurrence of foodborne outbreaks linked to foods in the United States (1996–2017) was investigated and suggests that STEC outbreaks peak from mid- to late July (Simpson et al., 2020). The study considered STEC outbreaks linked to all foods, and so results may not align with leafy green outbreaks data. Leonard et al. (2021) considered seasonality and other factors that may influence *E. coli* O157:H7 colonization of lettuce. The researchers included two lettuce cultivars with long and short shelf life harvested in the spring and the fall. The microbiome and STEC O157:H7 colonization of the fresh-cut lettuce was considered under storage in modified atmosphere packaging at cold and warm temperatures. The study demonstrated that the microbiome of lettuce grown in the fall and spring differed and evolved during storage. These changes particularly during cold storage influenced the survival of STEC. Potentially, these factors contribute to the seasonality of STEC outbreaks (Leonard et al., 2021). Considering produce is sourced globally year around to meet consumer demands, understanding the influence of season or climate on STEC persistence and survival in the environment and on crops is integral to the development of systems that will protect human health.

Changes in consumer preference for types of leafy greens and other produce and how they are produced may introduce different food safety considerations. Although controlled environment agriculture (CEA) was introduced >50 years ago, the practice has received increased interest in the past 10 years associated with sustainable agriculture. Now CEA is not confined to greenhouses and includes vertical farming or warehouse farming and typically without soil. The perception exists that produce grown in controlled environments have fewer microbial food safety concerns than field-grown commodities. In controlled environment systems, contamination of products with foodborne pathogens may be associated with water, seeds, substrate, nutrient solution, harvesting and processing equipment, and human handlers (Gomez et al., 2019). Recirculated water and nutrient solution are of particular concern since once contaminated they can rapidly spread human pathogens through the production system.

Microgreens are increasing in popularity as awareness of their potential health benefits increases. Microgreens are typically produced using CEA but may represent unique food safety challenges. Microgreens are immature shoots of spinach, beets, kale, and other vegetables and differ from sprouts and baby greens. Research on the microbial safety of microgreens is growing (Riggio et al., 2019). There have been no known foodborne illness outbreaks linked to microgreens, although recalls have occurred over concern with human pathogen contamination. Microgreens may be contaminated through workers' hands, contaminated water, seed contamination, and unsanitary conditions and tools in harvest and packing facilities. The rapid deterioration in the quality and microbial safety of microgreens may occur with temperature abuse during transportation and storage (Turner et al., 2020). Implementing measures that would improve quality and reduce food safety issues can represent challenges since many microgreen operations are small and lack sufficient resources (Misra and Gibson, 2021). Regardless of this, food safety measures must be developed and practiced to ensure microbial safety and public health.

6 Food safety regulations and future actions

Efforts to enhance the microbiological safety of leafy greens must be continuously evaluated (IFT, 2020). In the United States, despite the U.S.

Food and Drug Administration (FDA) implementing science- and risk-based preventative measures included in provisions of the Food Safety Modernization Act (FSMA) Produce Safety Rule and Good Agricultural Practices to prevent outbreaks linked to leafy greens, such outbreaks continue to occur. FMI, The Food Industry Association, provided a series of recommendations focused on food safety practices for leafy greens handled by food wholesalers and retailers (FMI, 2019). One recommendation encouraged shippers/handlers in Arizona and California to become certified through the Leafy Greens Marketing Agreement. In 2020, the FDA released the '2020 Leafy Greens STEC Action Plan' focused on three areas: prevention, response, and addressing knowledge gaps (FDA, 2020a). The FDA also released in 2020 the 'New Era of Smarter Food Safety' that represents a new approach to food safety (FDA, 2020c). The system is FSMA based, technology enabled, and people-led designed to protect a diverse food system using flexibility in approaches. Collectively, the pre-harvest environment is a focus of food safety reforms.

Recognition of approaches that countries have adopted and evaluated for improving the microbial safety of produce will facilitate the evolution of systems that mitigate microbial contamination and enhance produce safety. Fresh produce is exported/imported daily, and production practices in a country can impact human health in other countries (CFIA, 2020). There is no expectation that a single food safety system for produce will be adopted globally. Indeed, even within a country pre-harvest systems must be flexible to address, for example differences in geography and climate. Importantly, the collective efforts of growers, government agencies, and the food industry will improve the microbial safety of fresh produce regardless of the country in which it is grown.

7 Conclusion

A suite of approaches is required to address the complex issue of *E. coli* contamination of produce and produce safety. For example, the rapid surge in CEA requires real-time development of methods specific for CEA since those developed for open-field agriculture may not be relevant. Adding to the complexity of controlling *E. coli* across production systems is the introduction of new varieties (leafy greens, basil, and tomatoes) and products such as microgreens. One of the biggest challenges to controlling *E. coli* contamination is understanding mitigation practices used across the globe and how those practices may influence whether a product is safe and wholesome. The continued collective efforts to reduce microbial risks of produce will ensure that consumers can continue to enjoy a variety of fresh produce.

8 Where to look for further information

Subject overview:

- Machado-Moreira, B., Richards, K., Brennan, F., Abram, F. and Burgess, C. M. 2019. Microbial contamination of fresh produce: What, where, and how? *Comprehensive Reviews in Food Science and Food Safety*. 18:1727–1750.

Resources:

- EFSA Biohazard: https://www.efsa.europa.eu/en/science/scientific-committee-and-panels/biohaz.

9 References

Baker, C. A., Lee, S., De, J., Jeong, K. C. and Schneider, K. R. 2020. Survival of *Escherichia coli* O157 in autoclaved and natural sandy soil mesocosms. *PLoS ONE* 15(6):e0234562. https://doi.org/10.1371/journal.pone.0234562.

Berry, E. D., Wells, J. E., Durso, L. M., Friesen, K. M., Bono, J. L. and Suslow, T. V. 2019. Occurrence of *Escherichia coli* O157:H7 in cattle pest flies captured in leafy greens plots grown near a beef cattle feedlot. *Journal of Food Protection* 82(8):1300–1307.

Bhullar, M. S., Shaw, A., Mendonca, A., Monge, A., Nabwire. L., and Thomas-Popo, E. 2021. Shiga Toxin-Producing *Escherichia coli* in the Long-Term Survival Phase Exhibit Higher Chlorine Tolerance and Less Sublethal Injury Following Chlorine Treatment of Romaine Lettuce. *Foodborne Path. Dis.* https://doi.org/10.1089/fpd.2020.2873.

Bottichio, L., Keaton, A., Thomas, D., Fulton, T., Tiffany, A., Frick, A., Mattioli, M., Kahler, A., Murphy, J., Otto, M., Tesfai, A., Fields, A., Kline, K., Fiddner, J., Higa, J., Barnes, A., Arroyo, F., Salvatierra, A., Holland, A., Taylor, W., Nash, J., Morawski, B. M., Correll, S., Hinnenkamp, R., Havens, J., Patel, K., Schroeder, M. N., Gladney, L., Martin, H., Whitlock, L., Dowell, N., Newhart, C., Watkins, L. F., Hill, V., Lance, S., Harris, S., Wise, M., Williams, I., Basler, C. and Gieraltowski, L. 2020. Shiga toxin-producing *Escherichia coli* infections associated with romaine lettuce–United States, 2018. *Clinical Infectious Diseases* 71(8):e323–e330.

Carstens, C. K., Salazar, J. K. and Darkoh, C. 2019. Multistate outbreaks of foodborne illness in the United States associated with fresh produce from 2010 to 2017. *Frontiers in Microbiology* 10:2667. https://doi.org/10.3389/fmicb.2019.02667.

CDC 2020. *Outbreak of E. coli Infections Linked to Leafy Greens*. Available at: https://www.cdc.gov/ecoli/2020/o157h7-10-20b/index.html.

Ceuppens, S., Hessel, C. T., de Quadros, R. R., Bartz, S., Tondo, E. C. and Uyttendaele, M. 2014. Microbiological quality and safety assessment of lettuce production in Brazil. *International Journal of Food Microbiology* 181:67–76.

CFIA 2020. *Import Requirements for Romaine Lettuce from the United States*. Available at: https://www.inspection.gc.ca/importing-food-plants-or-animals/food-imports/food-specific-requirements/romaine-lettuce-from-the-united-states/eng/1601488215302/1601488215678 (Accessed June 20, 2022).

Detert, K. and Schmidt, H. 2021. Survival of enterohemorrhagic Escherichia coli O104:H4 Strain C227/11Fcu in Agricultural Soils Depends on *rpoS* and Environmental Factors. *Pathogens* 10(11):1443. https://doi.org/10.3390/pathogens10111443.

Durso, L. M., Gilley, J. E. and Miller, D. N. 2021. Differential survival of non-O157 Shiga toxigenic *Escherichia coli* in simulated cattle feedlot runoff. *Foodborne Pathogens and Disease* 18(11):771–777.

FDA 2020a. *Leafy Greens STEC Action Plan*. Available at: https://www.fda.gov/food/foodborne-pathogens/2020-leafy-greens-stec-action-plan (Accessed June 20, 2022).

FDA 2020b. *Investigation Report: Factors Potentially Contributing to the Contamination of Romaine Lettuce Implicated in the Three Outbreaks of E. coli O157:H7 During the Fall of 2019*. Available at: https://www.fda.gov/media/138238/download (Accessed June 20, 2022).

FDA 2020c. *New Era of Smarter Food Safety. FDA's Blueprint for the Future*. Available at: https://www.fda.gov/food/new-era-smarter-food-safety (Accessed June 20, 2022).

FDA 2022a. *Outbreak Investigation of E. coli O157H7 – Spinach*. Available at: https://www.fda.gov/food/outbreaks-foodborne-illness/outbreak-investigation-e-coli-o157h7-spinach-november-2021 (Accessed June 20, 2022).

FDA 2022b. *Outbreak Investigation of Listeria Monocytogenes: Dole Packaged Salad*. Available at: https://www.fda.gov/food/outbreaks-foodborne-illness/outbreak-investigation-listeria-monocytogenes-dole-packaged-salad-december-2021 (Accessed June 20, 2022).

FMI 2019. *FMI Recommended Food Safety Practices for Leafy Greens*. Available at: https://www.fmi.org/forms/store/ProductFormPublic/fmi-recommended-food-safety-practices-for-leafy-greens (Accessed June 20, 2022).

Fremaux, B., Prigent-Combaret, C., Delignette-Muller, M. L., Mallen, B., Dothal, M., Gleizal, A. and Vernozy-Rozand, C. 2008. Persistence of Shiga toxin-producing *Escherichia coli* O26 in various manure-amended soil types. *Journal of Applied Microbiology* 104(1):296–304.

Gobin, M., Hawker, J., Cleary, P., Inns, T., Gardiner, D., Mikhail, A., McCormick, J., Elson, R., Ready, D., Dallman, T., Roddick, L., Hall, I., Willis, C., Crook, P., Godbole, G., Tubin-Delic, D. and Oliver, I. 2018. National outbreak of Shiga toxin-producing *Escherichia coli* O157:H7 linked to mixed salad leaves, United Kingdom, 2016. *Euro Surveillance* 23(18):pii=17-00197. https://doi.org/10.2807/1560-7917.ES.2018.23.18.17-00197 (Accessed June 20, 2022).

Gomez, C., Currey, C. J., Dickson, R. W., Kim, H., Hernandez, R., Sabeh, N. C., Raudales, R. E., Brumfield, R. G., Laury-Shaw, A., Wilke, A. K., Lopez, R. G. and Burnett, S. E. 2019. Controlled environment food production for urban agriculture. *Hortscience* 54(9):1448–1458.

Gonzalez-Escalona, N. and Kase, J. A. 2019. Virulence gene profiles and phylogeny of Shiga toxin-positive *Escherichia coli* strains isolated from FDA regulated foods during 2010–2017. *PLoS ONE* 14(4):e0214620. https://doi.org/10.1371/journal.pone.0214620.

Han, Z., Huang, G., Liao, J., Li, J., Lyu, G. and Ma, J. 2020. Disentangling survival of *Escherichia coli* O157:H7 in soils: from a subpopulation perspective. *Science of the Total Environment* 749:141649. https://doi.org/10.1016/j.scitotenv.2020.141649.

Holmes, A., Rossez, Y., Wright, K. M., Hedley, P. E., Morris, J., Willats, W. G. T. and Holden, N. J. 2020. *Escherichia coli* O157:H7 F9 fimbriae recognize plant xyloglucan and elicit a response in *Arabidopsis thaliana*. *International Journal of Molecular Sciences*. MDPI 21(24), https://doi.org/10.3390/ijms21249720.

IFT 2020. (vol. 2020). Leafy Green Pilot Final Report. *Leafy Greens Action Plan Tech Enabled Traceability*. Available at: https://www.ift.org/press/press-releases/2020/december/2/leafy-green-traceability-pilots (Accessed June 20, 2022).

Jacob, C. and Melotto, M. 2019. Human pathogen colonization of lettuce dependent upon plant genotype and defense response activation. *Frontiers in Plant Science* 10:1769. https://doi.org/10.3389/fpls.2019.01769.

Jang, H. and Matthews, K. R. 2018a. Survival and interaction of *Escherichia coli* O104:H4 on *Arabidopsis thaliana* and lettuce (*Lactuca sativa*) in comparison to *E. coli* O157:H7: influence of plant defense response and bacterial capsular polysaccharide. *Food Research International* 108:35–41.

Jang, H. and Matthews, K. R. 2018b. Influence of surface polysaccharides of *Escherichia coli* O157:H7 on plant defense response and survival of the human enteric pathogen on *Arabidopsis thaliana* and lettuce (*Lactuca sativa*). *Food Microbiology* 70:254–261. https://doi.org/10.1016/j.fm.2017.10.013.

Jesser, K. J. and Levy, K. 2020. Updates on defining and detecting diarrheagenic Escherichia coli pathotypes. *Current Opinion in Infectious Diseases* 33(5):372–380.

Jung, Y., Jang, H. and Matthews, K. R. 2014. Effect of the food production chain from farm practices to vegetable processing on outbreak incidence. *Microbial Biotechnology* 7(6):517–527.

Irvin, K., Viazis, S., Fields, A., Seelman, S., Blickenstaff, K., Gee, E., Wise, M. E., Marshall, K. E., Gieraltowski, L. and Harris, S. 2021. An overview of traceback investigations and three case studies of recent outbreaks of *Escherichia coli* O157:H7 infections linked to Romaine lettuce. *Journal of Food Protection* 84(8):1340–1356.

Kim, J. S., Lee, M. S. and Kim, J. H. 2020. Recent updates on outbreaks of Shiga toxin-producing *Escherichia coli* and its potential reservoirs. *Frontiers in Cellular and Infection Microbiology* 10:273. https://doi.org/10.3389/fcimb.2020.00273.

Kintz, E., Byrne, L., Jenkins, C., McCarthy, N., Vivancos, R. and Hunter, P. 2019. Outbreaks of Shiga toxin–producing *Escherichia coli* linked to sprouted seeds, salad, and leafy greens: A systematic review. *Journal of Food Protection* 82(11):1950–1958.

Koutsoumanis, K., Allende, A., Alvarez-Ordóñez, A., Bover-Cid, S., Chemaly, M., Davies, R., De Cesare, A., Herman, L., Hilbert, F., Lindqvist, R., Nauta, M., Peixe, L., Ru, G., Simmons, M., Skandamis, P., Suffredini, E., Jenkins, C., Monteiro Pires, S., Morabito, S., Niskanen, T., Scheutz, F., da Silva Felício, M. T., Messens, W. and Bolton, D. 2020. Pathogenicity assessment of Shiga toxin-producing Escherichia coli (STEC) and the public health risk posed by contamination of food with STEC. *EFSA Journal* 18(1):5967, 105 pp. https://doi.org/10.2903/j.efsa.2020.5967.

Lee, K., Kusumoto, M., Iwata, T., Iyoda, S. and Akiba, M. 2017. Nationwide investigation of Shiga toxin-producing *Escherichia coli* among cattle in Japan revealed the risk factors and potentially virulent subgroups. *Epidemiology and Infection* 145(8):1557–1566.

Leonard, S. R., Simko, I., Mamel, M. K., Richter, T. K. S. and Brandl, M. T. 2021. Seasonality, shelf life and storage atmosphere are main drivers of the microbiome and *E. coli* O157:H7 colonization of post-harvest lettuce cultivated in a major production area in California. *Environmental Microbiome* 16(1):25. https://doi.org/10.1186/s40793-021-00393-y.

Luna-Guevara, J. J., Arenas-Hernandez, M. M. P., Martinez de la Pena, C., Silva, J. L. and Luna-Guevara, M. L. 2019. The role of pathogenic *E. coli* in fresh vegetables: behavior, contamination factors, and prevention measures. *International Journal of Microbiology* 2019:2894328. https://doi.org/10.1155/2019/2894328.

Machado-Moreira, B., Richards, K., Brennan, F., Abram, F. and Burgess, C. M. 2019. Microbial contamination of fresh Produce: what, where, and how? *Comprehensive Reviews in Food Science and Food Safety* 18(6):1727–1750.

Marshall, K. E., Hexemer, A., Seelman, S. L., Fatica, M. K., Blessington, T., Hajmeer, M., Kisselburgh, H., Atkinson, R., Hill, K., Sharma, D., Needham, M., Peralta, V., Higa, J., Blickenstaff, K., Williams, I. T., Jhung, M. A., Wise, M. and Gieraltowski, L. 2020. Lessons learned from a decade of investigations of Shiga toxin-producing *Escherichia coli* outbreaks linked to leafy greens, United States and Canada. *Emerging Infectious Diseases* 26(10):2319–2328.

Matthews, K. R. 2013. Sources of enteric pathogen contamination of fruits and vegetables: future directions of research. *Stewarts Post Har. Rev. ISSN* 9656:1945; https://doi.org/10.2212/spr2013.1.2.

Matthews, K. R. 2017. Enterohemorrhagic Escherichia coli. In: Matthews, K. R., Kniel, K. and Montville, T. J. (Eds). *Food Microbiology: an Introduction* (4th edn.). American Society for Microbiology Press, pp. 203–220. Washington D.C.

Merget, B., Forbes, K. J., Brennan, F., McAteer, S., Shepherd, T., Strachan, N. J. C. and Holden, N. J. 2019. Influence of plant species, tissue type, and temperature on the capacity of Shigatoxigenic *Escherichia coli* to colonize, grow, and be internalized by plants. *Applied and Environmental Microbiology* 85(11):e00123-19. https://doi.org/10.1128/AEM.00123-19.

Misra, G. and Gibson, K. E. 2021. Characterization of microgreen growing operations and associated food safety practices. *Food Protection Trends* 41(1):56–69.

Moriarty, M. J., Semmens, K., Bissonnette, G. K. and Jaczynski, J. 2019. Internalization assessment of *E. coli* O157:H7 in hydroponically grown lettuce. *LWT* 100:183–188.

Olimpi, E. M., Baur, P., Echeverri, A., Gonthier, D., Karp, D. S., Kremen, C., Sciligo, A. and De Master, K. T. 2019. Evolving food safety pressures in California's central coast region. *Frontiers in Sustainable Food Systems* 3:102. https://doi.org/10.3389/fsufs.2019.00102.

Peng, Z., Liang, W., Hu, Z., Li, X., Guo, R., Hua, L., Tang, X., Tan, C., Chen, H., Wang, X. and Wu, B. 2019. O-serogroups, virulence genes, antimicrobial susceptibility, and MLST genotypes of Shiga toxin-producing *Escherichia coli* from swine and cattle in central China. *BMC Veterinary Research* 15(1):427. https://doi.org/10.1186/s12917-019-2177-1.

Riggio, G. M., Wang, Q. , Kniel, K. E. and Gibson, K. E. 2019. Microgreens - A review of food safety considerations along the farm to fork continuum. *Inter, J. Food Micro.* 290:76-85.

Self, J. L., Conrad, A., Stroika, S., Jackson, A., Burnworth, L., Beal, J., Wellman, A., Jackson, K. A., Bidol, S., Gerhardt, T., Hamel, M., Franklin, K., Kopko, C., Kirsch, P., Wise, M. E. and Basler, C. 2016. Notes from the field: outbreak of listeriosis associated with consumption of packaged salad–United States and Canada, 2015–2016. *Morbidity and Mortality Weekly Reports* 65(33):879–881. https://doi.org/10.15585/mmwr.mm6533a6.

Seo, S. and Matthews, K. R. 2012. Influence of the plant defense response to *Escherichia coli* O157:H7 cell surface structures on survival of that pathogen on plant surfaces. *Applied and Environmental Microbiology* 78(16):5882–5889.

Simpson, R. B., Zhou, B. and Naumova, E. N. 2020. Seasonal synchronization of foodborne outbreaks in the United States, 1996–2017. Nature research. *Scientific Reports* 10(1):17500 https://doi.org/10.1038/s41598-020-74435-9.

Smith, C. R., Griffiths, A., Allison, S., Hoyano, D. and Hoang, L. 2022. *Escherichia coli* O103 outbreak associated with minced celery among hospitalized individuals in Victoria, British Columbia, 2021. *Canadian Community Disease Report* 48(1):46–50. https://doi.org/10.14745/ccdr.v48i01a07.

Solomon, E. B., Yaron, S. and Matthews, K. R. 2002. Transmission and internalization of *Escherichia coli* O157:H7 from contaminated manure and irrigation water into lettuce plant tissue. *Applied and Environmental Microbiology* 68(1):397–400.

Solomon, E. B., Pang, H. J. and Matthews, K. R. 2003. Persistence of *Escherichia coli* O157:H7 on lettuce plants following spray irrigation with contaminated water. *Journal of Food Protection* 66(12):2198–2202.

Tack, D. M., Kisselburgh, H. M., Richardson, L. C., Geissler, A., Griffin, P. M., Payne, D. C. and Gleason, B. L. 2021. Shiga toxin-producing *Escherichia coli* outbreaks in the United States, 2010–2017. *Microorganisms* 9(7):1529. https://doi.org/10.3390/microorganisms9071529.

Topalcengiz, Z., Jeamsripong, S., Spanninger, P., Persad, A., Aang, F., Buchanan, R. J., Lejeune, J., Kniel, K. E., Jay-Russell, M. T. and Danyluk, M. D. 2020. Survival of Shiga Toxin–producing Escherichia coli in various wild animal feces that may contaminate produce. *Journal of Food Protection* 83:1420–1429.

Turner, E. R., Luo, Y. and Buchanan, R. L. 2020. Microgreen nutrition, food safety, and shelf life: a review. *Journal of Food Science* 85(4):870–882.

Turner, K., Moua, C. N. O. U., Hajmeer, M., Barnes, A. and Needham, M. 2019. Overview of leafy greens–related food safety incidents with a California link: 1996 to 2016. *Journal of Food Protection* 82(3):405–414. https://doi.org/10.4315/0362-028X.JFP-18-316.

USDA 2012. *Risk Profile for Pathogenic Non-O157 Shiga Toxin-Producing Escherichia coli (Non-O157 STEC)*. Available at: https://www.fsis.usda.gov/sites/default/files/media_file/2020-07/Non_O157_STEC_Risk_Profile_May2012.pdf (Accessed June 20, 2022).

van Overbeek, L. S., Wichers, J. H., van Amerongen, A., van Roermund, H. J. W., van der Zouwen, P. and Willemsen, P. T. J. 2020. Circulation of Shiga toxin-producing *Escherichia coli* phylogenetic Group B1 strains between calve stable manure and pastureland with grazing heifers. *Frontiers in Microbiology* 11:1355. https://doi.org/10.3389/fmicb.2020.01355.

Wang, Y.-J., Deering, A. J. and Kim, H.-J. 2021. Effects of plant age and root damage on internalization of Shiga toxin-producing *Escherichia coli* in leafy vegetables and herbs. *Horticulturae* 7(4):68. https://doi.org/10.3390/horticulturae7040068.

Xing, J., Wang, H., Brookes, P. C., Falcao Salles, J. F. and Xu, J. 2019. Soil pH and microbial diversity constrain the survival of *E. coli* in soil. *Soil Biology and Biochemistry* 128:139–149.

Zhang, H., Yamamoto, E., Murphy, J., Carrillo, C. and Locas, A. 2021. Shiga toxin producing *Escherichia coli* (STEC) and STEC-associated virulence genes in raw ground pork in Canada. *Journal of Food Protection* 84(11):1956–1964.

Zhu, Q., Gooneratne, R. and Hussain, M. A. 2017. *Listeria monocytogenes* in fresh produce: outbreaks, prevalence and contamination levels. *Foods* 6(3):21; https://doi.org/10.3390/foods6030021.

Part 2

Detection and risk assessment

Chapter 4

Developments in rapid detection/high throughput screening techniques for identifying pathogens in food

*Kannappan Arunachalam and Chunlei Shi**, *Shanghai Jiao Tong University, China*

1 Introduction

In order to feed the global population, the production of adequate quantities of safe food, unmodified by the presence of foodborne pathogens, harmful substances and allergens, remains a major challenge in food production. Norovirus (NoV) alone has been responsible for the greatest number of foodborne illnesses, accounting for 125 million cases. In addition to NoV, *Camphylobacter jejuni*, *Salmonella*, *Escherichia coli*, *Listeria*, *Vibrio*, *Salmonella*, *Shigella* spp., *Cryptosporidium* spp. and *Giardia* spp. are also responsible for causing foodborne infections (WHO, 2020). A report by the World Health Organization (WHO) in 2020 stated that contaminated foods are responsible for more than 200 different diseases. The WHO also estimates that nearly 600 million people fall ill by eating unsafe food, resulting in the loss of 420 000 lives annually and 33 million healthy life years (WHO, 2020). The burden of infections caused

* Corresponding author: Shi Chunlei: clshi@sjtu.edu.cn.

http://dx.doi.org/10.19103/AS.2023.0121.08

by foodborne pathogens has a significant impact on the economy. According to a World Bank report in 2018, global foodborne illnesses are projected to cost US$95.2 billion in annual productivity loss in low- and middle-income nations, while treating foodborne illness costs US$15 billion (WHO, 2018).

Food deterioration due to microbial contamination is a major concern in terms of waste, health burden and economic loss (Zhuang et al., 2020). Visible growth, changes in texture and organoleptic characteristics are the most typical features of food spoilage caused by microorganisms. Given the prominence of microbial growth in processed food, it is imperative to implement real-time surveillance at certain important processing points to adequately control food quality and safety. Rapid and accurate analytical procedures are critical to ensuring product quality, label compliance and safety. Rapid screening of spoilage microorganisms in food during manufacturing and processing would facilitate their detection in a timely manner, thus controlling the spread of microbes and ensuring safe food supply (Saravanan et al., 2021).

Developing a method to analyse food samples is challenging, especially for fresh produce and raw foods, due to the complex composition of the food matrix, the nonhomogeneous distribution of pathogens, the stress experienced by microorganisms during food processing and the occurrence of non-targeted microbes (Macori et al., 2018). Traditional approaches, such as culture-based identification and immunological assays, have remained the primary methods of identifying microorganisms, even pathogenic ones (Rajapaksha et al., 2019). With the recent advent of new technologies, many approaches are being trialled to investigate the identification of microorganisms, and efforts to improve the sensitivity of microbial identification are on the rise (Rajapaksha et al., 2019). A variety of detection techniques are available, such as molecular, spectroscopic, spectrometry, biosensor and optical phenotyping methods, each with its own set of principles, advantages and limitations that are outlined in this chapter (Fig. 1). In order to better understand the steady development of detection systems, each technique is accompanied by appropriate examples. The purpose of this chapter is to present an overview of the available approaches and other recent techniques developed for detecting foodborne microorganisms, such as viruses [e.g. NoV, hepatitis A virus (HAV) and others], bacteria (e.g. *E. coli*, *Staphylococcus*, *Salmonella* and others), parasites (e.g. *Cyclospora* spp., *Cryptosporidium* spp. and others) and prions.

2 Conventional methods

2.1 Culture-based assay

Microorganisms are detected using conventional culturing techniques. Conventional methods are highly reliable, convenient, reproducible and

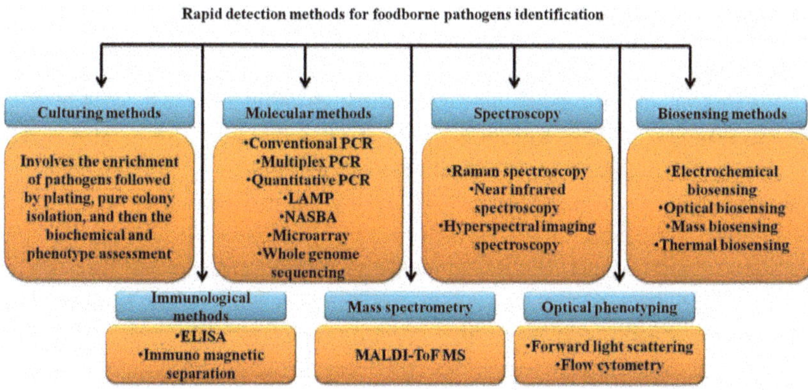

Figure 1 Summary of rapid detection methods for identifying foodborne pathogens in food.

sensitive and have a wide range of applications with reasonable working costs. Currently, most International Standards Organization (ISO) standards for the enumeration of microorganisms are based on the conventional culture-based method, with regulatory bodies defining acceptable values for microorganism categories such as hygiene indicators and pathogens (ISO, 2017). Culture-based and immune-based are two frequently used conventional methods. The culture-based method utilises the traditional way of identifying pathogens. Despite being simple and traditional, the main drawbacks of the biological culture-based detection method are (i) laborious work, as it involves preparation of samples, enrichment (either selective or non-selective), serial dilution, plating (either a selective or a general use agar), enumeration, single colony isolation and further characterisation that involves biochemical (e.g. catalase, oxidase and others) and physiological (Gram staining) characteristics and (ii) time consuming, as the incubation period for most microorganisms generally takes 24–48 h, which may vary up to 72 h depending on the target organism and further investigations using the combinations of biochemical and physiological characterisations (Rama Devi et al., 2016). Overall, the culture-based method takes 5-7 days to give exact results. Some modifications to the culture-based method can be carried out to reduce processing time, offering much faster detection. Modification includes the use of fluorogenic and chromogenic agar in rapid detection. For instance, the use of Rambach, SM-ID and CHROMagar agars are used to test *Salmonella*, while Sorbitol MacConkey and CHROMagar agars are used to test *E. coli* (Maciorowski et al., 2006; Hirvonen et al., 2012). However, in the event of a foodborne disease outbreak or a product recall, this improvement is deemed insufficient.

2.2 Immunoassays

Immune-based technology for detecting foodborne pathogens is well established. Immune-based methods work on the principle of binding a specific epitope of the antibody to the particular segment of the antigen present in the surface of the pathogen. The sensitivity and specificity of immune-based methods rely on the ability of the antibody to bind with the antigen. Immune-based detection approaches include enzyme-linked immunosorbent assays (ELISA, analysing colour production through the binding of antigen and antibody), enzyme-linked fluorescent assays (similar to ELISA, yet highly sensitive) and immune-magnetic separation (IMS, able to isolate and segregate cells directly from food samples). Among other detection methods, IMS is also known for the separation and identification of pathogenic microorganisms. In comparison to immunoassay approaches such as quantitative immunofluorescence, radioimmunoassay, immunodiffusion and immuno-electrophoretic assay, among others used in foodborne diagnostics and food poisoning, ELISA and lateral flow immunoassay are the most widely used and versatile immunoassay approaches (Välimaa et al., 2015; Nagaraj et al., 2016). Sandwich ELISA is the most commonly used approach for the detection of microbes and utilises two antibodies. The primary antibody is immobilised on the plate surface to which the antigen such as bacterial cells and toxin binds. The secondary antibody conjugated with the enzyme binds to the primary antibody–antigen complex and thus forms a sandwich model. Finally, the addition of a colourless substrate will be utilised by the enzyme conjugated with the secondary antibody to produce colour (Zhang, 2013). Despite the presence of several substrates, horseradish peroxidise, β-galactosidase and alkaline phosphatase are frequently used substrates in ELISA (Yeni et al., 2014). Commercially accessible ELISA kits are used to characterise the majority of foodborne pathogens. For instance, detection of *Salmonella* with BIOLINE ELISA kit and *Listeria monocytogenes*, *E. coli*, *Camphylobacter jejuni* and toxins like staphylotoxins with assurance enzyme immunoassay and Vitek immunodiagnostic assay system (VIDAS) kits have been commercialised for the use of foodborne pathogen detection (Bolton et al., 2000; Gómez-Govea et al., 2012; Fung, 2002).

ELISA is a sensitive and accurate method of detecting foodborne microorganisms. However, it necessitates the use of specialist equipment and qualified workers. In order to overcome these challenges and for rapid in-field recognition of foodborne pathogens, lateral flow immunoassays have been developed. Dipsticks and immunochromatographic strips are examples of this type of immunoassay. By this method, the sample will flow through four sections such as a sample pad, conjugate pad, nitrocellulose membrane and finally the absorbent pad via capillary force. As the sample passes through the four sections, it encounters coloured antibodies in the conjugate pad, then the nitrocellulose membranes immobilised with secondary antibodies and finally the absorbent pad. Visualisation of the result takes approximately 2–10 min, depending upon the target in the sample. This method can be performed either by competitive (sample with single antibody) or by sandwich (sample with multiple antibodies) approach (Zhao et al., 2014).

3 Molecular methods

Although the conventional method is reliable and simple for microbial detection, it is not feasible for microbes in a non-culturable state and for some emerging pathogens. There is therefore a need for the use of molecular methods that utilise nucleic acids to detect the same. This nucleic acid-based detection includes ribotyping, pulse field gel electrophoresis, plasmid typing and recent rapid detection methods such as polymerase chain reaction (PCR), loop-mediated isothermal detection, nucleic acid sequence-based amplification, DNA microarray and next-generation sequencing (NGS) (Lewis et al., 2020; Wang, 2021).

3.1 Polymerase chain reaction

PCR is a widely accepted approach for detecting microorganisms using target-specific DNA sequences. Due to its high accuracy, reliability and rapid detection outcomes, the PCR is considered the gold standard method in microbial identification and diagnostics. The idea behind PCR is that genes of particular interest from different pathogens will be amplified and investigated further. In a cyclic three-step process, PCR amplifies a particular target DNA sequence. Initially, the target double-stranded DNA will be denatured at a predefined high temperature. Then, the forward and reverse primers (oligonucleotides preferably below 20 bp in numbers, either synthetic or customised) will anneal to the denatured DNA strands. Thereafter, the forward and reverse primers complementary to the denatured DNA are extended by adding deoxyribonucleotides to the growing end by a thermostable DNA polymerase. Finally the amplified products are separated by agarose electrophoresis gel and visualised by ethidium bromide staining method (Zhao et al., 2014). The gene of particular interest includes 16S rDNA and 23S rDNA for bacterial and fungal pathogens, respectively, and microbe specific target DNA, such as *invA* (invasion protein) in *Salmonella* Enteritidis, *ipaH* (invasion plasmid antigen H) in *Shigella flexneri*, *hlyA* (listeriolysin O) in *L. monocytogenes*, *eaeA* (intimin) in *E. coli* and so on (Chen et al., 2012; Ricke et al., 2019; Shang et al., 2021). With a turnaround time of about 16 h from start to detection, PCR has shown to be an outstanding tool in diagnostic applications (Vinayaka et al., 2019).

PCR to confirm the identified microorganism is still in use and the technology continues to improve.

Genetic fingerprint methods are based on PCR amplification but are clone independent, making them simpler to conduct and less time demanding than PCR cloning and sequencing methodologies. Fingerprinting approaches work on the basis of determining the range of amplified genetic sequences via the size-dependent electrophoretic migration on gels agarose or polyacrylamide gels. Density-gradient gel electrophoresis, random amplified polymorphic DNA, and terminal restriction fragment length polymorphism are some of the genetic fingerprinting techniques that utilise PCR products for microorganism detection (Pinto et al., 2005; Lee et al., 2019b).

In contrast to conventional PCR, multiplex PCR (mPCR) offers rapid microbial detection through amplification of numerous genes in a single run. The core principle of mPCR is the same as traditional PCR. However, the mPCR test utilises 'n' number of primers sets against 'n' number of genes under consideration, while in a typical PCR assay, only one set of unique primers is used against the particular target gene. In order to achieve a successful mPCR result, primer designing is a critical step in which the designed primers must have the same annealing temperature (Zhao et al., 2014). In a study, Lei et al. performed mPCR analysis for the concurrent detection of six common foodborne pathogens utilising *stx* (verocytotoxin) gene of *E. coli*, *hly* (hemolysin) gene of *L. monocytogenes*, *tlh* (thermostable hemolysin) gene of *Vibrio parahaemolyticus*, *nuc* (thermostable nuclease) gene of *Staphylococcus aureus* and *ctx* (cholera toxin) gene of *Vibrio cholerae* as the target genes (Lei et al., 2008). It is worth noting that recent research by Wei et al. employed gene-specific DNA sequences of *rfb*E, *nuc* and *inv*A genes from *E. coli* O157:H7, *S. aureus* and *Salmonella* to perform mPCR analysis, respectively (Wei et al., 2018). The results revealed that 10^3 CFU/mL of bacterial load is sufficient to detect the pathogen of interest.

Quantitative PCR (qPCR) differs from traditional PCR and does not rely on agarose gel electrophoresis to visualise final results. The qPCR continuously monitors the result throughout the run time by analysing the fluorescent signal generated by certain fluorescent probes or intercalating dyes. The quantity of PCR amplicons is directly proportional to the amount of fluorescent signal generated during the run time (Omiccioli et al., 2009; Liu et al., 2019b). Fluorescent signals such as SYBR green (an intercalating dye that specifically binds to the double-stranded DNA) and TaqMan probes and molecular beacons (sequence-specific probes that can only bind to the target sequence) are the fluorescent signals often used in qPCR analysis to detect fluorescence (Chin et al., 2021). The TaqMan probes and molecular beacons contain fluorophore at the 5′ end and quencher at the 3′ end. It works on the fluorescence resonance energy transfer principle, in which the fluorophore is placed adjacent to the quencher in order to prevent the fluorescence of the fluorophore (Liming and Bhagwat, 2004). In general, qPCR is more responsive than traditional PCR and minimises the likelihood of cross-contamination. Table 1 contains further instances of qPCR identification of foodborne bacterial pathogens. Traditional PCR and mPCR that require agarose gel electrophoresis for the interpretation of results are strenuous and lengthy, making them unsuitable for high-throughput analysis, which can be avoided by qPCR analysis. The benefits of qPCR have resulted in the commercial development of several unique qPCR kits for the diagnosis of a wide range of foodborne pathogens (Maurer, 2011).

Foodborne viral infections are a major global public health problem, putting a significant burden on the economy of developed and developing nations, and have considerable detrimental effects on food safety. Viral illnesses may be transmitted via contaminated food, although, unlike bacteria, viruses cannot grow or replicate in food (Neethirajan et al., 2017). Viral foodborne pathogens are little known, and there is minimal awareness and monitoring, most likely due to the limitations of information and the availability and affordability of relevant technology and equipment. NoV, rotavirus and HAV have gained focus in foodborne viral research (Dirks et al., 2021; Victor et al., 2021). In addition, severe acute respiratory syndrome coronavirus, avian influenza virus and Nipah virus are other foodborne viruses to have recently emerged and are causing a significant amount of morbidity and fatality in humans (O'Shea et al., 2019).

NoV is a genus in the Caliciviridae family, having seven genogroups (GI-GVII), although only GI and GII cause the majority of human infections (Centers for Disease Control and Prevention, 2022). The GII.4 strain is ubiquitously prevalent, although GII.17 and the recombinant strain GII.P16.GII.2 have recently increased in popularity in several regions of Asia and Germany, respectively (De Graaf et al., 2015; Niendorf et al., 2017). Consumption of contaminated foods - especially fresh produce and ready-to-eat foods - is a frequent source of NoV foodborne outbreaks (Fino and Kniel, 2008; Müller et al., 2016). The dual typing method for the detection of NoV is being introduced and followed in CaliciNet labs. For GI and GII NoVs, dual typing involves reverse transcriptase (RT)-PCR amplification of a partial portion of both the polymerase and the capsid (regions B and C) genes in a single reaction (Centers for Disease Control and Prevention, 2022). In a recent study, Hernandez and co-workers utilised mPCR for the detection of human NoV GI strain from strawberry and found 10–100 plaque-forming units as the limit of detection (LoD) for mPCR analysis (Hernández Hernández et al., 2022). Similarly, Elmahdy et al. from Egypt recently employed the qPCR method for the detection of human NoV and HAV from green leafy vegetables and strawberry using ISO standards [ISO 15216-1:2017] (ISO, 2017; Elmahdy et al., 2022). The study results revealed that the occurrence of genome copies (GCs) of NoV and HAV was found to be 48% (6.1×10^3 GC/g) and 32.1% (9.2×10^4 GC/g), whilst the occurrence

Table 1 Summary of works studying the use of rapid detection methods in the identification of microbial pathogens in food systems

Detection method	Microorganism	Limit of detection	Food sample	Reference
	Conventional methods			
Culture-based detection	Different serovars of *Salmonella*	-	Raw poultry meat, raw red meat, fish, milk and vegetable salad	Siala et al. (2017)
ELISA	*Escherichia* coli O157: H7	1×10^4 CFU/mL	Commercial	Pang et al. (2018)
	Bacillus cereus	0.9×10^3 cells/mL	Artificial inoculation	Zhu et al. (2016)
	Nucleic acid-based detection			
Conventional PCR-based detection	*Salmonella*	-	Artificial inoculation in chicken	Vichaibun and Kanchanaphum (2020)
	Giardia intestinalis	20 cyst/g	Lettuce	Ramirez-Martinez et al. (2015)
	Staphylococcus aureus, *L. monocytogenes*, *Shigella flexneri*, *Yersinia enterocolitica* and *Clostridium difficile*	-	Artificially contaminated beef	Chen et al. (2021)
Multiplex PCR-based detection	*Salmonella*, *Y. enterocolitica*, *S. aureus*, *L. monocytogenes*, *E. coli* O157:H7	1×10^3 CFU/mL	Artificially contaminated pork	Guan et al. (2013)
	Vibrio parahaemolyticus, *Listeria monocytogenes* and *Salmonella*	10^2 CFU/g	Raw shrimp	Zhang et al. (2015)

	Different serovars of *Salmonella*	-	Raw poultry meat, raw red meat, fish, milk and vegetable salad	Siala et al. (2017)
	Salmonella spp., Shigella spp. and *S.* aureus	Less than 10 CFU/g	Pork samples	Ma et al. (2014)
Real-time PCR-based detection	NoV GI, human adenovirus, rotavirus group A and HAV	-	Lettuce, green onion, leek and watercress	Shaheen et al. (2019)
	Toxoplasma gondii, *Giardia duodenalis* and *Cryptosporidium parvum*	Three oocysts for *T. gondii* and *G. duodenalis*, and 1 oocyst for *Cryptosporidium parvum*	Basil	Hohweyer et al. (2016)
	Camphylobacter jejuni and *E. coli*	1-10 CFU/g	Artificially contaminated minced meat	Kreitlow et al. (2021)
Loop-mediated isothermal detection	*Vibrio parahaemolyticus*	0.3 CFU/25 g	Artificially contaminated squid	Kampeera et al. (2019)
	Hepatitis E virus	10^1 GC	-	Chen et al. (2014)
	T. gondii	0.5 oocyst/g	Baby lettuce	Zhang et al. (2019)
Nucleic acid sequence-based amplification	*Salmonella*	9.5×10^3 CFU/mL	Pork	Zhai et al. (2019)
Recombinase polymerase amplification-based detection	*S. aureus*	RPA = 7.9×10^2 copies/reaction Real-time RPA = 1.28×10^2 copies/reaction	Artificial inoculation in powdered infant formula, pork, steamed buns, codfish and orange juice	Geng et al. (2018)

(*Continued*)

Table 1 (*Continued*)

Detection method	Microorganism	Limit of detection	Food sample	Reference
DNA microarray-based detection	Shiga toxin producing *E. coli* O104:H7	-	-	Shridhar et al. (2018)
	L. monocytogenes	8 log CFU/mL	Artificial inoculation in milk	Bang et al. (2013)
	Astrovirus, NoV, rotavirus and HAV	-	Pure culture	Won et al. (2019)
Whole-genome sequencing-based detection	*Salmonella* typhimurium	-	Lettuce	Ogunremi et al. (2020)
	L. monocytogenes, *Camphylobacter* jejuni, *E.* coli, *Salmonella* enterica subsp. enterica serovar enteritidis, *E.* coli, *S. sonnei*, *S. aureus* and *Y. enterocolitica*	10^1 cells/g	Pure culture	Ferrario et al. (2017)
	Pantoea, *Erwinia*, *Klebsiella*, *Yersinia*, *Bacillus*, *Staphylococcus*, *Salmonella* and *Clostridium*	-	Chinese cabbage	Kim et al. (2018)
Spectroscopy-based detection				
Raman spectroscopy	*S. aureus*	10 cells/mL	Pure culture	Wang et al. (2016)
	Salmonella typhimurium, *Salmonella*, *Shigella boydii*, *Shigella sonnei*, *Shigella dysenteriae*, *Citrobacter freundii*, and *Enterobacter sakazakii*	-	Pure culture	Xie et al. (2013)

FT-IR spectroscopy	*Salmonella typhimurium*	10 CFU/mL	Pure culture	Gómez-Montaño et al. (2021)
	Listeria spp.	-	Raw milk	Al-Mariri et al. (2019)
Near-infrared spectroscopy	*Pseudomonas gessardii*, *Pseudomonas psychrophila*, *P. fragi* and *P. fluorescens*	-	Chicken	Xu et al. (2018)
	B. cereus, *B. thuringenesis*, *E. coli*, *P. aeruginosa* and *B. subtilis*	-	Apple juice	Rodriguez-Saona et al. (2004)
Hyperspectral imaging spectroscopy	*E. coli*	-	Artificially introduced fresh spinach	Siripatrawan et al. (2011)
	E. coli and *L. monocytogenes*	-	Milk and cheese	Unger et al. (2022)
	Spectrometry-based detection			
Matrix-assisted laser desorption/ionisation-time-of-flight mass spectrometry	*Clostridium aciditolerans*, *Clostridium aerotolerans*, *Clostridium argentinense*, *Clostridium beijerinckii*, *Clostridium bifermentans*, *Clostridium butyricum*, *Clostridium cochlearium*, *Clostridium difficile*, *Clostridium perfringens*, *Clostridium sporogenes* and *Clostridium subterminale*	-	Baby food and baby food supplement samples	Sulaiman et al. (2021)
	Optical phenotyping-based detection			
Forward light scattering	*S. aureus*	-	Chicken salad and bovine raw milk	Alsulami et al. (2018)

(Continued)

Table 1 (*Continued*)

Detection method	Microorganism	Limit of detection	Food sample	Reference
	S. aureus	7.50 cells/mL in milk; 8.30 cells/g in milk powder	Milk and milk powder	Liu et al. (2021)
Flow cytometry	*Salmonella*, *Shigella flexneri* and *E. coli*	3.1×10^3 cells/g for *Salmonella*, 2.1×10^3 cells/g for *E. coli* and 2.3×10^3 cells/g for *Shigella flexneri*	Ground beef	Wang et al. (2022)

CFU, colony-forming unit; ELISA, enzyme-linked immunosorbent assay; FT-IR, Fourier-transform infrared; GC, genome copy; GI, genogroup I; HAV, hepatitis A virus; NoV, norovirus; PCR, polymerase chain reaction; RPA, recombinase polymerase amplification.

of NoV GII was found to be 40% (2.4×10^3 GC/g) and 30% (2.03×10^3 GC/g) in fresh strawberry and green leafy vegetables, respectively. Occurrences of NoV and other viruses were consistent with previous global research, such as Terio and colleagues' discovery of NoV GII as the most frequent gastrointestinal virus in 74.8% of ready-to-eat salad samples (Terio et al., 2020). Furthermore, the authors used propidium monoazide, combined with real-time RT-PCR to differentiate infectious and inactivated viruses.

3.2 Loop-mediated isothermal detection

In 2000, Notomi et al. invented loop-mediated isothermal amplification/detection (LAMP), which is a unique nucleic acid-based molecular detection method with rapid, sensitive and specific microbial identification (Notomi et al., 2000). LAMP is based on auto-cycling strand displacement DNA synthesis performed by *Bst* DNA polymerase big fragment for 60 min at isothermal temperatures ranging from 59°C to 65°C. LAMP utilises four primers and two sets of inner and outer primers to mark six distinct DNA sequences. When compared to conventional PCR, LAMP may yield a huge number of amplicons in 60 min, which is often 10^3-fold or larger. The final LAMP results will be identified using agarose gel electrophoresis and further staining (Notomi et al., 2000; Zhao et al., 2014). Maruyama et al. used LAMP for the first time in 2003 to identify the *stxA* gene in *E. coli* O157:H7 (Maruyama et al., 2003). As a result, LAMP has now been used to identify a wide range of foodborne pathogens due to its speed and accuracy. LAMP has been found to be more precise and reliable than PCR analysis for identifying foodborne pathogens (Wang et al., 2008; Lakshmi and Kim, 2021b). Knowing its importance in high-throughput identification of foodborne microbes in food systems, several commercial kits have been made available for detecting foodborne pathogens such as *Salmonella*, *Camphylobacter*, *Listeria*, verotoxin producing *E. coli* and others (Mori and Notomi, 2009). Many kinds of LAMP tests for the identification of foodborne pathogens have been developed. For instance, in situ LAMP, multiplex LAMP, complementary-metal oxide-semiconductor-based LAMP, real-time LAMP and reverse transcription LAMP were developed for the identification of foodborne pathogens, which permits high-throughput screening along with rapid accuracy and specificity in detection (Lakshmi and Kim, 2021a). Similarly, reverse transcription LAMP (RT-LAMP) approaches including gold nanoparticles have been developed for the detection of hepatitis E viruses (Chen et al., 2014). Comparing current conventional detection methods, this one-step RT-LAMP, paired with the colorimetric labelling technique, may effectively identify hepatitis E viruses. This approach is highly recommended due to its ease of use, speed and high sensitivity as well as its detection limit of around 10^1 GC.

3.3 Nucleic acid sequence-based amplification

It was Compton (1991) who invented nucleic acid sequence-based amplification (NASBA), which uses isothermal conditions to amplify nucleic acids, as opposed to PCR, which requires thermocycling equipment (Compton, 1991). NASBA targets RNA, in which the RNA is converted into complementary DNA by reverse transcription enzyme. Avian myeloblastosis virus reverse transcriptase, T7 polymerase, and RNase H are the three enzymes along with two primers involved in the NASBA amplification process, which takes place at 41°C. The final amplicons of the NASBA approach can be visualised by agarose gel electrophoresis (Zhao et al., 2014). This approach may identify live bacteria in a sample and effectively remove the issues associated with PCR procedures. Detection of *Salmonella enterica*, *Camphylobacter jejuni*, *V. cholerae* and *S. aureus* using NASBA was shown to be effective and sensitive (Zhong and Zhao, 2018). Similar to LAMP, NASBA offers high-throughput screening analysis. Commercial high-throughput kits such as Nuclisens EasyQ® Basic Kit (bioMérieux, Marcy l'Etoile, France) can be used for the detection of several foodborne pathogens (Umesha and Manukumar, 2018).

3.4 Recombinase polymerase amplification

Another novel isothermal amplification technique, known as recombinase polymerase amplification (RPA), is capable of amplifying target nucleic acid sequences such as ssDNA, dsDNA, RNA and miRNA in a minimal time span (Lobato and O'Sullivan, 2018). This technique requires minimal sample preparation that excludes prior nucleic acid isolation and purification. This isothermal amplification technique is highly specific and sensitive and operates in the range of 37–42°C. Integration with other approaches will further increase the specificity and sensitivity of this technique. Research into the integration of approaches including microfluidic, lateral flow and real-time fluorescent clustered regularly interspaced short palindromic repeats was successful in the detection of foodborne pathogens in the samples. Due to its efficiency in detecting foodborne pathogens, the RPA is a suitable and practical alternative technique with minimal sample preparation, a fast turnaround time and a simple implementation process (Gao et al., 2017). More examples of the research studies utilising LAMP, NASBA and RPA in food samples are listed in Table 1.

3.5 DNA microarray

DNA microarray is a valuable tool for detecting and identifying pathogens on food matrices because it integrates nucleic acid- and biosensor-based techniques. DNA microarray is a multi-gene detection technique used initially for gene expression studies (Schena et al., 1995). The application of microarray was later extended to comparative genomics, sequence analysis and microbial identification. Microarray consists of either a glass slide or a chip coated with several hundreds of customised probes with a length of 25–80 bp that specifically targets the sequence of the gene of interest. To visualise the hybridisation, the single-stranded nucleic acid fragments from samples are labelled with a fluorescent dye. The fluorescent intensity of the probe–sample complex is directly proportional to the concentration of the labelled fragment (Severgnini et al., 2011). Serotypes with varying levels of toxicity, ranging from innocuous (strain K-12) to lethal (O157:H7), were reported in *E. coli*. Detection of a specific serotype is particularly important in the case of *E. coli* and other foodborne pathogens with multiple serotypes. Li and colleagues published the first report regarding the use of DNA microarray to identify pathogenic serotypes of *Shigella* and *E. coli* (Li et al., 2006). Reports on microarray-based detection of foodborne pathogens are also available (Won et al., 2019; Serrano et al., 2018). Wang and co-workers have developed a microarray method that is able to identify 22 different foodborne pathogens, including *S. aureus*, *L. monocytogenes*, *V. cholerae*, *Camphylobacter jejuni*, *Clostridium perfringens*, *Bacillus cerus*, *Shigella*, *V. parahaemolyticus* and *Salmonella*, among others (Wang et al., 2007). Table 1 lists further research using oligonucleotide DNA microarrays for the identification of foodborne pathogens.

3.6 Whole-genome sequencing

Another sophisticated molecular nucleic acid-based detection method is whole-genome sequencing (WGS), also known as NGS, which refers to highly automated and parallelised genome sequencers that annotate the whole genomes of bacterial pathogens. Rather than targeting specific genes with a DNA microarray, WGS works by sequencing whole fragments/genome of bacterial DNA, followed by sequence alignment and then comparing the obtained sequence with known genome sequence databases. The genomic database of the National Center for Biotechnology Information (https://www.ncbi.nlm.nih.gov/genome/, last accessed 7 March 2022), the Center for Food Safety and Applied Nutrition (https://github.com/CFSAN-Biostatistics/snp-pipeline) (Davis et al., 2015), the GenomeTrakr of US Food and Drug Administration (https://www.fda.gov/food/science-research-food/whole-genome-sequencing-wgs-program) and other genome databases are used to compare sequences and to identify target pathogens. The use of these genome databases is not limited to microbial identification; it also provides detailed information present in the sequenced nucleic acid fragments of genomic DNA, such as source tracking, antibiotic resistance, virulence factor production and so on (Rantsiou et al., 2018). A recent study linked with a foodborne-associated acute gastroenteritis outbreak in Southern Brazil resulted in the detection of uncommon NoV genotype GII.12[P16] in ice pops through qPCR and sequencing methods (Fumian et al., 2021).

4 Spectroscopic methods

4.1 Raman spectroscopy

Raman spectroscopy is a type of spectroscopy technique that relies on the inelastic scattering of monochromatic light to perform its analysis (preferably a laser source). Upon contact with the sample, the frequency of photons in monochromatic light varies due to the fact that they are absorbed by and subsequently re-emitted from the sample. The Raman effect is formed when the shifting frequency of the re-emitted photons from the sample is compared with the initial monochromatic frequency, which is described in detail elsewhere (Smith and Dent, 2019). Surface-enhanced Raman spectroscopy (SERS) is an improved form of the technique that allows for greater amplification of electromagnetic fields generated by the activation of localised surface plasmons. SERS is capable of detecting samples containing low concentrations of analytes while maintaining good sensitivity. This approach may be utilised to identify foodborne bacteria quickly (Zhao et al., 2018). An SERS-based aptasensor was used to detect *Salmonella typhimurium* in a real food sample by Duan and his colleagues (Duan et al., 2016). The study reported that the SERS-based aptasensor was effective in detecting the pathogen with 15 CFU/mL as its LoD.

4.2 Fourier-transform infrared spectroscopy

Fourier-transform infrared (FT-IR) spectroscopy is the most extensively used vibration spectroscopy method and is beginning to be recognised in microbiology due to its high throughput, high speed, low cost and simplicity. FT-IR spectroscopy is a physiochemical technique for measuring vibration from the functional groups present in the sample between the range of 400 cm^{-1} and 4000 cm^{-1} (Davis and Mauer, 2010). Vibration frequencies of the sample recorded using FT-IR spectral analysis provide both quantitative and qualitative analysis, which can differentiate bacterial cells at the metabolic level (Kannappan et al., 2019). Moreover, this approach can be used effectively to identify various bacterial cells at species and strain levels. With a combination of multivariate analysis, such as principal component analysis, hierarchical cluster analysis, canonical variate analysis, discriminate analysis, soft independent modelling by class analogy (SIMCA), artificial neural network and partial least squares regression, several pathogens have been successfully identified. Examples include the lactic acid bacterial from different food products such as milk, cheese and meat products (Lee et al., 2019a; Davis and Mauer, 2010).

4.3 Near-infrared spectroscopy

Near-infrared (NIR) spectroscopy is a spectroscopy method that uses a specific range of wavelengths known as the near infrared, ranging between 780 nm and 2526 nm (12 820 cm^{-1} to 3959 cm^{-1}). In this range, there are vibration transitions comprised of overtones and combination bands that are likely caused by changes in chemical bonds, such as those between O-H, C-H, N-H, C-O and other organic molecules, which can be used for microbial classification and detection (Ferone et al., 2020). Because pathogens' membrane structure is comprised of combinations of macromolecules such as nucleic acids, proteins, lipids and polysaccharides moieties particular to their individual species, the NIR spectrum of each species exhibits a highly specific absorption signature (Tian et al., 2021). NIR spectroscopy is reported to identify microorganisms directly from various kinds of food systems. NIR spectroscopy was shown to have the capacity to detect microbial presence in meat and cheese by Daskalov et al. (2010). NIR spectroscopy and the presence of *L. monocytogenes* in Bulgarian yellow cheese were shown to have a strong association. Consequently, NIR spectroscopy was able to distinguish between samples contaminated with *L. monocytogenes* and those free of *L. monocytogenes*. Furthermore, they found that, even at low bacterial contamination levels, NIR spectroscopy was able to differentiate between contaminated and uncontaminated products using the SIMCA model. Similarly, Pereira and co-workers reported the rapid detection of *Salmonella typhimurium* in a milk sample using NIR spectroscopy (Pereira et al., 2018). This approach allows for the measurement of the target in complicated sample matrices in a non-destructive, quick and precise manner.

4.4 Hyperspectral imaging spectroscopy

Hyperspectral imaging (HSI) is an emerging technology that offers great potential in identifying foodborne pathogens. HSI was originally designed for remote sensing, but its application has since been extended to medicine, food and pharmaceuticals (Gowen et al., 2015; Ferone et al., 2020). Rapid and non-destructive pathogen identification may be achieved with the use of HSI spectroscopy. Combining imaging and spectroscopy techniques, HSI presents detailed signatures regarding the spatial and spectral data of the sample being investigated (Bonah et al., 2019). HSI uses the sample's optical characteristics for detection and thereby focuses on the interactions between the specimen's molecular structure and visible or near-infrared light. During HSI analysis, reflectance, transmittance, absorbance and fluorescence are analysed throughout the

HSI's wavelength range to generate an HSI graph based on the molecular and chemical characteristics of the sample, and the spectral images of the sample are captured at predefined wavelengths. These results have been used to generate a database for known organisms and can be used for the rapid identification of the unknown (Gowen et al., 2015). In a study, Tao and Peng used the Gompertz function in conjunction with the HSI approach to rapidly determine the presence of *E. coli* in pig flesh (Tao and Peng, 2014). With a validation result (R_{CV}) of 0.939 for *E. coli* contamination, the Gompertz parameter surpassed the accuracy of other individual parameters. More examples of research studies utilising Raman, FT-IR, NIR and HSI spectroscopy in food samples are listed in Table 1.

5 Mass spectrometry methods

5.1 Matrix-assisted laser desorption/ionisation-time-of-flight mass spectrometry

Mass spectrometry is considered the best option for microbial diagnosis, as it has a wide detection range, high mass resolution and the capacity to analyse several components. Matrix-assisted laser desorption/ionisation-time-of-flight mass spectrometry (MALDI-TOF MS) has gained in popularity over the last decade in routine diagnostic laboratories and is now regarded as one of the most widely accepted gold standard approaches for identifying microorganisms. According to Holland et al. (1996), distinct spectral fingerprints may be produced from complete bacterial cells using MALDI-TOF MS analysis, opening the door for microbial identification using mass spectrometry (Holland et al., 1996). It has revolutionised the field of microbiology by gradually replacing all biochemical and phenotypic tests for microbial identification. MALDI-TOF MS can detect a wide range of microbial biomolecules, including carbohydrates, nucleic acids, peptides, proteins and other small compounds. The microbes isolated from various sources are placed directly on the target and coated with an acid reagent. The target is then evaluated using mass spectrometry, in which a laser shoots and ionises such biomolecules that are separated by their mass-to-charge ratio (*m/z*) and measured by a detector. The signal is converted into spectra, which is compared by commercial or in-house databases to deliver rapid and reliable identification at low cost and with high accuracy (e.g. relevance score) (Singhal et al., 2015). Besides microbial identification, other novel applications for MALDI-TOF MS are being investigated, including pathogen identification directly from samples, detection of drug resistance determinants, strain typing and detection of virulence factors (Florio et al., 2018).

6 Optical phenotyping methods

6.1 Forward light scattering

The use of a forward light scattering event to detect bacteria is an intriguing new method. This technology is contact free and non-destructive since it monitors microbe colonies that are allowed to grow on a culture plate to create scattering pictures directly, which are then compared to reference image resources of well-known microorganisms (Buzalewicz et al., 2019). Two methods used for microbial identification based on this technique are BARDOT (Bacterial Rapid Detection Using Optical Scattering Technology) and Bacteria Identification System by Light Diffraction. In a study, Abdelhaseib et al. (2019) used the BARDOT system in conjunction with a selective medium to detect multiple bacterial foodborne pathogens such as *Salmonella enterica*, *E. coli* and *L. monocytogenes* in a single experiment. Similarly, the BARDOT system, in combination with a laser optical sensor, was used by Singh et al. (2014) in a food sample to identify 36 distinct *Salmonella* serovars.

6.2 Flow cytometry

Flow cytometry (FCM) is one of the most powerful and popular techniques used in molecular biology to count and sort cells with different characteristics at single-cell level using laser-based technology. This technique allows the rapid and quantitative analysis of multi-parametric function of cells in heterogenic suspension. In addition to studies related to human cells and tumour cells, FCM applied to bacterial studies are on the increase. Initially discovered in the 1960s, FCM was used to detect the size of cells as a single parameter instrument. Advances in technology led to the development of a highly sophisticated instrument able to analyse 14 different parameters at a time (Wilkerson, 2012). FCM follows the basic principle of physics, which includes optics, fluidics and electronics. Through FCM, it is now possible to measure the optical and fluorescence characteristics of any particles at a single-cell level that pass in a fluid system through the light source. Analysis of cells and their differentiation is based on size, granular complexity and the type of fluorescent molecule (either probe or dye) being used. When the cell passes through the optics, the light gets scattered at slightly low angles depending upon the size of the sample, or at a perpendicular angle, depending upon the granular complexity. Moreover, the use of fluorescent dyes targeting different cells (either live or dead) and their components or fluorochrome-linked antibodies adds to the differentiation process, including cell types, membrane components, enzyme activity, DNA content and so on (Macey, 2007). Research reports indicate successful identification of the foodborne pathogen *E. coli* from different food sources such as apple, milk and ground beef (Yamaguchi et al., 2003; Hegde et al., 2012). However, the studies highlight the importance of a pre-treatment process in order to avoid the background noise from non-microbial particles in the sample for effective identification.

7 Biosensor methods

A biosensor is electronic equipment that turns biological signals into quantifiable and workable signals. As illustrated in Fig. 2, a biosensor is made up of three primary components - a biorecognition element, which is an amplifier that amplifies signals, a transducer component that processes the amplified signals and an electronic system that shows the final results in digital form (Xu et al., 2021). Following signal analysis, the electronic system converts the processed signal into an electrical signal, which is relative to the figure of analyte-bioreceptor contact events. Using different sensing approaches as mentioned below, the biosensor may analyse biological samples in a highly precise, rapid and consistent way.

Biosensors are classified based on two methods. The first is the biorecognition element-based detection, in which the bioreceptor analyses the analytes passing through it, where analytes are samples of interest such as whole microbial cells and other cellular components such as DNA, proteins, lipids and others. Based on the type of bioreceptor used, the different analytes can be identified. Examples include immunosensors in the case of antigen or antibody-based detection, phagosensors in the case of bacteriophage-based detection, antimicrobial peptide-based biosensors in the case of peptides with shorter amino acid residues and aptasensors and genosensors in the case of DNA- and RNA-based detection. The second is transducer-based detection, in which electrochemical-based (voltametric, potentiometric, ampherometric and others), optical-based (absorbance, reflectance, luminescence and others), mass-based (piezoelectric, crystalline and surface acoustic wave) and temperature-based (thermistor) transducers aid the detection of analytes, where the transducer converts the signal from the biorecognition element to a simple measurable signal (Bozal-Palabiyik et al., 2018; Xu et al., 2021). Table 2

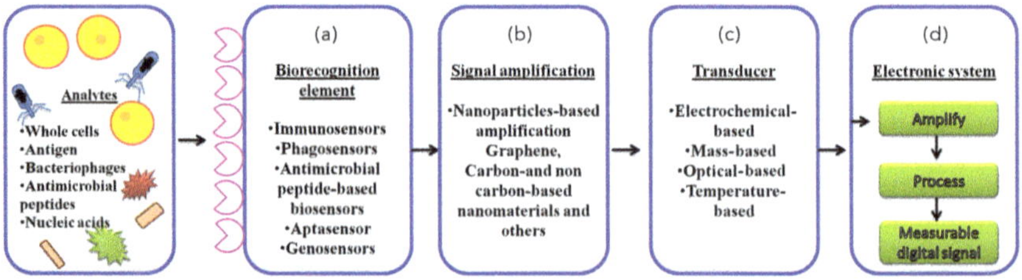

Figure 2 Schematic representation of biosensor with its major element responsible for microbial identification/detection.

Table 2 Summary of studies on the use of biosensor-based detection methods

Detection method	Method	Detection method; limit of detection	Food sample	Reference
		Electrochemical biosensor-based identification		
Potentiometric	Tris(2,2′-bipyridyl)ruthenium(II) fluorophore on gold electrode surfaces to enhance luminous potential. Changes in the bacterial concentration alter the fluorophore potential and thereby the luminescence signal	Potentiometric aptasensors to detect electrogenerated chemiluminescence from *Escherichia coli*; 2 CFU/mL	-	Gao et al. (2022)
Impedimetric	Graphene wrapped copper(II)-assisted cysteine molecules on gold electrode	Immunosensor for the detection of *E. coli*; 3.8 CFU/mL	Water samples	Pandey et al. (2017)
	Salmonella-specific antibody immobilisation	*Salmonella* concentration as low as 300 cells/mL	Ready-to-eat turkey matrix	Liu et al. (2019a)
Amperometric	Fabricated cadmium sulphide quantum dot particles as signal tags encapsulated in imidazolate framework-8	Immunobiosensor to detect *E.* coli; 3 CFU/mL	Milk	Zhong et al. (2019)
	Immobilisation of yeast-specific polyclonal antibody on the electrode via carbodiimide/*N*-hydroxysuccinimide coupling approach	Immunobiosensor to detect *S. cerevisiae*; 3 CFU/mL	White wine	Borisova et al. (2018)

(*Continued*)

Table 2 (*Continued*)

Detection method	Method	Detection method; limit of detection	Food sample	Reference
	Amine-labelled *Salmonella* enterica serovar typhi-specific single-strand capture probe on the SPE surface modified with gold nanoparticles and polycysteine	Genosensor using differential pulse voltammetry of anthraquinone-2-sulfonic acid monohydrate sodium salt as a signal indicator for detection; 6.8×10^{-25} mol L^{-1}	Human blood, raw milk, egg, and poultry faeces	Bacchu et al. (2022)
	The particular capture probe was functionalised on a disposable gold electrode and tested on complimentary ssDNA and HAV cDNA	0.65 pM for the complementary ssDNA and 6.94 fg/μL for HAV cDNA	-	Manzano et al. (2018)
Voltametric	Signal amplification based on hybridisation chain reaction with hemin/G-quadraplex DNAzyme	Immunosensor using differential pulse voltammetry based on hemin/G-quadraplex DNAzyme for prion detection; 0.38 pg/mL	-	Li et al. (2018)
	A microfluidic polydimethylsiloxane chip with screen-printed carbon electrode for electrochemical detection of NoV	Viral specific-based aptamer tagged with ferrocene molecule for detection; 100 pM	-	Chand and Neethirajan (2017)
	Gold nanoparticle-modified SPE for the detection of *Cryptosporidium Parvum*-bound aptamer	Aptamer using square-wave voltammetry; 100 oocysts	Mango and pineapple	Iqbal et al. (2015)
		Optical biosensor-based identification		
Electromagnetic piezoelectric acoustic sensor	Aptamer being covalently bound to the sensor surface by the anti-fouling linker, 3-(3-(trichlorosilylpropyloxy) propanoyl chloride	Aptamer-based sensor specific to *E. coli*; 8 CFU/mL	Milk	Spagnolo et al. (2022)

Surface plasmon resonance	Covalent binding	57 CFU/mL in *E. coli* 7.4×10^3 CFU/mL in *Salmonella*	Cucumber and hamburger extracts	Vaisocherová-Lísalová et al. (2016)
	Lectins as bioreceptor immobilised onto the surface of CM5 chip via an amide bond	3×10^3 CFU/mL for *E. coli*	Cucumber and ground beef	Wang et al. (2013)
Colorimetric	Gold nanorod-based multicolorimetric assay for pathogen detection via the combination of a magnetic bead-based sandwich immunoassay and 3,3′,5,5′-tetramethylbenzidine-mediated etching of gold nanorods	Aptamer to detect *L. monocytogenes*; 10/CFU mL	Spiked pork	Liu et al. (2019c)
	Cotton swab containing monoclonal antibody cocktails	Surface antigen as the target epitope to detect *Staphylococcus aureus*, *Salmonella typhimurium*, *S. enteritidis* and *Camphylobacter jejuni*; 10 CFU/mL	Poultry processing plants	Alamer et al. (2018)
	Antibody analyte–aptamer sandwich configuration	Internalin-specific aptamer to detect *L. monocytogenes*; 10^2 CFU/ 25 g	Sliced beef, chicken and turkey	Ohk et al. (2010)
Fibre optics	Detection based on Alexa-Fluor 647-labelled monoclonal antibodies	Antibody-based detection of *L. monocytogenes*, *E.* coli O157:H7 and *Salmonella* enterica: 10^3 CFU/mL	Ready-to-eat beef, chicken and turkey and meat samples	Ohk and Bhunia (2013)

CFU, colony-forming unit; HAV, hepatitis A virus; NoV, norovirus; SPE, screen-printed electrode.

lists several recent studies evaluating biosensor-based foodborne microbial identification.

Although the methods described earlier are useful and vigorous and provide accurate data, they are often time-consuming, complex and experimental procedures and necessitate the use of well-trained analysts and expensive instruments. Moreover, analysis of samples using the above-mentioned methods has to be performed in a centralised laboratory, and so it is not feasible for real-time analysis and data interpretation. Given these points, an analytical process that allows rapid screening of pathogens and spoilage indicators for short shelf-life food samples with on-site detection is essential.

Screen-printed electrodes (SPEs) in biosensors have attracted much interest recently. SPEs use working, counter and reference electrodes in a similar way to traditional electrodes. These electrodes are printed on stable, non-conductive substrates or a printed circuit board. The critical nature of SPEs is achieved via the deposit of layer-by-layer conductive ink on a solid surface with chance and amenability of using wide materials corresponding to the analyte of interest. Further introduction of different electrodes surfaces could improve sensor efficiency.

Similarly, particle-loaded electrodes have a significant impact during the electron transfer process with better working efficiency. With this knowledge, the use of nanomaterials in surface modification has become prominent. The use of nanomaterial (carbon based or non-carbon based)-loaded electrodes has advantages, including a larger surface area, biocompatibility to sense different biomolecules, superior electrical conductivity and electron transfer.

A lab-on-chip is an autonomous device to conduct on-site analysis that brings microfluidics integrated biosensors, the advantages of which include rapid detection, miniaturisation, less use of consumables, easy integration with other approaches and high-throughput screening. Furthermore, the introduction of CRISPR-Cas system in microfluidics biosensing platform enhances the sensitivity and specificity of the detection process.

8 Detection of parasites

Foodborne illnesses are also induced by parasites, although parasites are usually less harmful than those caused by bacterial and viral pathogens. Parasites such as fish-borne trematodes and tapeworms may infect people via food or by direct contact with infected animals. In addition, soil acts as the primary reservoir for parasite contaminations such as *Cryptosporidium*, *Entamoeba histolytica* and *Giardia* spp. (Ahmed and Karanis, 2018). Elution is a process used to remove parasites from plants and is vital since the efficacy of oocyst recovery and other subsequent steps are dependent upon it. Analysing different elution methods against *Cryptosporidium parvum* oocyst recovery from lettuce and raspberries, Cook et al. concluded that the type of extractants and the methods of extraction play a significant role in oocyst recovery (Cook et al., 2006). Following the elution process, the most common methods used for the detection of parasites are microscopic examinations and staining methods, such as the Ziehl–Neelsen staining method for *Cryptosporidium* spp. and Lugols' iodine staining method for *Giardia* spp. These methods are cost-effective and simple. However, the use of the PCR technique induces the sensitivity of detection compared to the above detection methods. Ramirez-Martinez and colleagues reported 20 ± 4.3% as the recovery efficacy with LoD of 20 oocysts of *Giardia*/g using light microscopy, while PCR targeting the β-giardin gene increased the sensitivity ten-fold compared to the microscopic analysis (Ramirez-Martinez et al., 2015). Another approach is the immunofluorescence assay (IFA), which is used to identify parasitic oocysts in diverse vegetable matrices (Indian Standards Organization, 2016). Immunofluorescence, size and shape helped recognition and differentiation of oocysts from other particles (Keserue et al., 2012). Although certain particles in food matrices have auto-fluorescence, this makes IFA inappropriate for analysing tagged oocysts. This procedure is also time consuming and costly and requires microscopy knowledge (Ahmed and Karanis, 2018). Rzezutka et al. (Rzeżutka et al., 2010) and Utaaker et al. (Utaaker et al., 2015) combined PCR with IFA to increase specificity and obtain genetic information. Other research studies utilising different methods to detect the presence of parasites in fresh produce are listed in Tables 1 and 2. However, the standardised method 18744 (Indian Standards Organization, 2016) is now the only one that can be used to identify and detect the presence of *Cryptosporidium* and *Giardia* in fresh leafy greens and berry fruits. Besides *Cryptosporidium* and *Giardia*, many harmful parasites such as *Entamoeba*, *Toxoplasma*, *Cyclospora* and others cannot be monitored or detected using this method. By developing standard methods and detection technologies against foodborne parasites, we may better understand the route of transmission, boost risk assessments and pinpoint crucial control points for implementation.

9 Advantages and limitation of microbial detection method

The pros and cons of rapid detection techniques for identifying microorganisms in food samples are shown in Table 3. Despite the fact that a variety of approaches have advantages in the pathogen detection process, the biosensor method has emerged as the most promising strategy due to its cost-effectiveness, user friendliness, real-time testing, precise detection and other qualities (Xu et al., 2021). Furthermore, the integration of approaches such as microfluidics enhances the rapid identification of pathogens (Fu et al., 2021).

Table 3 Overview of advantages and limitations of rapid detection methods involved in the identification of foodborne pathogens

Detection method	Advantages	Limitations
Culture-based detection	Necessitates the use of indicator strain Not complex Accurate and confirmatory results	Not able to identify the source of contamination Time consuming and laborious Requires sterile conditions throughout the experimental time Cannot be used on-site Does not provide information on microorganism in non-culturable state
Immune-based detection	Pathogen specific Use of automation reduces time and labour Also able to detect other virulence-like toxin	Low sensitivity Reactions may result in similar range antigen May lead to the generation of false-positive end results
PCR-based detection	Offers better specificity compared to conventional plating methods Decisive, robust and explicit results are provided with reliable counts qPCR eliminates laborious post-amplification processes Use of reverse transcriptase PCR can overcome the question of organism in culturable and non-culturable state Several organisms can be detected at a single run in mPCR	Apart from mPCR, other PCR methods lack differentiation between viable and dead cells Contamination may alter the final results Over-presence of template also alter the final results Prior to identification, PCR requires sequence data for primer design PCR can detect only the strain of interest
Next-generation sequencing	Interprets highly conserved sequence Public sequence database is available for comparison of known target Able to discriminate other parameters such as antibiotic resistance, virulence genes and others	High cost of equipment Requires highly skilled personnel to operate instrument and to interpret the data Copy number may interfere with the final result
Other molecular detections	Environmental factors may not influence the final results Ensure fast and sensitive detection Using new primers may improve efficiency Molecular subtyping may provide the details of strain differences within species identification	Complex structure of certain food matrixes may alter the final results Presence of PCR inhibitors may influence the efficacy of amplification and primer annealing

(Continued)

Table 3 (*Continued*)

Detection method	Advantages	Limitations
MALDI-TOF MS	High sensitivity and selectivity High throughput and quick Low operating cost per analysis Non-destructive in nature Can be used to analyse complex samples	Expensive instrument Limitations in accessing database Not feasible for on-site testing Requires well-trained personnel for operating
Raman spectroscopy	Easy device handling Non-destructive in nature Culture independent High sensitivity and selectivity	Requires fluorescent background Not feasible for on-site testing Expensive instrument
NIR spectroscopy	Non-destructive in nature Real-time detection High sensitivity	Signal saturation due to the presence of water Not feasible for on-site testing Expensive instrument Requires well-trained personnel for operating
HSI spectroscopy	Non-destructive in nature Real-time detection High sensitivity	Not feasible for on-site testing Expensive instrument Requires well-trained personnel for operating
Flow cytometry	Detects multiple parameters at an instant Offers single-cell analysis High specificity	Requires highly trained personnel for operating Relatively slow running time Expensive and need for specialist equipment
Forward light scattering	High specificity and selectivity Non-destructive in nature Real-time detection	Database limitations Not feasible for on-site testing Expensive instrument Requires highly trained personnel for operating
Electrochemical biosensor	High sensitivity and specificity Quick response, high-throughput, user friendly Low cost per testing Real-time detection Low detection limit Less prone to noise Feasible for on-site testing Miniaturisation potential	Expensive instrument Requires receptor-based sample preparation Interference of other electroactive substance may alter end result Probe variation results in signal instability Possibility to end up with batch variation

HSI, hyperspectral imaging; MALDI-TOF MS, matrix-assisted laser desorption/ionisation-time-of-flight mass spectrometry; mPCR, multiplex PCR; NIR, near infrared; PCR, polymerase chain reaction; qPCR, quantitative PCR.

In the near future, however, potential problems might be resolved if high-throughput screening technologies can overcome the constraints observed in the approaches discussed.

10 Conclusion

Progress in detection techniques, especially against foodborne microorganisms, increases year on year, highlighting the importance of rapid and precise detection methods to avoid morbidity and mortality. Although certain pathogens, such as *Salmonella*, have shown a decrease in infection rates in recent years, awareness of other foodborne pathogens, as well as their preparedness, must be improved to avoid outbreaks; hence, this chapter discusses on the progression of foodborne pathogen detection techniques, ranging from conventional to the latest biosensing methods. Conventional culture-based methods remain in use to identify pathogens of clinical origin, for which analytes have difficulty identifying pathogens through the latest high-throughput detection techniques. High-throughput screening based on conventional, molecular, technological, spectroscopic, spectrometry and optical phenotyping methods are more reliable as they have pathogen-specific niches for identifications. The biosensing system is undoubtedly the best detection method available for foodborne microorganisms, and the addition of nanomaterials in biosensing technology will enhance the sensitivity of this approach. Further developments in artificial intelligence technology and big data analytics will help take the identification of microorganisms in the food system to the next level.

11 Where to look for further information

The following articles provide a good overview of the subject:

1. Fung, D. Y. C. 2002. Rapid methods and automation in microbiology. *Comprehensive Reviews in Food Science and Food Safety* 1(1), 3–22.
2. Hameed, S., Xie, L. and Ying, Y. 2018. Conventional and emerging detection techniques for pathogenic bacteria in food science: A review. *Trends in Food Science & Technology*, 81, 61–73.
3. Rajapaksha, P., Elbourne, A., Gangadoo, S., Brown, R., Cozzolino, D. and Chapman, J. 2019. A review of methods for the detection of pathogenic microorganisms. *Analyst*, 144(2), 396–411.
4. Sun, Y., Li, J., Zhu, L. and Jiang, L. 2022. Cooperation and competition between CRISPR-and omics-based technologies in foodborne pathogens detection: a state of the art review. *Current Opinion in Food Science*, 44, 100813.

12 Acknowledgements

The authors gratefully acknowledge the financial support from the Shanghai Agriculture Applied Technology Development Program, China (grant number 2020-02-08-00-08-F01487), the Natural Science Foundation of China (Grant no. 32001798 and 31972169), and the Science and Technology Commission of Shanghai Municipality of China (Grant no. 21N31900200).

13 References

Abdelhaseib, M. U., Singh, A. K. and Bhunia, A. K. 2019. Simultaneous detection of *Salmonella enterica*, *Escherichia coli* and *Listeria monocytogenes* in food using a light scattering sensor. *Journal of Applied Microbiology* 126(5), 1496-1507.

Ahmed, S. A. and Karanis, P. 2018. An overview of methods/techniques for the detection of *Cryptosporidium* in food samples. *Parasitology Research* 117(3), 629-653.

Alamer, S., Eissa, S., Chinnappan, R. and Zourob, M. 2018. A rapid colorimetric immunoassay for the detection of pathogenic bacteria on poultry processing plants using cotton swabs and nanobeads. *Mikrochimica Acta* 185(3), 164.

Al-Mariri, A., Ramadan, L., Younes, A. and Al-Laham, A. 2019. Identification of Listeria species by Fourier-transform infrared spectroscopy. *Bulgarian Journal of Veterinary Medicine* 22(3), 285-296.

Alsulami, T. S., Zhu, X., Abdelhaseib, M. U., Singh, A. K. and Bhunia, A. K. 2018. Rapid detection and differentiation of *Staphylococcus* colonies using an optical scattering technology. *Analytical and Bioanalytical Chemistry* 410(22), 5445-5454.

Bacchu, M. S., Ali, M. R., Das, S., Akter, S., Sakamoto, H., Suye, S. I., Rahman, M. M., Campbell, K. and Khan, M. Z. H. 2022. A DNA functionalized advanced electrochemical biosensor for identification of the foodborne pathogen *Salmonella enterica* serovar Typhi in real samples. *Analytica Chimica Acta* 1192, 339332.

Bang, J., Beuchat, L. R., Song, H., Gu, M. B., Chang, H. I., Kim, H. S. and Ryu, J. H. 2013. Development of a random genomic DNA microarray for the detection and identification of *Listeria monocytogenes* in milk. *International Journal of Food Microbiology* 161(2), 134-141.

Bolton, F. J., Fritz, E., Poynton, S. and Jensen, T. 2000. Rapid enzyme-linked immunoassay for detection of *Salmonella* in food and feed products: Performance testing program. *Journal of AOAC International* 83(2), 299-303.

Bonah, E., Huang, X., Aheto, J. H. and Osae, R. 2019. Application of hyperspectral imaging as a nondestructive technique for foodborne pathogen detection and characterization. *Foodborne Pathogens and Disease* 16(10), 712-722.

Borisova, B., Sánchez, A., Soto-Rodríguez, P. E. D., Boujakhrout, A., Arévalo-Villena, M., Pingarrón, J. M., Briones-Pérez, A., Parrado, C. and Villalonga, R. 2018. Disposable amperometric immunosensor for *Saccharomyces cerevisiae* based on carboxylated graphene oxide-modified electrodes. *Analytical and Bioanalytical Chemistry* 410(30), 7901-7907.

Bozal-Palabiyik, B., Gumustas, A., Ozkan, S. A. and Uslu, B. 2018. Chapter 12: Biosensor-based methods for the determination of foodborne pathogens. In: Holban, A. M. and Grumezescu, A. M. (Eds.), *Foodborne Diseases*. Academic Press.

Buzalewicz, I., Suchwałko, A., Trzciński, P., Sas-Paszt, L., Sumorok, B., Kowal, K., Kozera, R., Wieliczko, A. and Podbielska, H. 2019. Integrated multi-channel optical system for bacteria characterization and its potential use for monitoring of environmental bacteria. *Biomedical Optics Express* 10(3), 1165–1183.

Centers for Disease Control and Prevention and U. C. 2022. CalciNet data - Norovirus US outbreak map. Available at: https://www.cdc.gov/norovirus/reporting/calicinet/data.html?CDC_AA_refVal=https%3A%2F%2Fwww.cdc.gov%2Fnorovirus%2Freporting%2Fcalicinet%2Fdata-tables.html [Accessed 5 May 2022].

Chand, R. and Neethirajan, S. 2017. Microfluidic platform integrated with graphene-gold nano-composite aptasensor for one-step detection of Norovirus. *Biosensors and Bioelectronics* 98, 47–53.

Chen, J., Tang, J., Liu, J., Cai, Z. and Bai, X. 2012. Development and evaluation of a multiplex PCR for simultaneous detection of five foodborne pathogens. *Journal of Applied Microbiology* 112(4), 823–830.

Chen, Q., Yuan, L., Wan, J., Chen, Y. and Du, C. 2014. Colorimetric detection of hepatitis E virus based on reverse transcription loop mediated isothermal amplification (RT-LAMP) assay. *Journal of Virological Methods* 197, 29–33.

Chen, Y., Wang, Z., Shi, Q., Huang, S., Yu, T., Zhang, L. and Yang, H. 2021. Multiplex PCR method for simultaneous detection of five pathogenic bacteria closely related to foodborne diseases. *3 Biotech* 11(5), 219.

Chin, N. A., Salihah, N. T., Shivanand, P. and Ahmed, M. U. 2021. Recent trends and developments of PCR-based methods for the detection of food-borne *Salmonella* bacteria and Norovirus. *Journal of Food Science and Technology*, 1–13.

Compton, J. 1991. Nucleic acid sequence-based amplification. *Nature* 350(6313), 91–92.

Cook, N., Paton, C. A., Wilkinson, N., Nichols, R. A., Barker, K. and Smith, H. V. 2006. Towards standard methods for the detection of *Cryptosporidium parvum* on lettuce and raspberries. Part 1: Development and optimization of methods. *International Journal of Food Microbiology* 109(3), 215–221.

Daskalov, H., Atanasova, S., Stoyanchev, T. and Chipilev, N. 2010. Microbiological and near infrared spectroscopy analysis of Bulgarian yellow cheese, contaminated with *Listeria monocytogenes*. *Bulgarian Journal of Veterinary Medicine* 13, 31.

Davis, R. and Mauer, L. J. 2010. Fourier transform infrared (FT-IR) spectroscopy: A rapid tool for detection and analysis of foodborne pathogenic bacteria. *Current Research, Technology and Education Topics in Applied Microbiology and Microbial Biotechnology* 2, 1582–1594.

Davis, S., Pettengill, J. B., Luo, Y., Payne, J., Shpuntoff, A., Rand, H. and Strain, E. 2015. CFSAN SNP pipeline: An automated method for constructing SNP matrices from next-generation sequence data. *PeerJ Computer Science* 1, e20.

De Graaf, M., Van Beek, J., Vennema, H., Podkolzin, A., Hewitt, J., Bucardo, F., Templeton, K., Mans, J., Nordgren, J. and Reuter, G. 2015. Emergence of a novel GII. 17 Norovirus–end of the GII era? *Eurosurveillance* 4, 21178.

Dirks, R. A. M., Jansen, C. C. C., Hägele, G., Zwartkruis-Nahuis, A. J. T., Tijsma, A. S. L. and Boxman, I. L. A. 2021. Quantitative levels of Norovirus and hepatitis A virus in bivalve molluscs collected along the food chain in the Netherlands, 2013-2017. *International Journal of Food Microbiology* 344, 109089.

Duan, N., Chang, B., Zhang, H., Wang, Z. and Wu, S. 2016. *Salmonella* Typhimurium detection using a surface-enhanced Raman scattering-based aptasensor. *International Journal of Food Microbiology* 218, 38–43.

Elmahdy, E. M., Shaheen, M. N. F., Mahmoud, L. H. I., Hammad, I. A. and Soliman, E. R. S. 2022. Detection of Norovirus and hepatitis A virus in strawberry and green leafy vegetables by using RT-qPCR in Egypt. *Food and Environmental Virology* 14(2), 178–189.

Feng, W., K., Jiang, A., Xiu, Z., Lao, Y., Li, Y. and Long, Y. 2020. An in situ-synthesized gene chip for the detection of food-borne pathogens on fresh-cut cantaloupe and lettuce. *Frontiers in Microbiology* 10.

Ferone, M., Gowen, A., Fanning, S. and Scannell, A. G. M. 2020. Microbial detection and identification methods: Bench top assays to omics approaches. *Comprehensive Reviews in Food Science and Food Safety* 19(6), 3106–3129.

Ferrario, C., Lugli, G. A., Ossiprandi, M. C., Turroni, F., Milani, C., Duranti, S., Mancabelli, L., Mangifesta, M., Alessandri, G., Van Sinderen, D. and Ventura, M. 2017. Next generation sequencing-based multigene panel for high throughput detection of food-borne pathogens. *International Journal of Food Microbiology* 256, 20–29.

Fino, V. R. and Kniel, K. E. 2008. Comparative recovery of foodborne viruses from fresh produce. *Foodborne Pathogens and Disease* 5(6), 819–825.

Florio, W., Tavanti, A., Barnini, S., Ghelardi, E. and Lupetti, A. 2018. Recent advances and ongoing challenges in the diagnosis of microbial infections by MALDI-TOF mass spectrometry. *Frontiers in Microbiology* 9, 1097.

Fu, X., Sun, J., Liang, R., Guo, H., Wang, L. and Sun, X. 2021. Application progress of microfluidics-integrated biosensing platforms in the detection of foodborne pathogens. *Trends in Food Science and Technology* 116, 115–129.

Fumian, T. M., Ferreira, F. C., De Andrade, J. D. S. R., Canal, N., Silva Gomes, G., Teixeira, L. B. and Miagostovich, M. P. 2021. Norovirus foodborne outbreak associated with the consumption of ice pop, Southern Brazil, 2020. *Food and Environmental Virology* 13(4), 553–559.

Fung, D. Y. C. 2002. Rapid methods and automation in microbiology. *Comprehensive Reviews in Food Science and Food Safety* 1(1), 3–22.

Gao, W., Huang, H., Zhang, Y., Zhu, P., Yan, X., Fan, J. and Chen, X. 2017. Recombinase polymerase amplification-based assay for rapid detection of *Listeria monocytogenes* in food samples. *Food Analytical Methods* 10(6), 1972–1981.

Gao, X., Jiang, T. and Qin, W. 2022. Potentiometric aptasensing of *Escherichia coli* based on electrogenerated chemiluminescence as a highly sensitive readout. *Biosensors and Bioelectronics* 200, 113923.

Geng, Y., Liu, S., Wang, J., Nan, H., Liu, L., Sun, X., Li, D., Liu, M., Wang, J. and Tan, K. 2018. Rapid detection of *Staphylococcus aureus* in food using a recombinase polymerase amplification-based assay. *Food Analytical Methods* 11(10), 2847–2856.

Gómez-Govea, M., Solís-Soto, L., Heredia, N., García, S., Moreno, G., Tovar, O. and Isunza, G. 2012. Analysis of microbial contamination levels of fruits and vegetables at retail in Monterrey, Mexico. *Journal of Food, Agriculture and Environment* 10, 152–156.

Gómez-Montaño, F. J., Orduña-Díaz, A., Avelino-Flores, M. D. C. G., Avelino-Flores, F., Reyes-Betanzo, C. and Rojas-López, M. 2021. Determination of *Salmonella typhimurium* by Fourier transform infrared (FTIR) spectroscopy using a hydrogenated amorphous silicon carbide (a-SiC:H) thin film. *Analytical Letters* 54(16), 2655–2665.

Gowen, A. A., Feng, Y., Gaston, E. and Valdramidis, V. 2015. Recent applications of hyperspectral imaging in microbiology. *Talanta* 137, 43–54.

Guan, Z. P., Jiang, Y., Gao, F., Zhang, L., Zhou, G. H. and Guan, Z. J. 2013. Rapid and simultaneous analysis of five foodborne pathogenic bacteria using multiplex PCR. *European Food Research and Technology* 237(4), 627–637.

Hegde, N. V., Jayarao, B. M. and Debroy, C. 2012. Rapid detection of the top six non-O157 Shiga toxin-producing *Escherichia coli* O groups in ground beef by flow cytometry. *Journal of Clinical Microbiology* 50(6), 2137–2139.

Hernández Hernández, O., Gutiérrez-Escolano, A. L., Cancio-Lonches, C., Iturriaga, M. H., Pacheco-Aguilar, J. R., Morales-Rayas, R. and Arvizu-Medrano, S. M. 2022. Multiplex PCR method for the detection of human Norovirus, *Salmonella* spp., *Shigella* spp., and shiga toxin producing *Escherichia coli* in blackberry, coriander, lettuce and strawberry. *Food Microbiology* 102, 103926.

Hirvonen, J. J., Siitonen, A. and Kaukoranta, S. S. 2012. Usability and performance of CHROMagar STEC medium in detection of Shiga toxin-producing *Escherichia coli* strains. *Journal of Clinical Microbiology* 50(11), 3586–3590.

Hohweyer, J., Cazeaux, C., Travaillé, E., Languet, E., Dumètre, A., Aubert, D., Terryn, C., Dubey, J. P., Azas, N., Houssin, M., Loïc, F., Villena, I. and La Carbona, S. 2016. Simultaneous detection of the protozoan parasites *Toxoplasma*, *Cryptosporidium* and *Giardia* in food matrices and their persistence on basil leaves. *Food Microbiology* 57, 36–44.

Holland, R. D., Wilkes, J. G., Rafii, F., Sutherland, J. B., Persons, C. C., Voorhees, K. J. and Lay, J. O. 1996. Rapid identification of intact whole bacteria based on spectral patterns using matrix-assisted laser desorption/ionization with time-of-flight mass spectrometry. *Rapid Communications in Mass Spectrometry: RCM* 10(10), 1227–1232.

International Standards Organization. 2016. ISO 18744. Microbiology of the food chain - Detection and enumeration of Cryptosporidium and Giardia in fresh leafy green vegetables and berry fruits. ISO. Available at: https://www.iso.org/standard/63252.html [Accessed 5 March 2022].

International Standards Organization. 2017a. ISO/IEC 17025:2017 - General requirements for the competence of testing and calibration laboratories [Online]. ISO. Available at: https://www.iso.org/standard/66912.html [Accessed 5 March 2022].

International Standards Organization (ISO). 2017b. *Microbiology of the Food Chain—Horizontal Method for Determination of Hepatitis A Virus and Norovirus Using Real-Time RT-PCR—Part 1: Method for Quantification*. International Organization for Standardization, pp. 15216–15211.

Iqbal, A., Labib, M., Muharemagic, D., Sattar, S., Dixon, B. R. and Berezovski, M. V. 2015. detection of *Cryptosporidium parvum* oocysts on fresh produce using DNA aptamers. *PLoS ONE* 10(9), e0137455.

Kampeera, J., Pasakon, P., Karuwan, C., Arunrut, N., Sappat, A., Sirithammajak, S., Dechokiattawan, N., Sumranwanich, T., Chaivisuthangkura, P., Ounjai, P., Chankhamhaengdecha, S., Wisitsoraat, A., Tuantranont, A. and Kiatpathomchai, W. 2019. Point-of-care rapid detection of *Vibrio parahaemolyticus* in seafood using loop-mediated isothermal amplification and graphene-based screen-printed electrochemical sensor. *Biosensors and Bioelectronics* 132, 271–278.

Kannappan, A., Balasubramaniam, B., Ranjitha, R., Srinivasan, R., Packiavathy, I. A. S. V., Balamurugan, K., Pandian, S. K. and Ravi, A. V. 2019. *In vitro* and *in vivo* biofilm inhibitory efficacy of geraniol-cefotaxime combination against *Staphylococcus* spp.

Food and Chemical Toxicology: An International Journal Published for the British Industrial Biological Research Association 125, 322–332.

Keserue, H.-A., Füchslin, H. P., Wittwer, M., Nguyen-Viet, H., Nguyen, T. T., Surinkul, N., Koottatep, T., Schürch, N. and Egli, T. 2012. Comparison of rapid methods for detection of *Giardia* spp. and *Cryptosporidium* spp. (oo)cysts using transportable instrumentation in a field deployment. *Environmental Science & Technology* 46, 8952–8959.

Kim, D., Hong, S., Kim, Y. T., Ryu, S., Kim, H. B. and Lee, J. H. 2018. Metagenomic approach to identifying foodborne pathogens on Chinese cabbage. *Journal of Microbiology and Biotechnology* 28(2), 227–235.

Kreitlow, A., Becker, A., Ahmed, M. F. E., Kittler, S., Schotte, U., Plötz, M. and Abdulmawjood, A. 2021. Combined loop-mediated isothermal amplification assays for rapid detection and one-step differentiation of *Campylobacter jejuni* and *Campylobacter coli* in meat products. *Frontiers in Microbiology* 12, 668824–668824.

Lakshmi, B. A. and Kim, S. 2021a. Chapter 27: Recent trends in the utilization of LAMP for the diagnosis of viruses, bacteria, and allergens in food. In: Viswanath, B. (Ed.), *Recent Developments in Applied Microbiology and Biochemistry*. Academic Press.

Lakshmi, B. A. and Kim, S. 2021b. Recent trends in the utilization of LAMP for the diagnosis of viruses, bacteria, and allergens in food. In: *Recent Developments in Applied Microbiology and Biochemistry*. Elsevier.

Lee, J., Ahn, M. S., Lee, Y. L., Jie, E. Y., Kim, S. G. and Kim, S. W. 2019a. Rapid tool for identification of bacterial strains using Fourier transform infrared spectroscopy on genomic DNA. *Journal of Applied Microbiology* 126(3), 864–871.

Lee, S. I., Kim, S. A., Park, S. H. and Ricke, S. C. 2019b. Molecular and new-generation techniques for rapid detection of foodborne pathogens and characterization of microbial communities in poultry meat. In: Venkitanarayanan, K., Thakur, S. and Ricke, S. C. (Eds.), *Food Safety in Poultry Meat Production*. Cham: Springer International Publishing.

Lei, I.-F., Roffey, P., Blanchard, C. and Gu, K. 2008. Development of a multiplex PCR method for the detection of six common foodborne pathogens. *Journal of Food and Drug Analysis* 16(4).

Lewis, E., Hudson, J. A., Cook, N., Barnes, J. D. and Haynes, E. 2020. Next-generation sequencing as a screening tool for foodborne pathogens in fresh produce. *Journal of Microbiological Methods* 171, 105840.

Li, X., Li, J., Zhu, C., Zhang, X. and Chen, J. 2018. A new electrochemical immunoassay for prion protein based on hybridization chain reaction with hemin/G-quadruplex DNAzyme. *Talanta* 182, 292–298.

Li, Y., Liu, D., Cao, B., Han, W., Liu, Y., Liu, F., Guo, X., Bastin, D. A., Feng, L. and Wang, L. 2006. Development of a serotype-specific DNA microarray for identification of some *Shigella* and pathogenic *Escherichia coli* strains. *Journal of Clinical Microbiology* 44(12), 4376–4383.

Liming, S. H. and Bhagwat, A. A. 2004. Application of a molecular beacon–Real-time PCR technology to detect *Salmonella* species contaminating fruits and vegetables. *International Journal of Food Microbiology* 95(2), 177–187.

Liu, J., Jasim, I., Shen, Z., Zhao, L., Dweik, M., Zhang, S. and Almasri, M. 2019a. A microfluidic based biosensor for rapid detection of *Salmonella* in food products. *PLoS ONE* 14(5), e0216873.

Liu, Y., Cao, Y., Wang, T., Dong, Q., Li, J. and Niu, C. 2019b. Detection of 12 common food-borne bacterial pathogens by TaqMan Real-Time PCR using a single set of reaction conditions. *Frontiers in Microbiology* 10, 222.

Liu, Y., Wang, J., Zhao, C., Guo, X., Song, X., Zhao, W., Liu, S., Xu, K. and Li, J. 2019c. A multicolorimetric assay for rapid detection of *Listeria monocytogenes* based on the etching of gold nanorods. *Analytica Chimica Acta* 1048, 154–160.

Liu, S., Wang, B., Sui, Z., Wang, Z., Li, L., Zhen, X., Zhao, W. and Zhou, G. 2021. Faster detection of *Staphylococcus aureus* in milk and milk powder by flow cytometry. *Foodborne Pathogens and Disease* 18(5), 346–353.

Lobato, I. M. and O'Sullivan, C. K. 2018. Recombinase polymerase amplification: Basics, applications and recent advances. *Trends in Analytical Chemistry: TRAC* 98, 19–35.

Ma, K., Deng, Y., Bai, Y., Xu, D., Chen, E., Wu, H., Li, B. and Gao, L. 2014. Rapid and simultaneous detection of *Salmonella*, *Shigella*, and *Staphylococcus aureus* in fresh pork using a multiplex real-time PCR assay based on immunomagnetic separation. *Food Control* 42, 87–93.

Macey, M. G. 2007. Principles of flow cytometry. In: *Flow Cytometry*. Springer.

Maciorowski, K. G., Herrera, P., Jones, F. T., Pillai, S. D. and Ricke, S. C. 2006. Cultural and immunological detection methods for *Salmonella* spp. in animal feeds - A review. *Veterinary Research Communications* 30(2), 127–137.

Macori, G., Gilardi, G., Bellio, A., Bianchi, D. M., Gallina, S., Vitale, N., Gullino, M. L. and Decastelli, L. 2018. Microbiological parameters in the primary production of berries: A pilot study. *Foods* 7(7), 105.

Manzano, M., Viezzi, S., Mazerat, S., Marks, R. S. and Vidic, J. 2018. Rapid and label-free electrochemical DNA biosensor for detecting hepatitis A virus. *Biosensors and Bioelectronics* 100, 89–95.

Maruyama, F., Kenzaka, T., Yamaguchi, N., Tani, K. and Nasu, M. 2003. Detection of bacteria carrying the *stx2* gene by in situ loop-mediated isothermal amplification. *Applied and Environmental Microbiology* 69(8), 5023–5028.

Maurer, J. J. 2011. Rapid detection and limitations of molecular techniques. *Annual Review of Food Science and Technology* 2, 259–279.

Mori, Y. and Notomi, T. 2009. Loop-mediated isothermal amplification (LAMP): A rapid, accurate, and cost-effective diagnostic method for infectious diseases. *Journal of Infection and Chemotherapy: Official Journal of the Japan Society of Chemotherapy* 15(2), 62–69.

Müller, L., Rasmussen, L. D., Jensen, T., Schultz, A. C., Kjelsø, C., Barnadas, C., Sigsgaard, K., Larsen, A. R., Jensen, C. W., Jeppesen, S., Uhrbrand, K., Hove, N., Mølbak, K. and Ethelberg, S. 2016. Series of Norovirus outbreaks caused by consumption of green coral lettuce, Denmark, April 2016. *PLoS Currents* 8.

Nagaraj, S., Ramlal, S., Kingston, J. and Batra, H. V. 2016. Development of IgY based sandwich ELISA for the detection of staphylococcal enterotoxin G (SEG), an EGC toxin. *International Journal of Food Microbiology* 237, 136–141.

Neethirajan, S., Ahmed, S. R., Chand, R., Buozis, J. and Nagy, É. 2017. Recent advances in biosensor development for foodborne virus detection. *Nanotheranostics* 1(3), 272–295.

Niendorf, S., Jacobsen, S., Faber, M., Eis-Hübinger, A. M., Hofmann, J., Zimmermann, O., Höhne, M. and Bock, C. T. 2017. Steep rise in Norovirus cases and emergence of a new recombinant strain GII. P16-GII. 2, Germany, winter 2016. *Eurosurveillance* 22(4), 30447.

Notomi, T., Okayama, H., Masubuchi, H., Yonekawa, T., Watanabe, K., Amino, N. and Hase, T. 2000. Loop-mediated isothermal amplification of DNA. *Nucleic Acids Research* 28(12), E63.

Ogunremi, D., Dupras, A. A., Naushad, S., Gao, R., Duceppe, M. O., Omidi, K., Márquez, I. G., Huang, H., Goodridge, L., Lévesque, R. C., Hasan, N. A., Dadlani, M., Dixon, B., Magierowski, S. and Masson, L. 2020. A new whole genome culture-independent diagnostic Test (WG-CIDT) for rapid detection of *Salmonella* in lettuce. *Frontiers in Microbiology* 11, 602.

Ohk, S. H. and Bhunia, A. K. 2013. Multiplex fiber optic biosensor for detection of *Listeria monocytogenes*, *Escherichia coli* O157:H7 and *Salmonella enterica* from ready-to-eat meat samples. *Food Microbiology* 33(2), 166–171.

Ohk, S. H., Koo, O. K., Sen, T., Yamamoto, C. M. and Bhunia, A. K. 2010. Antibody-aptamer functionalized fibre-optic biosensor for specific detection of *Listeria monocytogenes* from food. *Journal of Applied Microbiology* 109(3), 808–817.

Omiccioli, E., Amagliani, G., Brandi, G. and Magnani, M. 2009. A new platform for real-time PCR detection of *Salmonella* spp., *Listeria monocytogenes* and *Escherichia coli* O157 in milk. *Food Microbiology* 26(6), 615–622.

O'Shea, H., Blacklaws, B. A., Collins, P. J., McKillen, J. and Fitzgerald, R. 2019. Viruses associated with foodborne infections. *Reference Module in Life Sciences*, B978–B970-12-809633-8.90273-5.

Pandey, C. M., Tiwari, I., Singh, V. N., Sood, K. N., Sumana, G. and Malhotra, B. D. 2017. Highly sensitive electrochemical immunosensor based on graphene-wrapped copper oxide-cysteine hierarchical structure for detection of pathogenic bacteria. *Sensors and Actuators, Part B: Chemical* 238, 1060–1069.

Pang, B., Zhao, C., Li, L., Song, X., Xu, K., Wang, J., Liu, Y., Fu, K., Bao, H., Song, D., Meng, X., Qu, X., Zhang, Z. and Li, J. 2018. Development of a low-cost paper-based ELISA method for rapid *Escherichia coli* O157: H7 detection. *Analytical Biochemistry* 542, 58–62.

Pereira, J. M., Leme, L. M., Perdoncini, M. R. F. G., Valderrama, P. and Março, P. H. 2018. Fast discrimination of milk contaminated with *Salmonella* sp. via near-infrared spectroscopy. *Food Analytical Methods* 11(7), 1878–1885.

Pinto, B., Chenoll, E. and Aznar, R. 2005. Identification and typing of food-borne *Staphylococcus aureus* by PCR-based techniques. *Systematic and Applied Microbiology* 28(4), 340–352.

Rajapaksha, P., Elbourne, A., Gangadoo, S., Brown, R., Cozzolino, D. and Chapman, J. 2019. A review of methods for the detection of pathogenic microorganisms. *Analyst* 144(2), 396–411.

Rama Devi, K., Srinivasan, R., Kannappan, A., Santhakumari, S., Bhuvaneswari, M., Rajasekar, P., Prabhu, N. M. and Veera Ravi, A. 2016. In vitro and in vivo efficacy of rosmarinic acid on quorum sensing mediated biofilm formation and virulence factor production in *Aeromonas hydrophila*. *Biofouling* 32(10), 1171–1183.

Ramirez-Martinez, M. L., Olmos-Ortiz, L. M., Barajas-Mendiola, M. A., Giono Cerezo, S., Avila, E. E. and Cuellar-Mata, P. 2015. A PCR procedure for the detection of *Giardia intestinalis* cysts and *Escherichia coli* in lettuce. *Letters in Applied Microbiology* 60(6), 517–523.

Rantsiou, K., Kathariou, S., Winkler, A., Skandamis, P., Saint-Cyr, M. J., Rouzeau-Szynalski, K. and Amézquita, A. 2018. Next generation microbiological risk assessment: Opportunities of whole genome sequencing (WGS) for foodborne pathogen surveillance, source tracking and risk assessment. *International Journal of Food Microbiology* 287, 3-9.

Ricke, S. C., Feye, K. M., Chaney, W. E., Shi, Z., Pavlidis, H. and Yang, Y. 2018. Developments in rapid detection methods for the detection of foodborne *Campylobacter* in the United States. *Frontiers in Microbiology* 9, 3280.

Rodriguez-Saona, L. E., Khambaty, F. M., Fry, F. S., Dubois, J. and Calvey, E. M. 2004. Detection and identification of bacteria in a juice matrix with Fourier transform-near infrared spectroscopy and multivariate analysis. *Journal of Food Protection* 67(11), 2555–2559.

Rzeżutka, A., Nichols, R. A., Connelly, L., Kaupke, A., Kozyra, I., Cook, N., Birrell, S. and Smith, H. V. 2010. *Cryptosporidium* oocysts on fresh produce from areas of high livestock production in Poland. *International Journal of Food Microbiology* 139(1–2), 96–101.

Saravanan, A., Kumar, P. S., Hemavathy, R. V., Jeevanantham, S., Kamalesh, R., Sneha, S. and Yaashikaa, P. R. 2021. Methods of detection of food-borne pathogens: A review. *Environmental Chemistry Letters* 19(1), 189–207.

Schena, M., Shalon, D., Davis, R. W. and Brown, P. O. 1995. Quantitative monitoring of gene expression patterns with a complementary DNA microarray. *Science* 270(5235), 467–470.

Serrano, N. S., Zweifel, C., Corti, S. and Stephan, R. 2018. Microbiological quality and presence of foodborne pathogens in raw milk cheeses and raw meat products marketed at farm level in Switzerland. *Italian Journal of Food Safety* 7(2), 7337–7337.

Severgnini, M., Cremonesi, P., Consolandi, C., De Bellis, G. and Castiglioni, B. 2011. Advances in DNA microarray technology for the detection of foodborne pathogens. *Food and Bioprocess Technology* 4(6), 936–953.

Shaheen, M. N. F., Elmahdy, E. M. and Chawla-Sarkar, M. 2019. Quantitative PCR-based identification of enteric viruses contaminating fresh produce and surface water used for irrigation in Egypt. *Environmental Science and Pollution Research International* 26(21), 21619–21628.

Shang, D., Zhao, H., Xu, X., Arunachalam, K., Chang, J., Bai, L. and Shi, C. 2021. Conjugative IncHI2 plasmid harboring novel class 1 integron mediated dissemination of multidrug resistance genes in *Salmonella typhimurium*. *Food Control* 122, 107810.

Shridhar, P. B., Patel, I. R., Gangiredla, J., Noll, L. W., Shi, X., Bai, J., Elkins, C. A., Strockbine, N. and Nagaraja, T. G. 2018. DNA microarray-based assessment of virulence potential of Shiga toxin gene-carrying *Escherichia coli* O104:H7 isolated from feedlot cattle feces. *PLoS ONE* 13(4), e0196490.

Siala, M., Barbana, A., Smaoui, S., Hachicha, S., Marouane, C., Kammoun, S., Gdoura, R. and Messadi-Akrout, F. 2017. Screening and detecting *Salmonella* in different food matrices in Southern Tunisia using a combined enrichment/real-time PCR method: Correlation with conventional culture method. *Frontiers in Microbiology* 8, 2416.

Singh, A. K., Bettasso, A. M., Bae, E., Rajwa, B., Dundar, M. M., Forster, M. D., Liu, L., Barrett, B., Lovchik, J. and Robinson, J. P. 2014. Laser optical sensor, a label-free on-plate *Salmonella enterica* colony detection tool. *mBio* 5, e01019-13.

Singhal, N., Kumar, M., Kanaujia, P. K. and Virdi, J. S. 2015. MALDI-TOF mass spectrometry: An emerging technology for microbial identification and diagnosis. *Frontiers in Microbiology* 6, 791.

Siripatrawan, U., Makino, Y., Kawagoe, Y. and Oshita, S. 2011. Rapid detection of *Escherichia coli* contamination in packaged fresh spinach using hyperspectral imaging. *Talanta* 85(1), 276–281.

Smith, E. and Dent, G. 2019. *Modern Raman Spectroscopy: A Practical Approach*. John Wiley & Sons.

Spagnolo, S., De La Franier, B., Davoudian, K., Hianik, T. and Thompson, M. 2022. Detection of *E. coli* bacteria in milk by an acoustic wave aptasensor with an anti-fouling coating. *Sensors* 22(5), 1853.

Sulaiman, I. M., Miranda, N. and Simpson, S. 2021. MALDI-TOF mass spectrometry and 16S rRNA gene sequence analysis for the identification of foodborne *Clostridium* spp. *Journal of AOAC International* 104(5), 1381–1388.

Tao, F. and Peng, Y. 2014. A method for nondestructive prediction of pork meat quality and safety attributes by hyperspectral imaging technique. *Journal of Food Engineering* 126, 98–106.

Terio, V., Lorusso, P., Mottola, A., Buonavoglia, C., Tantillo, G., Bonerba, E. and Di Pinto, A. 2020. Norovirus detection in ready-to-eat salads by propidium monoazide real time RT-PCR assay. *Applied Sciences* 10(15), 5176.

Tian, Y., Gao, X., Qi, W. L., Wang, Y., Wang, X., Zhou, J., Lu, D. and Chen, B. 2021. Advances in differentiation and identification of foodborne bacteria using near infrared spectroscopy. *Analytical Methods: Advancing Methods and Applications* 13(23), 2558–2566.

Umesha, S. and Manukumar, H. M. 2018. Advanced molecular diagnostic techniques for detection of food-borne pathogens: Current applications and future challenges. *Critical Reviews in Food Science and Nutrition* 58(1), 84–104.

Unger, P., Sekhon, A. S., Chen, X. and Michael, M. 2022. Developing an affordable hyperspectral imaging system for rapid identification of *Escherichia coli* O157:H7 and *Listeria monocytogenes* in dairy products. *Food Science and Nutrition* 10(4), 1175–1183.

Utaaker, K. S., Huang, Q. and Robertson, L. J. 2015. A reduced-cost approach for analyzing fresh produce for contamination with *Cryptosporidium* oocysts and/or *Giardia* cysts. *Food Research International* 77, 326–332.

Vaisocherová-Lísalová, H., Víšová, I., Ermini, M. L., Špringer, T., Song, X. C., Mrázek, J., Lamačová, J., Scott Lynn, N., Šedivák, P. and Homola, J. 2016. Low-fouling surface plasmon resonance biosensor for multi-step detection of foodborne bacterial pathogens in complex food samples. *Biosensors and Bioelectronics* 80, 84–90.

Välimaa, A.-L., Tilsala-Timisjärvi, A. and Virtanen, E. 2015. Rapid detection and identification methods for *Listeria monocytogenes* in the food chain – A review. *Food Control* 55, 103–114.

Vichaibun, V. and Kanchanaphum, P. 2020. Quantitative LAMP and PCR detection of *Salmonella* in chicken samples collected from local markets around Pathum Thani Province, Thailand. *International Journal of Food Science* 2020, 8833173.

Victor, C. P., Ellis, K., Lamar, F. and Leon, J. S. 2021. Agricultural detection of Norovirus and hepatitis A using fecal indicators: A systematic review. *International Journal of Microbiology* 2021, 6631920.

Vinayaka, A. C., Ngo, T. A., Kant, K., Engelsmann, P., Dave, V. P., Shahbazi, M. A., Wolff, A. and Bang, D. D. 2019. Rapid detection of *Salmonella enterica* in food samples by a novel approach with combination of sample concentration and direct PCR. *Biosensors and Bioelectronics* 129, 224–230.

Wang, J., Wu, X., Wang, C., Rong, Z., Ding, H., Li, H., Li, S., Shao, N., Dong, P., Xiao, R. and Wang, S. 2016. Facile synthesis of Au-coated magnetic nanoparticles and their application in bacteria detection via a SERS method. *ACS Applied Materials and Interfaces* 8(31), 19958–19967.

Wang, L., Shi, L., Alam, M. J., Geng, Y. and Li, L. 2008. Specific and rapid detection of foodborne *Salmonella* by loop-mediated isothermal amplification method. *Food Research International* 41(1), 69–74.

Wang, P. 2021. Nucleic acid-based rapid methods for the detection of foodborne pathogens. *Journal of Physics: Conference Series* 1759(1), 012023.

Wang, X. W., Zhang, L., Jin, L. Q., Jin, M., Shen, Z. Q., An, S., Chao, F. H. and Li, J. W. 2007. Development and application of an oligonucleotide microarray for the detection of food-borne bacterial pathogens. *Applied Microbiology and Biotechnology* 76(1), 225–233.

Wang, Y., Ye, Z., Si, C. and Ying, Y. 2013. Monitoring of *Escherichia coli* O157: H7 in food samples using lectin based surface plasmon resonance biosensor. *Food Chemistry* 136(3–4), 1303–1308.

Wang, Z., Xu, Q., Liu, S., Liu, Y., Gao, Y., Wang, M., Zhang, L., Chang, H., Wei, Q. and Sui, Z. 2022. Rapid and multiplexed quantification of *Salmonella*, *Escherichia coli* O157:H7, and *Shigella flexneri* in ground beef using flow cytometry. *Talanta* 238(1), 123005.

Wei, C., Zhong, J., Hu, T. and Zhao, X. 2018. Simultaneous detection of *Escherichia coli* O157:H7, *Staphylococcus aureus* and *Salmonella* by multiplex PCR in milk. *3 Biotech* 8(1), 76.

Wilkerson, M. J. 2012. Principles and applications of flow cytometry and cell sorting in companion animal medicine. *Veterinary Clinics of North America. Small Animal Practice* 42(1), 53–71.

Won, Y.-J., Lee, S.-G., Paik, S.-Y. and Lyoo, K.-S. 2019. Development of an oligonucleotide-based microarray for the detection of foodborne viruses. *Journal of Consumer Protection and Food Safety* 14(3), 287–291.

World Health Organization (WHO). 2018. The safe food imperative: Accelerating progress in low- and middle-income countries [Online]. Available at: https://www.worldbank.org/en/topic/agriculture/publication/the-safe-food-imperative-accelerating-progress-in-low-and-middle-income-countries [Accessed 23 October 2018].

World Health Organization (WHO). 2020. Food safety. Available at: https://www.who.int/en/news-room/fact-sheets/detail/food-safety [Accessed 30 April 2020].

Xie, Y., Xu, L., Wang, Y., Shao, J., Wang, L., Wang, H., Qian, H. and Yao, W. 2013. Label. Label-free detection of the foodborne pathogens of Enterobacteriaceae by surface-enhanced Raman spectroscopy. *Analytical Methods* 5(4), 946–952.

Xu, L., Bai, X. and Bhunia, A. K. 2021. Current state of development of biosensors and their application in foodborne pathogen detection. *Journal of Food Protection* 84(7), 1213–1227.

Xu, Y., Kutsanedzie, F. Y. H., Sun, H., Wang, M., Chen, Q., Guo, Z. and Wu, J. 2018. Rapid *Pseudomonas* species identification from chicken by integrating colorimetric sensors with near-infrared spectroscopy. *Food Analytical Methods* 11(4), 1199–1208.

Yamaguchi, N., Sasada, M., Yamanaka, M. and Nasu, M. 2003. Rapid detection of respiring *Escherichia coli* O157: H7 in apple juice, milk, and ground beef by flow cytometry. *Cytometry. Part A: The Journal of the International Society for Analytical Cytology* 54(1), 27–35.

Yeni, F., Acar, S., Polat, Ö. G., Soyer, Y. and Alpas, H. 2014. Rapid and standardized methods for detection of foodborne pathogens and mycotoxins on fresh produce. *Food Control* 40, 359–367.

Zhai, L., Liu, H., Chen, Q., Lu, Z., Zhang, C., Lv, F. and Bie, X. 2019. Development of a real-time nucleic acid sequence-based amplification assay for the rapid detection of *Salmonella* spp. from food. *Brazilian Journal of Microbiology: [Publication of the Brazilian Society for Microbiology]* 50(1), 255–261.

Zhang, G. 2013. Foodborne pathogenic bacteria detection: An evaluation of current and developing methods. *The Meducator* 1.

Zhang, J., Cao, J., Zhu, M., Xu, M. and Shi, F. 2019. Loop-mediated isothermal amplification-lateral-flow dipstick (LAMP-LFD) to detect *Mycoplasma ovipneumoniae*. *World Journal of Microbiology and Biotechnology* 35(2), 31.

Zhang, Z., Xiao, L., Lou, Y., Jin, M., Liao, C., Malakar, P. K., Pan, Y. and Zhao, Y. 2015. Development of a multiplex real-time PCR method for simultaneous detection of *Vibrio parahaemolyticus*, *Listeria monocytogenes* and *Salmonella* spp. in raw shrimp. *Food Control* 51, 31–36.

Zhao, X., Li, M. and Xu, Z. 2018. Detection of foodborne pathogens by surface enhanced Raman spectroscopy. *Frontiers in Microbiology* 9, 1236.

Zhao, X., Lin, C. W., Wang, J. and Oh, D. H. 2014. Advances in rapid detection methods for foodborne pathogens. *Journal of Microbiology and Biotechnology* 24(3), 297–312.

Zhong, J. and Zhao, X. 2018. Isothermal amplification technologies for the detection of foodborne pathogens. *Food Analytical Methods* 11(6), 1543–1560.

Zhong, M., Yang, L., Yang, H., Cheng, C., Deng, W., Tan, Y., Xie, Q. and Yao, S. 2019. An electrochemical immunobiosensor for ultrasensitive detection of *Escherichia coli* O157: H7 using CdS quantum dots-encapsulated metal-organic frameworks as signal-amplifying tags. *Biosensors and Bioelectronics* 126, 493–500.

Zhu, L., He, J., Cao, X., Huang, K., Luo, Y. and Xu, W. 2016. Development of a double-antibody sandwich ELISA for rapid detection of *Bacillus cereus* in food. *Scientific Reports* 6, 16092.

Zhuang, S., Hong, H., Zhang, L. and Luo, Y. 2021. Spoilage-related microbiota in fish and crustaceans during storage: Research progress and future trends. *Comprehensive Reviews in Food Science and Food Safety* 20(1), 252–288.

Chapter 5

Advances in modelling pathogen behaviour in fresh produce

Panagiotis N. Skandamis, Agricultural University of Athens, Greece

1 Introduction

Consumption of fresh produce has significantly increased in recent years mainly because of the increased consumer awareness of the benefits of a healthy diet (Anon, 2007; Heaton and Jones, 2008). In addition, changes in lifestyles have led to an increased demand for packed ready-to-eat (RTE) produce and out-of-season fruit and vegetables. This trend can be attributed partly to the increase in demands for convenience in handling and preparation before consumption. These foods of plant origin, usually called 'fresh produce', are low-calorie, nutritious foods which help to ensure an adequate intake of vitamins, fibres and antioxidants (Giugliano and Esposito, 2008). It has been established that diets which are low in fat and high in fibre and include high consumption of fresh produce help protect against many cancers and lessen the risk of coronary heart disease (De Roever, 1998). Leafy vegetables and RTE salads have a relatively short shelf-life and a composition that is favourable for the growth of pathogenic microorganisms. RTE salads are considered more perishable than whole vegetables because the cutting and chopping process breaks the plant tissue and results in the release of nutrients from the interior (Beuchat, 1996; De Roever, 1998). Damaging plant tissue also creates safety issues. The need for large quantities of fresh, often even exotic, produce is achieved by harvesting and production in many different countries with very diverse practices relating to cultivation, irrigation, harvesting cleaning and disinfection, with consequently variable levels of quality and safety.

http://dx.doi.org/10.19103/AS.2023.0121.09

The conditions of these foods are almost ideal for the survival and growth of many microorganisms. Most of the microorganisms that are initially observed on whole fruit or vegetable surfaces are soil inhabitants, members of a very large and diverse microbial community. As a result, bacteria, yeasts and fungi become an integral part of the epiphytic flora of fruits and vegetables and many are present at the time of consumption. Bacteria generally present in fruits and vegetables include species of *Pseudomonas*, *Bacillus*, *Enterobacter*, *Staphylococcus, Streptococcus, Lactobacillus, Leuconostoc*, etc. Of these, Pseudomonads, Enterobacteriaceae and lactic acid bacteria are the dominant groups (Caponigro et al., 2010). The analysis of various RTE salads showed bacterial counts of up to 7.0 $\log_{10}$ cfu/g (Ragaert et al., 2007). The level of bacteria in final products is influenced by the microbiological quality of raw material, processing practices, storage and handling. In addition to bacteria, many different yeast species of comparable quantitative importance have been identified in fresh produce, including species of *Candida, Cryptococcus, Rhodotorula, Trichosporon, Pichia* and *Torulaspora* (Ragaert et al., 2007). In contrast to bacteria and yeasts, moulds are more rarely present in vegetables. Common fungi identified in fruits are *Botrytis cinerea*, *Alternaria*, *Cladosporium*, *Penicillium*, *Fusarium* and *Rhizopus* (Tournas and Katsoudas, 2005). Finally, certain parasites, such as cryptosporidium, and viruses, such as hepatitis A, have also been linked to outbreaks associated with fresh produce.

Under natural conditions (e.g. pre-harvest or post-harvest storage of non-injured fruits), the outer layer of the plant tissue consists of a hydrophobic surface providing a natural barrier against microorganisms (Lund, 1992; Beuchat, 2002; Brackett, 1999). The surface of plants may, however, be subjected to external stress before, during and after harvest due to weather, insects, birds, rodents, farm implements, harvesting, and post-harvest handling and processing (e.g. shredding chopping and washing). This kind of damage may lead to physical damage to the tissue, which may subsequently permit microorganisms to reach nutrients within the plant and multiply. In addition, when shredded products pass through a washing tank, they may be exposed to a cumulative microbial load in the water, which may or may not have been re-circulated between different processing lines within the same plant. This may cause a substantial increase in the microbial population unless proper water cleaning and disinfection interventions are in place. Furthermore, some of the microorganisms inhabiting the surfaces of fruits and vegetables produce pectinolytic enzymes, which degrade the cell structure, thus providing more nutrients for microbiological activity (Ragaert et al., 2007). Surface decontamination treatments with chemicals, most commonly chlorine, reduce the microbial load of fruits and vegetables and prevent the accumulation of microbial loads during washing processes. The shelf-life of these foods ranges from 4 days to 2 weeks, depending on the application of a surface decontamination treatment, the type of packaging and, in particular, the packaging atmosphere.

Microbiological safety risks from the consumption of fruit and vegetables are linked to their contamination with enteric pathogens, among which Norovirus, *Escherichia coli* O157:H7, *Salmonella* sp. and *Listeria monocytogenes* have been the most common cause of recent enteric disease outbreaks (FAO/WHO, 2008; EFSA, 2014, 2020; Schierstaedt et al., 2019). Other pathogens (including emerging pathogens), which are less frequently associated with fresh produce outbreaks, include *Cyclospora cayetanensis*, *Echinococcus multilocularis*, hepatitis A, *Arcobacter*, *Bacillus* and *Campylobacter* (Bartosova et al., 2021; Bell et al., 2021; Pradhan et al., 2018; Victor et al., 2021). *Salmonella* spp. are often isolated from lettuce, tomatoes, spinach, endive, cauliflower, sprouted seeds and mushrooms (Thunberg et al., 2002; Doran et al., 2005). *E. coli* O157:H7 can be found in the faeces of livestock and therefore is very likely to contaminate soil and leafy vegetables that are commonly associated with *E. coli* infections (Heaton and Jones, 2008). *Listeria* spp. are ubiquitous in the environment (vegetation, soil, water, faeces of livestock) and are often isolated from leafy vegetables, though root vegetables have a higher prevalence due to the closest contact with the soil (EFSA, 2020). Overall, pathogens may find their way onto fresh produce through faecal or water contamination. However, their establishment on the plant tissue strongly depends on their ability to attach and form biofilms.

Given the perishable character of fresh produce, predictive microbiology and quantitative microbial risk assessment (QMRA) tools are essential to assess the growth potential of key microorganisms in fresh-cut salads which compromise consumer health and/or render the products unfit for consumption. These tools contribute to the timely assessment of the microbiological status of products under different packaging and storage conditions in the food supply chain and help producers to re-assess their safety management systems. The next section provides an overview of advances in modelling the behaviour of spoilage and pathogenic microorganisms in packaged fresh produce. Modelling the growth of spoilage organisms is an integral part of fresh produce safety, since spoilage organisms and pathogens interact in determining the quality and safety of fresh produce at the time of consumption (Carlin et al., 1995, 1996).

2 Modelling spoilage of fresh produce

Significant amounts of fresh produce are discarded annually due to microbial spoilage during storage. According to EC Regulation 178/2002, foods unfit for human consumption may still be considered unsafe after a certain date even though they may not carry pathogens or harmful contaminants. The majority of commercial fresh-cut salads packaged under modified atmosphere packaging

(MAP) conditions have a common date marking of no more than 7 days from the point of packaging to ensure the product maintains its freshness and fresh-like appearance. There is a particular need to model the growth of spoilage indigenous organisms in RTE fresh-cut salads under commercial packaging conditions and assess the spoilage potential of different packaging solutions under variable chill chain conditions.

The optimum modified atmosphere packaging recommended for fresh-cut salads of leafy greens is so-called 'equilibrium modified atmosphere packaging' (EMAP), where the CO_2 and O_2 levels in the packaging headspace maintain the optimal appearance of the product, as determined by factors such as minimum respiratory activity, ripening rate, senescence fermentation, initial gas flushing and film permeability (Jacxsens et al., 2000, 2002; Fonseca et al., 2002; Gontard and Guillaume, 2009; Oliveira et al., 2015). Recommended concentrations of O_2 in MAP during storage of fresh-cut vegetables should range between 1% and 5% (Sandhya, 2010), whilst CO_2 concentrations should not exceed 10–15% (Zagory and Kader, 1988), in order to reduce respiration and deterioration rates and the growth of aerobic bacteria, and consequently extend the shelf-life of various fresh-cut vegetables (Cagnon et al., 2013; Kapetanakou et al., 2019).

Establishing EMAP requires mathematical models simulating the impact of temperature, O_2 and CO_2 on the growth of spoilage bacteria in RTE fresh-cut salads. One approach is the use of generic models, i.e. trained with data from solid growth media to simulate the composition of fresh-cut salads. These models depend on experimental conditions with variations, e.g. of headspace gas concentration or temperature on microbial response. The 'trade-off' of this simplified approach is that model predictions do not account for any impact of variability in food substrate characteristics (e.g. cavities or variability in moisture distribution) on microbial interactions. The accuracy of these models can be judged on the basis of their agreement with actual microbial growth in target foods.

Geysen et al. (2005a) described a multiple linear secondary model ($\sqrt{\mu_{max}} = 0.375 - 0.096 \cdot O_2 - 0.637 \cdot CO_2$) expressing the dependence of the maximum specific growth rate (μ_{max}) and lag time of *P. fluorescence* on super-atmospheric O_2 (20–100%) and CO_2 (0–25%), when grown on nutrient agar plates at 7°C. Lag time (λ) was found to be inversely proportional to μ_{max} and given by the equation $\lambda = 1.732/\mu_{max}$. This was one of the earliest *in vitro* models in this area which aimed to exclude the impact of uncontrolled factors on pseudomonads, such as food structure and microbial interactions, in order to study the effect of headspace gases in isolation. The experimental setup involved a constant humidified gas flow of 10 L per h in the headspace of the jars containing the nutrient agar plates. This assumed that initial gas flushing is maintained at equilibrium in the packaging headspace throughout storage, taking into account the respiratory activity of the plant and the permeability characteristics of a commercial packaging film. The model was then validated in lettuce using the same experimental setup in jars and isothermal storage at 7°C (Geysen et al., 2006). Validation suggested that the model could be used to predict growth of pseudomonads under isothermal conditions in modified atmosphere (MA)-packaged fresh-cut salads and proved the usefulness of generic modelling approaches in simulating microbial growth in foods based on their training in laboratory media.

However, along with gas composition, other factors such as control of storage temperature are also critical for minimizing the effect of rapid respiratory metabolism of fresh-cut vegetables on deterioration in quality (Koseki and Itoh, 2001; Zhu et al., 2018). Kapetanakou et al. (2019) have proposed a MA-packaged fresh-cut salad model to simulate the impact of temperature and CO_2, either initially flushed or dynamically equilibrated (from respiratory activity and gas release from the packaging film), on the growth of indigenous spoilage organisms, particularly pseudomonads and lactic acid bacteria. The model system requires a number of steps (Kapetanakou et al., 2019):

1 to mechanically destroy the plant tissue and convert it to a pulp, thereby minimizing the respiratory activity of the plant;
2 to package the pulp into a low permeability film flushed with the desired percentage of CO_2 in the headspace, so that the CO_2 hardly changes throughout storage; and
3 to use an increased headspace volume to the size of packaged pulp, e.g. 10 g of pulp in a plastic bag of 23 cm × 26 cm, in order to delay any change in gas concentration caused by microbial activity.

The model is applicable to all types of fresh-cut salads, regardless of botanic origin, plant respiratory activity and film permeability. It captures the growth of mixed spoilage organisms in response to levels of CO_2 in the headspace, regardless of flushing or respiratory activity. However, application of predictive models in simulating growth of spoilage organisms requires knowing the initial microbial load and the time-temperature history. Microbiological analysis takes on average 2–3 days, which is almost 20–30% of the total shelf-life. Time-temperature data can be translated into a log increase but is not directly correlated to the local enzymatic activity.

An alternative approach is provided by instrumental advances and automation in vibrational spectroscopy, coupled with the development of bioinformatics and machine learning algorithms. These have paved the way for multi/hyperspectral methods in detecting food adulteration as well as in freshness characterization and authentication. Supported by advanced statistical processing, these techniques correlate changes in colour and chemistry of the food surfaces with the freshness and microbiological status of foods, regardless of time-temperature history and packaging atmosphere. The combined effect of microbial growth and accelerated enzymatic activity (due to shredding, dicing or chopping) results in off-odours, texture breakdown, softening and browning of the tissue of the packaged fresh-cut produce. The use of FTIR, VIS- or NIR-spectroscopy and multi-spectral imaging can rapidly collect large amounts of data on changes in the quality of a packaged product, stored at various temperatures and packaging conditions. Processing of spectral data with linear, e.g. partial least squares regression, or nonlinear algorithms, e.g. artificial neural networks and supporting vector regression, can lead to a rapid and accurate assessment of freshness and microbiological status of packaged products in a non-invasive/non-destructive manner (Manthou et al., 2022). These methods performed well when targeting high microbial populations, e.g. >6 log CFU/g, but their applicability in detecting pathogens that commonly occur in low numbers (e.g. <100 CFU/g) is yet to be explored. In addition, even though some of these changes are visually detectable, the roughness of the plant surface and the complex appearance of fresh-cut salads can limit the reliable detection of decay.

3 Modelling pathogen behaviour in fresh produce

Fresh produce associated with disease outbreaks in the US includes greens-based salads, lettuce, potatoes, some fruits and sprouts (DeWaal and Bhuiya, 2007). Similar trends in vegetable outbreaks can also be found in Europe (EFSA, 2020). A WHO (2008) report on microbial safety in fresh produce prioritized leafy green vegetables as a global safety risk. The proportion of all foodborne disease outbreaks associated with fresh produce increased from 0.7% in the 1970s to 12% in the 1990s. In a compilation of USA outbreak data for the period 1990–2005, fresh produce has become even more dominant as a vehicle in foodborne outbreaks, accounting for 13% (713/5416) of outbreaks with an identified food source (DeWaal and Bhuiya, 2007). Sivapalasingam et al. (2004) identified 190 produce-associated outbreaks that were associated with 16.058 illnesses, 598 hospitalizations and 8 deaths from 1973 to 1997 in 32 USA states. In autumn 2006, a number of outbreaks (within a period of two months) of foodborne illness in the US were associated with the consumption of fresh produce. Sources included spinach (205 cases - with 103 hospitalizations and 3 deaths), lettuce (152 cases - 79 hospitalizations) and tomatoes (283 cases - 22 hospitalizations) (Doyle and Erickson, 2008). Significant outbreaks, such as one associated with spinach contaminated with *E. coli* O157:H7, resulted in >US$300 m market losses (Warriner et al., 2009).

Since 2010, the US has experienced a number of listeriosis outbreaks associated with the consumption of fresh produce products that are routinely considered as 'moderate' to 'low risk', compared to other food commodities based on QMRAs (Buchanan et al., 2017). These included the outbreak in 2011 associated with cantaloup, which affected 147 people and caused 33 deaths and the multi-state caramel apple outbreak in 2014–2015 resulting in 35 cases of illness, including 11 pregnant women, with 34 requiring hospitalization. Both outbreaks were due to the growth of the pathogen at high numbers. In the case of the caramel apple outbreak, the disease was attributed to an increase in the levels of *L. monocytogenes* in growth-supporting micro-environments on the flesh of apples that were accessed by the pathogen upon penetration of the stick in the apples from the stem side (Glass et al., 2015; Salazar et al., 2016). A multi-country listeriosis outbreak (2015–2018) in Europe was linked to the consumption of blanched frozen vegetables providing a vehicle for *L. monocytogenes*, causing 53 human cases and 10 deaths. The outbreak was attributed to *Listeria* growth during thawing and storage of contaminated products at chill temperatures until consumption (EFSA, 2020).

3.1 Growth modelling

The growth potential of pathogens on fresh-cut salads depends on factors such as the organism, plant product characteristics, packaging headspace, the competitive effect of epiphytic microbiota, temperature and the permeability of the packaging film.

Epiphytic (indigenous) microbiota of fresh-cut salads can have a remarkable antagonistic effect against pathogens, causing sometimes total growth arrest, depending on product pH and spatial distribution of microbial cells on the vegetable tissue (Geysen et al., 2005b, 2006; Kakiomenou et al., 1998; Manios et al., 2013). Other key factors responsible for the competitive advantage of epiphytic biota over pathogens include better acclimatization to vegetable surface characteristics and nutritional composition, allowing them to colonize surfaces before pathogens. The inhibition of pathogen growth by epiphytic microorganisms can be seen in the reduced growth rate of pathogens compared to that in laboratory media and lower maximum population density compared to a pure culture in broth (Koseki and Isobe, 2005b; Puerta-Gomez et al., 2013; Table 1). This is most likely the reason why the growth of pathogens on fresh-cut salads is often overestimated by predictive modelling tools (e.g. PMP and ComBase) which do not account for epiphytic biota (Sant'Ana et al., 2012; Kataoka et al., 2017).

To overcome such limitations, models need to be fitted to data reflecting the behaviour of pathogens directly on the food matrix of concern. A comprehensive list of currently available growth models for pathogens in fresh produce is provided in Table 1. In some studies, the constraints posed by the competitive growth of the epiphytic microbiota on the maximum population density (MPD) achieved by the pathogens are addressed by expressing MPD as a function of temperature via an additional secondary model describing the

Table 1 Growth models for pathogens on fresh-cut salads

Vegetable	Pathogen	Headspace composition	Temperature range	Packaging material	Secondary model	Reference
Nutrient agar plates, validation in lettuce	*Listeria innocua*	9 conditions expressed as $\%O_2$: $\%CO_2$: $\%N_2$: 100 87.5:12.5 75:25 60:0:40 53.75:12.5:33.75 47.5:25:27.5 20:0:80 20:12.5:67.5 20:25:55	7°C	Jars flushed with gases at 10 L/h	$\mu_{max} = 0.00204\ (31.5^{a} - CO_2)$ $\lambda = \sqrt{118(CO_2 + 0.007^{b})}$	Geysen et al. (2005b)
Lettuce	*Salmonella* Typhimurium	5% O_2:15% CO_2:80% N_2	7, 10, 15, 20, 25, 30°C	Low-density polyethylene (LDPE)	$\sqrt{\mu_{max}} = 0.0178\ (T - 6.65^{c})$ $\ln(\lambda^{d}) = -0.118\ (T + 34.84)$	Sant'Ana et al. (2012b)
	Listeria monocytogenes				$\sqrt{\mu_{max}} = 0.0144\ (T - 1.96)$ $\ln(\lambda) = -0.1445\ (T + 35.3)$	
	L. monocytogenes	Air	4, 10, 15, 20, 25, 30, 35°C	Barrier bag	$\sqrt{\mu_{max}} = 0.014\ (T + 10.2)$	Ding et al. (2010)
Lettuce treated with alkaline electrolyzed water for 3 min					$\sqrt{\mu_{max}} = 0.012\ (T + 12.9)$	

Iceberg lettuce	*S.* Typhimurium	Air		5, 10, 15, 25°C	Stomacher bag	$\sqrt{\mu_{max}} = 0.033\,(T - 4.96)$ $y_{max}{}^{e} = 0.257T + 8.641$	Koseki and Isobe (2005)
	Escherichia coli O157:H7					$\sqrt{\mu_{max}} = 0.033\,(T - 4.54)$ $y_{max} = 0.218T + 10.320$	
	L. monocytogenes					$\sqrt{\mu_{max}} = 0.016\,(T + 4.26)$ $y_{max} = 0.037T + 12.434$	
Mixture of lettuce and cabbage	Mixture of EHEC, EPEC, EIEC, ETEC, EAEC *E. coli* strains	Air		4, 10, 12, 15, 25, 30, 37°C	Sterilized container	$\ln(\lambda) = 5.52 - 0.267T + 0.0037T^2$ $\mu_{max} = -0.014 - 0.0014T + 0.000441T^2$	Kim et al. (2021)
Baby spinach	*E. coli*	Air		10, 20, 30, 37°C	Stomacher bag	$\sqrt{\mu_{max}} = 0.000605\,(T + 4.76)$ $y_{max}{}^{f} = 0.0075T^2 - 0.2059T + 5.6769$	Puerta-Gomez et al. (2013)
	S. Typhimurium					$\sqrt{\mu_{max}} = 0.000296\,(T + 5.88)$ $y_{max} = 0.009T^2 - 0.2369T + 5.2411$	

(*Continued*)

Table 1 (*Continued*)

Vegetable	Pathogen	Headspace composition	Temperature range	Packaging material	Secondary model	Reference
Data[g] from various leafy greens: romaine lettuce, iceberg lettuce, heat treated lettuce at 50°C, cilantro, chopped tomatoes, jalapeño peppers, cilantro, baby spinach, shredded carrots,	*Salmonella choleraesuis* and *Salmonella* enterica (various serovars: Typhimurium, Thomson, Enteritidis, Heidelberg)	Variable, including air, or MAP (for lettuce: passive MAP, for shredded carrots and lettuce: 2.1% O_2:4.9% CO_2:93% N_2)	7–30°C	Variable (stomacher bags, polypropylene, polyethylene)	$\sqrt{\mu_{max}}$ =0.020 $(T + 0.571)$[h]	Mishra et al. (2017)
Data[i] from various leafy greens (fresh endive, coleslaw, stir fry, rutabaga, butternut squash, carrots, onion, lettuce, swedes, bean sprouts, diced celery	*L. monocytogenes*	Variable, including air (for diced celery), passive MAP (for lettuce, swedes, bean sprouts), or MAP (for endives: 10% O_2:10% CO_2:80% N_2, 10% O_2:30% CO_2:60% N_2, 10% O_2:50% CO_2:40% N_2, for iceberg lettuce: 2.1–4.3% O_2: 4.65–6.2% CO_2: balance of N_2, for shredded lettuce: 100% N_2, for the rest: 0.5% (onion) to 29.4% (carrots) CO_2 – other gases not specified)	3–30°C	Variable (polypropylene, perforated film with 1.25 perforations/cm^2, polyethylene)	$\sqrt{\mu_{max}}$ = 0.023 $(T + 0.599)$	

Corn	*L. monocytogenes*		Frozen (−18°C) for 7 days and then thawed[j] at 4, 8, 12, 20°C	Stomacher bag	$\sqrt{EGR}^{k} = 0.0168(T + 0.916)$ $\sqrt{\lambda} = 21.41 - 2.3T + 0.0628T^2$	Kataoka et al. (2017)
Green peas					$\sqrt{EGR} = 0.0225(T+0.79)$ $\sqrt{\lambda} = 19.77 - 2.164T + 0.0673T^2$	

[a]The number represents the theoretical maximum CO_2 concentration that is needed for total growth arrest.
[b]The number represents the theoretical minimum CO_2 concentration that induces a significant lag time.
[c]The number represents the minimum temperature allowing growth.
[d]Lag time.
[e]Maximum population density (ln CFU/g).
[f]Maximum population density (log CFU/g).
[g]Growth and inactivation data were retrieved from the following studies at various packaging headspace compositions and storage temperatures: Brandl and Amundson (2008); Brandl and Mandrell (2002); Chang and Fang (2007); Koseki and Isobe (2005a); Ma et al. (2010); Oliveira et al. (2010); Puerta-Gomez et al. (2013); Kakiomenou et al. (1998).
[h]In dynamic simulations, no growth or first-order inactivation was assumed at $T < 7$°C. Thus, the rate model was only used at $T > 7$°C.
[i]Growth and inactivation data were retrieved from the following studies at various packaging headspace compositions and storage temperatures: Carlin et al. (1995, 1996); Farber et al. (1998); Francis and O'Beirne (2001); Kaminski et al. (2014); Li et al. (2002).
[j]Growth was monitored by the time that the samples were transferred to any temperature above 0°C.
[k]Exponential growth rate (log CFU/g/h).

dependency of μ_{max} to temperature (Table 1). Integrating models describing both the μ_{max} and MPD change in response to temperature in a primary model may provide reliable predictions of the growth of pathogens under non-isothermal conditions along the food supply chain.

Significant variation has been reported in the log change (from <1 to 3.2 log_{10}) of *Salmonella* and *L. monocytogenes* on various fresh-cut vegetables, including escarole, collard green, cabbage, spinach, arugula and carrot, stored at 7°C or 15°C for 6 days or sequentially for 1.8 days (30% of a week shelf life) at 7°C and then for 3.2 days (70% of a week shelf-life) at 15°C (Sant'Ana et al., 2012a). Notably, cabbage and carrot were the only vegetables that prohibited the growth of both pathogens, yet their pH, i.e. 7.2 and 6.7, respectively, was similar to that of the other vegetables (6.2–6.8) that supported growth, suggesting the existence of pH-independent factors responsible for growth inhibition in cabbage and carrots (Manios et al., 2013). Film perforation may lead to a lower increase of pathogens compared to MAP with elevated CO_2 or N_2 (e.g. 5% O_2: 15% CO_2: 80% N_2 or 100% N_2), because oxygen influx enhances the growth of aerobic organisms that compete with pathogens, whereas MAP inhibits indigenous microbiota and moderates their competitive advantage against pathogens (Carlin et al., 1996; Francis and O'Beirne, 1997; Sant'Ana et al., 2012b; Tarlak et al. 2020; Farber et al. 1998).

In general, pathogen growth is suppressed by CO_2 at concentrations higher than 30% (Farber et al. 1996, 1998). However, this is not applicable in MA-packaged fresh produce because CO_2 levels above 20% accelerate senescence and compromise sensory quality (Carlin et al. 1996). A similar growth-enhancing effect for *L. monocytogenes*, i.e. due to inhibition of competing microbiota, has also been attributed to mild heat treatment (50°C, 90 s) of lettuce, aiming to prevent discoloration by enzymatic browning (Li et al. 2002).

3.2 Modelling pathogen inactivation

The main types of fresh produce commonly involved in foodborne outbreaks are products that can support the growth of pathogens, including fresh-cut EMA-packaged and frozen-thawed products. These products are commonly characterized by high pH (>4.5) and a_w (>0.980), as is the case for most vegetables and a few fruits, e.g. cantaloupes and coconut. They are not expected to cause microbial inactivation, as opposed to most fruits of low pH (3.4–4.2), such as avocados, mangos, pineapple, apples, grapefruits, ananas, pomegranate and oranges (Cabrera-Díaz et al., 2022; Ziegler et al., 2018) and cucumbers (Possas et al., 2021). Improving the safety of fresh produce partly relies on water and/or product decontamination (-disinfection) technologies, which reduce the level and growth potential of pathogens and may extend the shelf-life of the packaged products (e.g. RTE salads).

A critical mitigation factor affecting pathogen-related risk is temperature control during serving, e.g. in catering settings, as outlined in relevant QMRAs (Franz et al., 2010). Efficient maintenance of the chill chain, especially at temperatures below <7°C, i.e. the lowest threshold value for the growth of most enteric pathogens, may cause the inactivation of *Salmonella*, even in produce with growth-permitting pH, e.g. iceberg lettuce, due to the cumulative stress posed by low temperatures and growth of competing microbiota (Tarlak et al., 2020).

Quantifying the reduction of pathogens in fresh produce requires knowledge of their survival traits from pre-harvest to post-harvest stages. The survival of pathogens in the phyllosphere (the above-ground parts of plants) is strongly dependent on the capability of the strains to grow and form biofilms in specific environments. In biofilms, cells are enclosed into exopolymeric matrices that protect cells embedded in the biofilm against environmental stress such as desiccation and lack of nutrients (Monier and Lindow, 2005). The potential mechanisms explaining the attachment of bacteria to the surface of fresh produce include:

(i) the presence of extracellular polymeric substances;
(ii) the presence or absence of fimbriae;
(iii) cell surface hydrophobicity;
(iv) divalent cationic bridges; and
(v) bacterial surface charge.

Attachment is also dependent on the variable capacity of different strains to adhere to surfaces, which may be further impacted by the structure of the surface (rough, intact or cut) and the presence of nutrients (Merget et al., 2019). The presence of wounds or structures such as lenticels or stomata can also provide an entry point for enteric pathogens. Bacteria can also be entrapped in the plant tissue due to internalization and translocation (Solomon and Matthews, 2005). Internalization is mostly a passive event: bacteria are taken up from irrigation water by root systems (Wachtel et al., 2002). Bacteria in internalized sites are likely more resistant to sanitizing agents (Grivokostopoulos et al., 2022). More than 50% of bacterial populations of leafy vegetables will grow in biofilms on the surface of the leaf, using leaf veins and glandular trichomes as nutrient sources (Chahar et al., 2021; Merget et al., 2019; Zhou et al., 2018).

Leaf microbiota also contains epiphytic bacteria that will compete with enteric pathogens for nutrient sources (Cooley et al., 2006; Grivokostopoulos et al., 2022). The associations between pathogens and leaf microbiota (especially Enterobacteriaceae) can sometimes lead to pathogen death or a decrease in their initial loads (Carlin et al., 1996; Janisiewicz et al., 1999; Cooley et al., 2006), which make naturally occurring microbiota a good candidate for biocontrol of enteric pathogens on vegetable surfaces (Heaton and Jones, 2008). The establishment of bacteria on vegetables is also affected by the production of signalling molecules such as acyl-homoserine lactones which mediate metabolism regulation.

3.2.1 Decontamination interventions for fresh produce

The fresh produce industry relies heavily on pre- or post-harvest, interventions, e.g. surface decontamination, to ensure microbiological safety and quality. The surface of fresh produce is often treated with disinfectants. However, these treatments often have limited efficacy due to the presence of biofilms on the plant surfaces and the internalization of pathogens below the plant surface. Pathogens, such as *E. coli* O157:H7 and *Salmonella*, can become internalized within growing plants and are thereby protected from post-harvest washing treatments (Jablasone et al., 2005). Disinfection efficacy is also limited by the accumulation of organic load in the water used in a washing tank expressed via the increase in chemical oxygen demand (COD; mg/L), which quenches disinfectants, especially free chlorine (Zhou et al., 2015; Banach et al., 2020; Luo et al., 2018) Chemical decontamination treatments can also result in contamination of produce with potentially harmful biocide residues (Aharoni et al., 1997; Gómez-López et al., 2014). The health and environmental risks associated with the formation of carcinogenic halogenated disinfection by-products have raised concerns about the excessive use of chlorine.

Alternative approaches include chemical-based washing treatments using organic acids, electrolyzed water, peroxyacetic acid, ozone, water-assisted UV applications and chlorine dioxide (Benarde et al., 1965; Rico et al., 2007; Ölmez and Temur, 2010; Huang et al., 2018; López-Gálvez et al., 2021). Ozone processing minimizes contamination of the treated surface as it produces oxygen as the major end product. The cost-effectiveness and sustainable nature of ozone technology is also due to sanitizing effects on process water which can be recycled and reused (Novak et al., 2008). There are advantages to each of these technologies; however, they are not widely used in industry as many still require optimization, meaning that chlorine remains the industry standard due to cost and efficacy benefits.

Existing predictive models simulate the impact of disinfection treatments on the initial reduction of pathogens on treated produce and their subsequent growth potential (Table 1). A systematic study by Cuggino et al. (2020) has modelled the impact of lettuce cut size (8–22 cm^2), contact time (s) and concentration (mg/L) of two disinfectants (benzyl-isothiocyanate (BITS) and sodium hypochlorite (expressed as free chlorine from 0 ppm to 200 ppm) on the time needed for log reductions of *Salmonella*. The optimum combination of process parameter values yielding the highest *Salmonella* reduction was a lettuce cut size of 15 cm^2 washed for 110 s with a solution of 160 mg/L free chlorine and 40 mg/L BITC. In general, disinfectants may cause sub-lethal injury and extended lag time upon subsequent re-growth of pathogens under favourable conditions in packaged products at retail.

4 Modelling transfer of pathogens during fresh produce processing

Contamination of fresh produce may occur either in the field through the rhizosphere (below ground) or phyllosphere (above ground), e.g. via contaminated soil, irrigation water, aerosols or penetration of microorganisms through tissue openings, e.g. wounds or physical openings in leaves, e.g. stomata. At the post-harvest stage, contamination is further mediated by processes like cutting, chopping and shredding or through washing processes. In the processing environment, there is a bi-directional movement of bacteria from equipment surfaces, personnel, water and plant material. Microorganisms detach from fresh produce, become suspended in water and may re-attach to the same or different plant tissue, depending on the contact time between water and fresh produce, the accumulation of microbial contamination and organic load (i.e. COD increase) in the washing tank and the chemical or physical disinfection of water and its re-circulation. This leads to the following transfer patterns of microbial contamination (Holvoet et al., 2014; Munther and Wu, 2013; Munther et al., 2015; Mokhtari et al., 2018) (Fig. 1):

1 produce-to-water;
2 water-to-produce; and
3 cross-contamination of produce by produce via the same or re-circulated water.

The complexity of the processing environment makes modelling microbiological risks challenging.

Agent-based simulation (modelling) can assess the relative efficacy of various corrective actions that may control pathogen dynamics in a processing-packing plant and the final product (Barnett-Neefs et al., 2022; Madamba et al., 2022). The approach identifies all possible routes of (cross-)contamination that may occur within a fresh produce processing and packaging plant, including water, equipment, floor, personnel and packaging materials and stochastically expresses their contribution to the prevalence and spread of microbial contamination. Finally, sensitivity analysis helps identify the main factors controlling the dynamics of pathogens in the plant and the product.

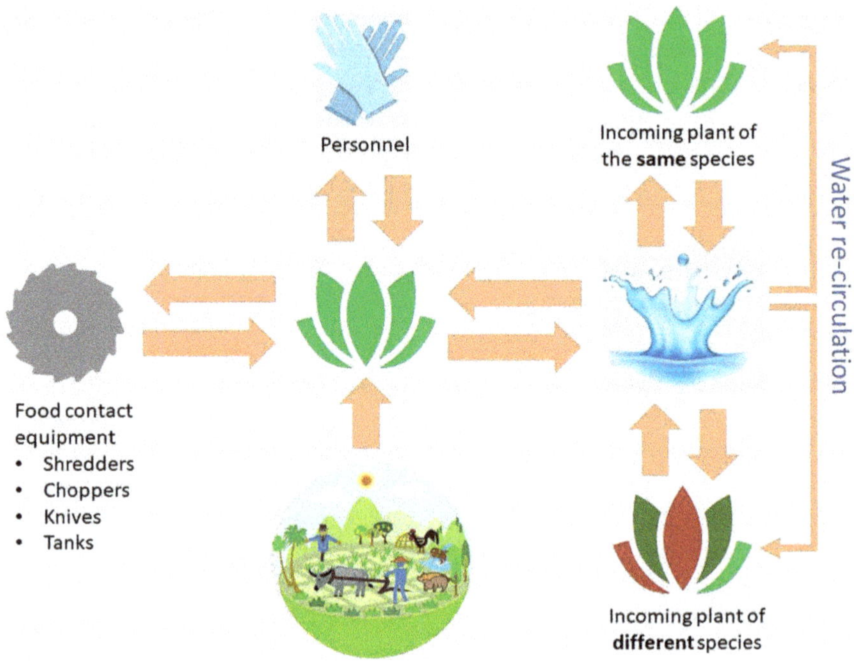

Figure 1 Sources and transfer routes of microbial contamination in minimally processed fresh produce.

Modelling can be used to estimate the impact and rank different mitigation strategies, i.e. cleaning and sanitation, zoning, flow of incoming materials, etc., in controlling *Listeria* occurrence in the processing environment and the final product (Barnett-Neefs et al., 2022).

The capability of enteropathogenic bacteria to grow and develop in biofilms on vegetables and equipment surfaces related to their production, handling, washing and packaging is now acknowledged. However, the specific ability of different species and strains to grow on vegetables and equipment surfaces is less explored. Additionally, the associations between pathogens and the microbiota commonly contaminating vegetables and equipment surfaces are currently unknown. This last aspect could play a key role in the establishment of the biofilm due to possible nutrient competition dynamics that could have a significant influence on the development and persistence of the pathogens on vegetables and inert equipment surfaces. There is also evidence that pathogens and background microbiota form multi-species biofilms exhibiting different levels of resistance to current sanitation treatments. The accumulation of microbial load in water or plant tissues, as well as the contamination of fresh produce by equipment and personnel in the processing environment contamination, can be quantified via the so-called transfer models.

4.1 Modelling microbial transfer between processing equipment and foods

The recent deeper insight into bacterial transfer phenomena has led to new types of model: bacterial transfer models. Modelling bacterial transfer is a challenge for predictive microbiologists. The significance of choosing the right model to describe a microbiological process is well known as is the choice of the appropriate parameters as model inputs. As Pérez-Rodríguez et al. (2008) state, not only do unknown factors govern the bacterial transfer, but even the known factors (contact time, bacterial strain properties, pressure, etc.) involved in cross-contamination events cannot be easily controlled. The challenge in modelling bacterial transfer in food processing environments, whether domestic kitchens or factories, is the dynamic nature of the environment. As an example, a recipient surface can become the donor surface in terms of contamination. The contamination status changes constantly and each new generated pathway becomes the new starting point for more transfer scenarios.

The first step in modelling transfer is the calculation of transfer rates as the percentage of cells transferred from the source to the destination of contamination. The multitude of factors which govern the bacterial transfer lead to a high variability which affects the reliability of experimental data. This variability can be captured through probability distributions. Probability distributions can be constructed based on the log-transformed transfer rates instead of transfer rates. Normal distribution is considered as most suitable to best fit log-transformed transfer rates. Hoelzer et al. (2012) used available scientific data to produce probability distributions and models in order to describe the transfer of *L. monocytogenes* between foods and surfaces or during slicing, as well as the impact of sanitation methods on subsequent cross-contamination events. Normal distributions were also chosen to fit the log-transformed transfer data. In a similar study carried out by Chen et al. (2001) to determine transfer rates of *Enterobacter aerogenes* between food and surfaces used for food preparation, normal distribution was chosen as the most suitable and statistically convenient to represent the data. Similarly, Kusumaningrum et al. (2004) used the Anderson-Darling criteria and decided that normal distribution gives the best fit to log-transformed transfer rates. Logistic distributions were ranked first when normal distributions did not give the best results.

Models describing cross-contamination scenarios are usually empirical due to the lack of knowledge on mechanisms of bacterial transfer. Interest has mainly been focused on pathogen transfer that takes place between cutting, shredding or slicing equipment and foods. Models represent transfer as a function of cut/slice number or inoculation level. Pérez-Rodríguez et al. (2011) found that, when different transfer scenarios were investigated, the logarithmic decrease that was observed in the numbers of the microorganisms could be efficiently described by a log-linear model. The goodness of fit of the log-linear model to the experimental data was comparable to that of the Weibull model.

The empirical models used so far have been proven sufficiently accurate for risk assessment studies (Aarnisalo et al., 2007; Sheen, 2008; Sheen and Hwang, 2010). A more mechanistic approach was introduced by Møller et al. (2012). They developed a model simulating the transfer of *Salmonella* Typhimurium during the grinding of pork. The model was an improved version of the model presented by Nauta et al. (2005), taking into account the tailing phenomenon that is observed after sequential slices during a grinding process. The authors claim that this type of model, though similar to empirical models such as that of Sheen and Hwang (2010), can more efficiently describe all the events that take place in a grinder. It also introduces the idea of the two environments existing in the grinder, similarly to a chopping or shredding machine, which is responsible for two different transfer rates occurring during the process.

This approach was successfully applied by Zilelidou et al. (2015) in modelling the transfer of *S.* Typhimurium and *L. monocytogenes* between lettuce and knives and vice versa in a simulated process of fresh-cut salad preparation in a domestic or catering environment. This involved 50 consecutive cuts with a knife that was first contaminated by cutting a contaminated batch of lettuce leaves. Several authors (Hoelzer et al., 2012; Kusumaningrum et al., 2004; van Asselt et al., 2008) have made some of the following assumptions for the setup of the experiment and subsequent model development:

- No bacteria are lost during processes;
- No cleaning of the equipment takes place between consecutive uses;
- No new contamination is introduced in the processing equipment for a certain number of operations; and
- The time after which the transfer event will take place does not have any impact on the event.

Such assumptions seem to be necessary for the development of bacterial transfer models since the complexity and inherent variability of the phenomenon require some simplification and sometimes a more linear representation of transfer pathways (van Asselt et al., 2008).

The probabilistic models by Mokhtari et al. (2018) and Mokhtari et al. (2022) represent a more pragmatic view of bacterial transfer in a fresh produce packing plant. These models map the various routes of equipment and edible material contamination, assuming bi-directional transfer, and describe the variability and uncertainty of factors controlling this transfer and the residual viable pathogen load on the packaged product. Data used to populate the appropriate probability distributions were collected from the literature (Munther and Wu, 2013; Munther et al., 2015), or assumed by the authors, based on experimental observations. The simulations representatively captured the variability in the transferring of pathogens among equipment and products that may occur in a fresh produce processing plant. These and other studies have modelled the change in pathogen load as a function of available free chlorine, which in turn depends on natural chlorine decay and its reaction (neutralization) rate with the accumulating organic load in the washing tank (e.g. flume) (Azimi et al., 2017; Abnavi et al., 2019; Munther et al., 2015; Mokhtari et al., 2022). Such studies together with agent-based simulations (Barnett-Neefs et al., 2022; Madamba et al., 2022) may be used to follow pathogen dynamics in a processing plant and be integrated in the exposure assessment from the plant to consumer.

5 Conclusion and future trends

The microbial safety of fresh produce is closely linked to the pre-harvest occurrence of pathogens and their ability to colonize plant surfaces and become internalized in tissue sublayers. Depending on post-harvest processing, pre-harvest microbial contamination, in combination with the contamination existing in the processing environment, may be re-distributed, increased or accumulated in the processing environment. The efficacy of disinfection and decontamination practices for water and produce surfaces, respectively, determines the prevalence and final levels of pathogens in packaged fresh-cut products by the time the product leaves the manufacturing plant.

Agent-based modelling may assist in mapping and parametrizing the impact of plant hygiene, through handling and processing practices (including zoning and sanitation) on pathogen dynamics in food processing environments (Barnett-Neefs et al., 2022). Modelling the subsequent behaviour of pathogens in retail environments helps estimate the final level of microbial hazard at the time of consumption. This requires the knowledge of pathogens' response to different plant surfaces, storage temperatures, packaging atmosphere and the impact of endogenous competing microbiota.

A variety of existing growth models are available for quantifying the impact of packaging atmosphere and temperature on the growth of pathogens in fresh-cut salads. Transfer-rate models may be modularly combined with growth models by assessing the prevalence and concentration of pathogens in packaged products, as the starting population for growth models at retail (Franz et al., 2010). Recent advances in next-generation sequencing may reveal new routes of microbial contamination of fresh produce and shed more light on the microbial interactions within these plant ecosystems, which in turn may drive new or more effective (targeted) decontamination strategies (Jacob et al., 2021). Of note are also recent insights on the induction by sanitizers of a viable-but-non-culturable state in pathogens by sanitizers, e.g. peroxyacetic acid and chlorine, even at the recommended doses, which may lead to overestimation of sanitation efficacy. Overestimation of efficacy significantly impacts the initial population introduced into growth models at retail which are an integral part of QMRA.

6 Where to look for further information

The readers may consult other relevant publications by the Research Group on Quality, Safety and Bioactivity of Plant Foods, The Centre of Edafology and Applied Biology of Segura, Spanish National Research Council (CEBAS-CSIC) in Murcia, Spain. This group has thoroughly investigated fresh produce safety from different perspectives, including the impact of water, disinfection agents, such as peroxyacetic acid and chlorine, ClO_2, the factors affecting their efficacy, e.g. COD, as well as the transfer of pathogens between water and fresh produce surface.

An additional source of relevant information comprise the QMRA reports by FAO/WHO and particularly the documents generated by the Joint FAO/WHO Expert Meeting on Microbiological Risk Assessment (JEMRA) on the Prevention and Control of Microbiological Hazards in Fresh Fruits and Vegetables.

7 References

Aarnisalo, K., Sheen, S., Raaska, L. and Tamplin, M. (2007). Modelling transfer of *Listeria monocytogenes* during slicing of 'Gravad' salmon. *International Journal of Food Microbiology* 118(1), 69-78.

Abnavi, M. D., Alradaan, A., Munther, D., Kothapalli, C. R. and Srinivasan, P. (2019). Modeling of free chlorine consumption and *Escherichia coli* O157:H7 cross-contamination during fresh-cut produce wash cycles. *Journal of Food Science* 84(10), 2736–2744. doi:10.1111/1750-3841.14774.

Aharoni, Y., Fallik, E., Copel, A., Gil, M., Grinberg, S. and Klein, J. D. (1997). Sodium bicarbonate reduces postharvest decay development on melons. *Postharvest Biology and Technology* 10(3), 201–206.

Anon (2007). *Consumer Attitudes to Food Standards Report, Wave 7*. London, UK: Food Standard Agency.

Azimi, V., Munther, D., Fakoorian, S. A., Nguyen, T. T. and Simon, D. (2017). Hybrid extended Kalfman filtering and noise statistics optimization for produce wash state estimation. *Journal of Food Engineering* 212, 136–145.

Banach, J. L., van Bokhorst-van de Veen, H., van Overbeek, L. S., van der Zouwen, P. S., Zwietering, M. H. and van der Fles-Klerx, H. J. (2020). Effectiveness of peracetic acid solution on *Escherichia coli* reduction during fresh-cut lettuce processing at

the laboratory and industrial scales. *International Journal of Food Microbiology* 321, 108537.
Barnett-Neefs, C., Sullivan, G., Zoellner, C., Wiedmann, M. and Ivanek, R. (2022). Using agent-based modeling to compare corrective actions for *Listeria contamination* in produce packinghouses. *PLoS ONE* 17(3), e0265251.
Bartosova, B., Koudela, B. and Slana, I. (2021). Detection of *Cyclospora cayetanensis, Echinococcus multilocularis, Toxocara* spp. and Microsporidia in fresh produce using molecular methods: - a review. *Food and Waterborne Parasitology* 23, e00124.
Bell, R. L., Kase, J. A., Harrison, L. M., Balan, K. V., Babu, U., Chen, Y., Mascarisin, D., Kwon, H. J., Zheng, J., Stevens, E. L., Meng, J. and Brown, E. W. (2021). The persistence of bacterial pathogens in surface water and its impact on global food safety. *Pathogens* 10(11), 1391. doi:10.3390/pathogens10111391.
Benarde, M. A., Israel, B. M., Oliveri, V. P. and Granstrom, M. L. (1965). Efficiency of chlorine dioxide as a bactericide. *Applied Microbiology* 13(5), 776–780.
Beuchat, L. R. (1996). Pathogenic microorganisms associated with fresh produce. *Journal of Food Protection* 59(2), 204–216.
Beuchat, L. R. (2002). Ecological factors influencing survival and growth of human pathogens on raw fruits and vegetables. *Microbes and Infection* 4(4), 413–423.
Brackett, R. E. (1999). Incidence, contributing factors, and control of bacterial pathogens in produce. *Postharvest Biology and Technology* 15(3), 305–311.
Brandl, M. T. and Amundson, R. (2008). Leaf age as a risk factor in contamination of lettuce with *Escherichia coli* O157:H7 and Salmonella enterica. *Applied and Environmental Microbiology* 74(8), 2298–2306.
Brandl, M. T. and Mandrell, R. E. (2002). Fitness of *Salmonella enterica* serovar Thompson in the cilantro phyllosphere. *Applied and Environmental Microbiology* 68(7), 3614–3621.
Buchanan, R. L., Gorris, L. G. M., Hayman, M. M., Jackson, T. C. and Whiting, R. C. (2017). A review of *Listeria monocytogenes*: an update on outbreaks, virulence, dose-response, ecology, and risk assessments. *Food Control* 75, 1–13.
Cabrera-Díaz, E., Martínez-Chávez, L., Gutiérrez-González, P., Pérez-Montaño, J. A., Rodríguez-García, M. O. and Martínez-Gonzáles, N. E. (2022). Effect of storage temperature and time on the behavior of *Salmonella, Listeria monocytogenes*, and background microbiota on whole fresh avocados (*Persea americana* var Hass). *International Journal of Food Microbiology* 369, 109614.
Cagnon, T., Méry, A., Chalier, P., Guillaume, C. and Gontard, N. (2013). Fresh food packaging design: a requirement driven approach applied to strawberries and agro-based materials. *Innovative Food Science and Emerging Technologies* 20, 288–298.
Caponigro, V., Ventura, M., Chiancone, I., Amato, L., Parente, E. and Piro, F. (2010). Variation of microbial load and visual quality of ready-to-eat salads by vegetable type, season, processor and retailer. *Food Microbiology* 27(8), 1071–1077.
Carlin, F., Nguyen-the, C. and Abreu da Silva, A. (1995). Factors affecting the growth of Listeria monocytogenes on minimally processed fresh endive. *Journal of Applied Bacteriology* 78(6), 636–646.
Carlin, F., Nguyen-the, C., Abreu Da Silva, A. and Cochet, C. (1996). Effects of carbon dioxide on the fate of Listeria monocytogenes, of aerobic bacteria and on the

development of spoilage in minimally processed fresh endive. *International Journal of Food Microbiology* 32(1–2), 159–172.

Chahar, M., Kroupitski, Y., Gollop, R., Belausov, E., Melotto, M. and Sela-Saldlinger, S. (2021). Determination of *Salmonella enterica* leaf internalization varies substantially according to the method and conditions used to assess bacterial localization. *Frontiers in Microbiology* 12, 622068. doi:10.3389/fmicb.2021.622068.

Chang, J. M. and Fang, T. J. (2007). Survival of *Escherichia coli* O157:H7 and *Salmonella enterica* serovars Typhimurium in iceberg lettuce and the antimicrobial effect of rice vinegar against E. coli O157:H7. *Food Microbiology* 24(7–8), 745–751.

Chen, Y., Jackson, K. M., Chea, F. P. and Schaffner, D. W. (2001). Quantification and variability analysis of bacterial cross-contamination rates in common food service tasks. *Journal of Food Protection* 64(1), 72–80.

Cooley, M. B., Chao, D. and Mandrell, R. E. (2006). *Escherichia coli* O157:H7 survival growth in lettuce is altered by the presence of epiphytic bacteria. *Journal of Food Protection* 69(10), 2329–2335.

Cuggino, S. G., Bascón-Villegas, I., Rincón, F., Pérez, M. A., Posada-Izquierdo, G., Marugán, J., Pablos Carro, C. and Pérez-Rodríguez, F. (2020). Modelling the combined effect of chlorine, benzyl isothiocyanate, exposure time and cut size on the reduction of *Salmonella* in fresh-cut lettuce during washing process. *Food Microbiology* 86, 103346.

De Roever, C. (1998). Microbiological safety evaluations and recommendations on fresh produce. *Food Control* 9(6), 321–347.

DeWaal, C. S. and Bhuiya, F. (2007). Outbreaks by the numbers: fruits and vegetables 1990–2005. International Association for Food Protection Poster Presentation P3-03, July 8–11.

Ding, T., Jin, Y.-G. and Oh, D.-H. (2010). Predictive model for growth of *Listeria monocytogenes* in untreated and treated lettuce with alkaline electrolyzed water. *World Journal of Microbiology and Biotechnology* 26(5), 863–869.

Doran, G., Sheridan, F., DeLappe, N., O'Hare, C., Anderson, W., Corbett-Feeney, G. and Cormican, M. (2005). *Salmonella enterica* serovar Kedougou contamination of commercially grown mushrooms. *Diagnostic Microbiology and Infectious Disease* 51(1), 73–76.

Doyle, M. P. and Erickson, M. C. (2008). Summer meeting 2007 – The problems with fresh produce: An overview. *Journal of Applied Microbiology* 105(2), 317–330.

EFSA BIOHAZ Panel (EFSA Panel on Biological Hazards) (2014). Scientific Opinion on the risk posed by pathogens in food of non-animal origin. Part 2 (Salmonella and Norovirus in leafy greens eaten raw as salads). *EFSA Journal* 12(3), 3600, 118 pp. doi:10.2903/j.efsa.2014.3600.

EFSA BIOHAZ Panel (EFSA Panel on Biological Hazards), Koutsoumanis, K., Alvarez-Ordóñez, A., Bolton, D., Bover-Cid, S., Chemaly, M., Davies, R., De Cesare, A., Herman, L., Hilbert, F., Lindqvist, R., Nauta, M., Peixe, L., Ru, G., Simmons, M., Skandamis, P., Suffredini, E., Jordan, K., Sampers, I., Wagner, M., Da Silva Felicio, M. T., Georgiadis, M., Messens, W., Mosbach-Schulz, O. and Allende, A. (2020). Scientific Opinion on the public health risk posed by *Listeria monocytogenes* in frozen fruit and vegetables including herbs, blanched during processing. *EFSA Journal* 18(4), 6092, 102 pp. doi:10.2903/j.efsa.2020.6092.

Farber, J. M., Cai, Y. and Ross, W. H. (1996). Predictive modeling of the growth of Listeria monocytogenes in CO_2 environments. *International Journal of Food Microbiology* 32(1-2), 133-144.

Farber, J. M., Wang, S. L., Cai, Y. and Zhang, S. (1998). Changes in populations of *Listeria monocytogenes* inoculated on packaged fresh-cut vegetables. *Journal of Food Protection* 61(2), 192-195.

Fonseca, S. C., Oliveira, F. A. R. and Brecht, J. K. (2002). Modelling respiration rate of fresh fruits and vegetables for modified atmosphere packages: a review. *Journal of Food Engineering* 52(2), 99-119.

Food and Agricultural Organization-World Health Organization. (2008). *Microbiological Hazards in Fresh Leafy Vegetables and Herbs*. Microbiological risk assessment series (Vol 14). 138 pp.

Francis, G. A. and O'Beirne, D. O. (1997). Effect of gas atmosphere, antimicrobial dip and temperature on the fate of *Listeria innocua* and *Listeria monocytogenes* on minimally processed lettuce. *International Journal of Food Science and Technology* 32(2), 141-151.

Franz, E., Tromp, S. O., Rijgersberg, H. and van der Fels-Klerx, H. J. (2010). Quantitative microbial risk assessment for *Escherichia coli* O157:H7, *Salmonella*, and *Listeria monocytogenes* in leafy green vegetables consumed at salad bars. *Journal of Food Protection* 73(2), 274-285.

Geysen, S., Escalona, V. H., Verlinden, B. E., Aertsen, A., Geeraerd, A. H., Michiels, C. W., Van Impe, J. F. and Nicolaï, B. M. (2006). Validation of predictive growth models describing superatmospheric oxygen effects on *Pseudomonas fluorescens* and *Listeria innocua* on fresh-cut lettuce. *International Journal of Food Microbiology* 111(1), 48-58.

Geysen, S., Geeraerd, A. H., Verlinden, B. E., Michiels, C. W., Van Impe, J. F. and Nicolaï, B. M. (2005a). Predictive modelling and validation of *Pseudomonas fluorescens* growth at superatmospheric oxygen and carbon dioxide concentrations. *Food Microbiology* 22(2-3), 149-158.

Geysen, S., Verlinden, B. E., Geeraerd, A. H., Van Impe, J. F. and Michiels, N., B. M. (2005b). Predictive modelling and validation of *Listeria innocua* growth at superatmospheric oxygen and carbon dioxide concentrations. *International Journal of Food Microbiology* 102, 333-345.

Giugliano, D. and Esposito, K. (2008). Mediterranean diet and metabolic diseases. *Current Opinion in Lipidology* 19(1), 63-68.

Glass, K. A., Golden, M. C., Wanless, B. J., Bedale, W. and Czuprynski, C. (2015). Growth of *Listeria monocytogenes* within a caramel-coated apple microenvironment. *mBio* 6(5), e01232-e01215. doi:10.1128/mBio.01232-15.

Gómez-López, V. M., Lannoo, A. S., Gil, M. I. and Allende, A. (2014). Minimum free chlorine residual level required for the inactivation of *Escherichia coli* O157: H7 and trihalomethane generation during dynamic washing of fresh-cut spinach. *Food Control* 42, 132-138.

Gontard, G. and Guillaume, C. (2009). Food packaging and the shelf life of fruits and vegetables. In: *Food Packaging and Shelf Life*, Robertson, G. L. (Ed.), pp. 297-315. Boca Raton: CRC Press, Taylor and Francis Group.

Grivokostopoulos, N. C., Makariti, I. P., Hilaj, N., Apostolidou, Z. and Skandamis, P. N. (2022). Internalization of *Salmonella* in leafy greens and impact on acid tolerance. *Applied and Environmental Microbiology* 88(6), e0224921.

Heaton, J. C. and Jones, K. (2008). Microbial contamination of fruit and vegetables and the behaviour of enteropathogens in the phyllosphere: a review. *Journal of Applied Microbiology* 104(3), 613–626.

Hoelzer, K., Pouillot, R., Gallagher, D., Silverman, M. B., Kause, J. and Dennis, S. (2012). Estimation of *Listeria monocytogenes* transfer coefficients and efficacy of bacterial removal through cleaning and sanitation. *International Journal of Food Microbiology* 157(2), 267–277.

Holvoet, K., De Keuckelaere, A., Sampers, I., Van Haute, S., Stals, A. and Uyttendaele, M. (2014). Quantitative study of cross-contamination with *Escherichia coli*, *E. coli* O157, MS2 phage and murine Norovirus in a simulated fresh-cut lettuce wash process. *Food Control* 37, 218–227.

Huang, R., de Vries, D. and Chen, H. (2018). Strategies to enhance fresh produce decontamination using combined treatments of ultraviolet, washing and disinfectants. *International Journal of Food Microbiology* 283, 37–44.

Jablasone, J., Warriner, K. and Griffiths, M. (2005). Interactions of *Escherichia coli* O157:147, *Salmonella* typhimurium and *Listeria monocytogenes* plants cultivated in a gnotobiotic system. *International Journal of Food Microbiology* 99(1), 7–18.

Jacob, C., Velásquez, A. C., Josh, N. A., Settles, M., He, S. Y. and Melotto, M. (2021). Dual transcriptomic analysis reveals metabolic changes associated with different persistence of human pathogenic bacteria in leaves of Arabidopsis and lettuce. G3 (Bethesda). 2021 Dec 8;11(12):jkab331. doi: 10.1093/g3journal/jkab331. PMID: 34550367; PMCID: PMC8664426.

Jacxsens, L., Devlieghere, F. and Debevere, J. (2002). Predictive modelling for packaging design: equilibrium modified atmosphere packages of fresh–cut vegetables subjected to a simulated distribution chain. *International Journal of Food Microbiology* 73(2–3), 331–341.

Jacxsens, L., Devlieghere, F., De Rudder, T. and Debevere, J. (2000). Designing equilibrium modified atmosphere packages for fresh–cut vegetables subjected to changes in temperature. *LWT - Food Science and Technology* 33(3), 178–187.

Janisiewicz, W. J., Conway, W. S., Brown, M. W., Sapers, G. M., Fratamico, P. and Buchanan, R. L. (1999). Fate of *Escherichia coli* O157:H7 on fresh-cut apple tissue and its potential for transmission by fruit flies. *Applied and Environmental Microbiology* 65(1), 1–5.

Kakiomenou, K., Tassou, C. and Nychas, G.-J. (1998). Survival of *Salmonella enteritidis* and *Listeria monocytogenes* on salad vegetables. *World Journal of Microbiology and Biotechnology* 14(3), 383–387.

Kapetanakou, A. E., Taoukis, P. and Skandamis, P. N. (2019). Model development for microbial spoilage of packaged fresh-cut salad products using temperature and in-package CO_2 levels as predictor variables. *LWT* 113, 108285.

Kataoka, A., Wang, H., Elliott, P. H., Whiting, R. C. and Hayman, M. M. (2017). Growth of *Listeria monocytogenes* in thawed frozen foods. *Journal of Food Protection* 80(3), 447–453.

Kim, Y. J., Park, J. Y., Suh, S. H., Kim, M. G., Kwak, H. S., Kim, S. H. and Hoe, E. J. (2021). Development and validation of a predictive model for pathogenic *Escherichia coli* in fresh-cut produce. *Food Science and Nutrition* 9(12), 6866–6872.

Koseki, S. and Isobe, S. (2005a). Growth of *Listeria monocytogenes* on iceberg lettuce and solid media. *International Journal of Food Microbiology* 101(2), 217–225.

Koseki, S. and Isobe, S. (2005b). Prediction of pathogen growth on iceberg lettuce under real temperature history during distribution from farm to table. *International Journal of Food Microbiology* 104(3), 239–248.

Koseki, S. and Itoh, K. (2001). Prediction of microbial growth in fresh-cut vegetables treated with acidic electrolyzed water during storage under various temperature conditions. *Journal of Food Protection* 64(12), 1935–1942.

Kusumaningrum, H. D., van Asselt, E. D., Beumer, R. R. and Zwietering, M. H. (2004). A quantitative analysis of cross-contamination of *Salmonella* and *Campylobacter* spp. *via* domestic kitchen surfaces. *Journal of Food Protection* 67(9), 1892–1903.

Li, Y., Brackett, R. E., Chen, J. and Beuchat, L. R. (2002). Mild heat treatment of lettuce enhances growth of *Listeria monocytogenes* during subsequent storage at 5°C or 15°C. *Journal of Applied Microbiology* 92(2), 269–275.

López-Gálvez, F., Allende, A. and Gil, M. I. (2021). Recent progress on the management of the industrial washing of fresh produce with a focus on microbiological risks. *Current Opinion in Food Science* 38, 46–51.

Lund, B. M. (1992). Ecosystem in vegetable foods. *Journal of Applied Bacteriology* 73, 115S–126S.

Luo, Y., Zhou, B., Van Haute, S., Nou, X., Zhang, B., Teng, Z., Turner, E. R., Wang, Q. and Millner, P. D. (2018). Association between bacterial survival and free chlorine concentration during commercial fresh-cut produce wash operation. *Food Microbiology* 70, 120–128.

Ma, L., Zhang, G., Gerner-Smidt, P., Tauxe, R. V. and Doyle, M. P. (2010). Survival and growth of *Salmonella* in salsa and related ingredients. *Journal of Food Protection* 73(3), 434–444.

Madamba, T., Moreira, R. G., Castell-Perez, E., Benerjee, A. and da Silva, D. (2022). Agent-based simulation of cross-contamination of *Escherichia coli* O157:H7 on lettuce during processing and temperature fluctuations during storage in a produce facility. Part 2: Model implementation. *Journal of Food Process Engineering* 45(4), e13983.

Manios, S. G., Konstantinidis, N., Gounadaki, A. S. and Skandamis, P. N. (2013). Dynamics of low (1–4 cells) vs high populations of *Listeria monocytogenes* and *Salmonella* Typhimurium in fresh-cut salads and their sterile liquid or solidified extracts. *Food Control* 29(2), 318–327.

Manthou, E., Karnavas, A., Fengou, L. C., Bakali, A., Lianou, A., Tsakanikas, P. and Nychas, G.-J. E. (2022). Spectroscopy and imaging technologies coupled with machine learning for the assessment of the microbiological spoilage associated to ready-to-eat leafy vegetables. *International Journal of Food Microbiology* 361, 109458.

Merget, B., Forbes, K. J., Brennan, F., McAteer, S., Shepherd, T., Strachan, N. J. C. and Holden, N. J. (2019). Influence of plant species, tissue type, and temperature on the capacity of Shiga-toxigenic *Escherichia coli* to colonize, grow, and be internalized by plants. *Applied and Environmental Microbiology* 85(11), e00123-19.

Mishra, A., Guo, M., Buchanan, R. L., Schaffner, D. W. and Pradhan, A. K. (2017). Development of growth and survival models for Salmonella and Listeria monocytogenes during non-isothermal time-temperature profiles in leafy greens. *Food Control* 71, 32–41.

Mokhtari, A., Oryang, D., Chen, Y., Pouillot, R. and Van Doren, J. (2018). A mathematical model for pathogen cross-contamination dynamics during the postharvest processing of leafy greens. *Risk Analysis* 38(8), 1718–1737. doi:10.1111/risa.12960.

Mokhtari, A., Pang, H., Santillana Farakos, S., Davidson, G. R., Williams, E. N. and Van Doren, J. M. (2022). Evaluation of potential impacts of free chlorine during washing of fresh-cut leafy greens on *Escherichia coli* O157:H7 cross-contamination and risk of illness. *Risk Analysis* 42(5), 966–988. doi:10.1111/risa.13818.

Møller, C. O. A., Nauta, M. J., Christensen, B. B., Dalgaard, P. and Hansen, T. B. (2012). Modelling transfer of *Salmonella* Typhimurium DT104 during simulation of grinding of pork. *Journal of Applied Microbiology* 112(1), 90–98.

Monier, J. M. and Lindow, S. E. (2005). Aggregates of resident bacteria facilitate survival of immigrant bacteria on leaf surfaces. *Microbial Ecology* 49(3), 343–352.

Munther, D., Luo, Y., Wu, J., Magparntay, F. M. G. and Srinivasan, P. (2015). A mathematical model for pathogen cross-contamination dynamics during produce wash. *Food Microbiology* 51, 101–107.

Munther, D. and Wu, J. (2013). Enhanced surveillance on food-borne disease outbreaks: dynamics of cross-contamination in biocidal wash procedure. *Journal of Theoretical Biology* 321, 28–35.

Nauta, M., van der Fels-Klerx, I. and Havelaar, A. (2005). A poultry-processing model for quantitative microbiological risk assessment. *Risk Analysis* 25(1), 85–98.

Novak, J., Demirci, A. and Han, Y. (2008). Novel chemical processes: ozone, supercritical CO_2, electrolyzed oxidizing water, and chlorine dioxide gas. *Food Science and Technology International* 14(5), 437–441.

Oliveira, M., Abadias, M., Usall, J., Torres, R., Teixidó, N. and Viñas, I. (2015). Application of modified atmosphere packaging as a safety approach to fresh–cut fruits and vegetables - a review. *Trends in Food Science and Technology* 46(1), 13–26. doi:10.1016/j. tifs.2015.07.017.

Oliveira, M., Usall, J., Solsona, C., Alegre, I., Viñas, I. and Abadias, M. (2010). Effects of packaging type and storage temperature on the growth of foodborne pathogens on shredded "Romaine" lettuce. *Food Microbiology* 27(3), 375–380.

Ölmez, H. and Temur, S. D. (2010). Effects of different sanitizing treatments on biofilms and attachment of *Escherichia coli* and *Listeria monocytogenes* on green leaf lettuce. *LWT - Food Science and Technology* 43(6), 964–970.

Pérez-Rodríguez, F., Valero, A., Carrasco, E., García, R. M. and Zuera, G. (2008). Understanding and Modelling Bacterial Transfer to Foods: a Review. Trends in Food Science & Technology 19(3), 131–144.

Pérez-Rodríguez, F. P., Campos, D., Ryser, E. T., Buchholz, A. L., Posada-Izquierdo, G. D., Marks, B. P., Zurera, G. and Todd, E. (2011). A mathematical risk model for *Escherichia coli* O157:H7 cross-contamination of lettuce during processing. *Food Microbiology* 28(4), 694–701.

Possas, A., Posada-Izquierdo, G. D., Zurera, G. and Pérez-Rodríguez, F. (2021). Evaluating the fate of *Escherichia coli* O157:H7 and *Salmonella* spp. on cucumbers. *Food Microbiology* 99, 103830.

Pradhan, A., Mishra, A. and Pang, H. (2018). Chapter 3. Relevant pathogenic and spoilage microorganisms in vegetable products. In: *Quantitative Methods for Food Safety and Quality in the Vegetable Industry*, Pérez-Rodríguez Skandamis, F., P. and Valdramidis, V. (Eds), pp. 29–58. Springer International Publishing.

Puerta-Gomez, A. F., Moreira, R. G., Kim, J. and Castell-Perez, E. (2013). Modeling the growth rates of *Escherichia coli* and *Salmonella* Typhimurium LT2 in baby spinach leaves under slow cooling. *Food Control* 29(1), 11–17.

Ragaert, P., Devlieghere, F. and Debevere, J. (2007). Role of microbiological and physiological spoilage mechanisms during storage of minimally processed vegetables. *Postharvest Biology and Technology* 44(3), 185–194.

Rico, D., Martin-Diana, A. B., Barat, J. M. and Barry-Ryan, C. (2007). Extending and measuring the quality of fresh-cut fruit and vegetables: a review. *Trends in Food Science and Technology* 18(7), 373–386.

Salazar, J. K., Carstens, C. K., Bathija, V. M., Narula, S. S., Parish, M. and Lou Tortorello, M. L. (2016). Fate of *Listeria monocytogenes* in fresh applies and caramel apples. *Journal of Food Protection* 79(5), 696–702.

Sandhya (2010). Modified atmosphere packaging of fresh produce: current status and future needs. *LWT – Food Science and Technology* 43(3), 381–392.

Sant'Ana, A. S., Barbosa, M. S., Destro, M. T., Landgraf, M. and Franco, B. D. G. M. (2012a). Growth potential of *Salmonella* spp. and *Listeria monocytogenes* in nine types of ready-to-eat vegetables stored at variable temperature conditions during shelf-life. *International Journal of Food Microbiology* 157(1), 52–58.

Sant'Ana, A. S., Franco, B. D. G. M. and Schaffner, D. W. (2012b). Modeling the growth rate and lag time of different strains of *Salmonella enterica* and *Listeria monocytogenes* in ready-to-eat lettuce. *Food Microbiology* 30(1), 267–273.

Schierstaedt, J., Grosch, R. and Schikora, A. (2019). Agricultural production systems can serve as reservoir for human pathogens. *FEMS Microbiology Letters* 366(23), fnaa016. doi:10.1093/femsle/fnaa016.

Sheen, S. (2008). Modeling surface transfer of *Listeria monocytogenes* on salami during slicing. *Journal of Food Science* 73(6), E304–E311.

Sheen, S. and Hwang, C. A. (2010). Mathematical modeling the cross-contamination of *Escherichia coli* O157:H7 on the surface of ready-to-eat meat product while slicing. *Food Microbiology* 27(1), 37–43.

Sivapalasingam, S. F., Friedman, C. R., Cohen, L. and Tauxe, R. V. (2004). Fresh produce: a growing cause of outbreaks of foodborne illness in the United States, 1973 through 1997. *Journal of Food Protection* 67(10), 2342–2353.

Solomon, E. B. and Matthews, K. R. (2005). Use of fluorescent microspheres as a tool to investigate bacterial interactions with growing plants. *Journal of Food Protection* 68(4), 870–873.

Tarlak, F., Johannessen, G., Bascón Villegas, I., Bolívar, A., Posada-Izquierdo, G. D. and Pérez-Rodríguez, F. (2020). Modelling of the behaviour of *Salmonella enterica serovar* Reading on commercial fresh-cut iceberg lettuce stored at different temperatures. *Foods* 9(7), 946. doi:10.3390/foods9070946.

Thunberg, R. L., Tran, T. T., Bennett, R. W., Mattews, R. N. and Belay, N. (2002). Research Note. Microbial evaluation of selected fresh produce obtained at retail markets. *Journal of Food Protection* 65(4), 677–682.

Tournas, V. H. and Katsoudas, E. (2005). Mould and yeast flora in fresh berries, grapes and citrus fruits. *International Journal of Food Microbiology* 105(1), 11–17.

Van Asselt, E. D., de Jong, A. E. I., de Jonge, R. and Nauta, M. J. (2008). Cross-contamination in the kitchen: estimation of transfer rates for cutting boards, hands and knives. *Journal of Applied Microbiology* 105(5), 1392–1401.

Victor, C. P., Ellis, K., Lamar, F. and Leon, J. S. (2021). Agricultural detection of Norovirus and hepatitis A using fecal indicators: a systematic review. *International Journal of Microbiology* 2021, Article ID 6631920. doi:10.1155/2021/6631920.

Wachtel, M. R., Whitehand, L. C. and Mandrell, R. E. (2002). Association of Escherichia coli O157:H7 with preharvest leaf lettuce upon exposure to contaminated irrigation water. *Journal of Food Protection* 65(1), 18–25.

Warriner, K., Huber, A., Namvar, A., Fan, W. and Dunfield, K. (2009). Chapter 4. Recent advances in the microbial safety of fresh fruits and vegetables. In: *Advances in Food and Nutrition Research* (vol. 57), pp. 155–208. doi: 10.1016/S1043-4526(09)57004-0. PMID: 19595387.

Zagory, D. and Kader, A. A. (1988). Modified atmosphere packaging of fresh produce. *Food Technology* 42, 70–77.

Zhou, B., Luo, Y., Nou, X., Lyu, S. and Wang, Q. (2015). Inactivation dynamics of *Salmonella enterica*, *Listeria monocytogenes*, and *Escherichia coli* O157:H7 in wash water during simulated chlorine depletion and replenishment processes. *Food Microbiology* 50, 88–96.

Zhou, B., Luo, Y., Bauchan, G. R., Feng, H. and Stommel, J. R. (2018). Visualizing pathogen internalization pathways in fresh tomatoes using MicroCT and confocal laser scanning microscopy. *Food Control* 85, 276–282.

Zhu, Z., Wu, X., Geng, Y., Sun, D., Chen, H., Zhao, Y., Zhou, W., Li, X. and Pan, H. (2018). Effects of modified atmosphere vacuum cooling (MAVC) on the quality of three different leafy cabbages. *LWT* 94, 190–197.

Ziegler, M., Rüegg, S., Stephan, R. and Guldimann, C. (2018). Growth potential of *Listeria monocytogenes* in six different RTE fruit products: impact of food matrix, storage temperature and shelf life. *Italian Journal of Food Safety* 7(3), 7581.

Zilelidou, E. A., Tsourou, V., Poimenidou, S., Loukou, A. and Skandamis, P. N. (2015). Modelling transfer of *Escherichia coli* O157:H7 and *Listeria monocytogenes* during preparation of fresh-cut salads: impact of cutting and shredding practices. *Food Microbiology* 45(B), 254–265.

Chapter 6

Advances in quantitative microbiological risk assessment for pathogens in fresh produce

Donald W. Schaffner, Rutgers University, USA; Marina Girbal, University of Barcelona, Spain; Matt Igo, Henry M. Jackson Foundation for the Advancement of Military Medicine, USA; and Kaitlyn Casulli, University of Georgia, USA

1 Introduction

Microbiological risk assessment is an increasingly valuable tool to evaluate and manage microbial food safety (Haas et al., 2014; Vose, 2008; Schaffner, 2008; Council for Agricultural Science Technology, 2006). Expert guidance on microbial risk assessment is available from the Food and Agriculture Organization (FAO) and the World Health Organization (WHO) (2008, 2009, 2021). Risk assessment is often considered one of three parts of risk analysis (Codex Alimentarius Commission, 1999). Those three components are:

- risk assessment (scientific evaluation of known or potential risks);
- risk management (weighting and selecting various policy alternatives); and
- risk communication (exchange of information about risk).

http://dx.doi.org/10.19103/AS.2023.0121.10

Risk assessment is commonly thought to have four components (National Research Council, 1983):

- Hazard identification;
- Exposure assessment;
- Hazard characterization; and
- Risk characterization.

These four components have been defined by Codex (1999) for microbial food hazards as a qualitative process to identify hazards (hazard identification); a qualitative or quantitative evaluation of the chance that a hazard causes an adverse health effect (exposure assessment); the adverse effects of ingestion, using a dose-response function if available (hazard characterization); and the integration of these three steps for overall risk estimate (risk characterization).

Codex's recognition and definition of microbial risk assessment in 1999 opened the door for the development of microbiological risk assessment by food safety and public health agencies around the world. While some of the earliest risk assessments were performed by federal agencies, the field has received a great deal of attention from academic researchers as well as those in the food industry as tools and techniques have become more widely available.

Because different microorganisms have different characteristics, the remainder of this chapter will be organized by the pathogen. Within the discussion of a given pathogen, risk assessments for similar commodities will be grouped and discussed together and presented in chronological order to show the development of the field.

2 Microbial risk assessment of *Listeria*

Listeria monocytogenes occurs commonly in both agricultural and food processing environments. While the number of reported cases of listeriosis is generally low compared to other pathogens, the hospitalization and fatality rate can be very high (Scallan et al., 2011). Risk assessments for this organism must consider a variety of unique attributes, including its ability to grow under refrigeration conditions. In the exposure assessment any *L. monocytogenes* risk assessment must also consider the unique aspects of victims who are seriously affected (pregnant women, the elderly and the immunocompromised), and the means by which these victims encounter the organism.

2.1 Multi-food risk ranking

One of the earliest examples of a risk assessment for *L. monocytogenes* was the 2003 report entitled 'Quantitative assessment of the relative risk to public health from foodborne *Listeria monocytogenes* among selected categories of ready-to-eat foods' issued by the FDA Center for Food Safety and Applied Nutrition and the USDA Food Safety and Inspection Service (U.S. Department of Health and Human Services, 2003). While most quantitative microbial risk assessments focus on a single pathogen in a single food product, this report evaluated more than 20 ready-to-eat (RTE) foods for their ability to support the growth of *L. monocytogenes* and cause listeriosis in consumers. The reason for this unique approach was to examine current science and models at the time to inform the effectiveness of current and future policies, programs, and regulatory practices. Only two categories of foods discussed in this report are relevant to the content of this chapter (fruits and vegetables), but a brief review of the entire risk assessment is needed to place these two foods in the proper context.

Foods were grouped according to risk per serving as well as risk per annum. Risk per serving means the probability of foodborne disease resulting from the consumption of a single serving of the food in question. Risk per annum estimates the total number of cases in the US food supply arising from a particular food which considers not only the risk per serving but the total number of servings consumed by the US population. Risk per serving foods was grouped into high, moderate, and low-risk categories. Risk per annum foods were grouped into four categories: high very, high, moderate, and low risk.

The relative risk ranking of all foods predicted deli meats as the most probable causative agent of listeriosis cases when analyzed both from a per serving and a per annum basis. The high-risk per serving category also included unheated frankfurters, meat spreads, unpasteurized milk, smoked seafood, and RTE crustaceans. Deli meats were the only very high-risk per annum food, but other high-risk per annum foods included pasteurized milk fat and other dairy products and unheated frankfurters.

Fruits and vegetables fell into the low-risk categories both on a risk per serving as well as a risk per annum basis. The report estimates the risk per serving of fruit to be 1.9×10^{-11} and the risk per serving of vegetables to be about an order of magnitude lower at 2.8×10^{-12}. Risk per serving of fruits and vegetables ranked 14th and 18th, respectively, out of the list of 23 foods evaluated. Fruits and vegetables were ranked higher in the risk per annum ranking, at 10th and 12th places, respectively. The estimated number of cases per year arising from fruit was 0.9, while cases arising from vegetables were 0.2.

2.2 Vegetables and leafy greens

Crépet et al. (2007) published one of the first detailed studies assessing the risk of *L. monocytogenes* in fresh vegetables. These authors used 165 prevalence studies (15 with concentration data) in their analysis. They categorized source data into five groups:

- Minimally processed vegetables;
- Minimally processed leafy salad vegetables;
- Sprouts and germinated seed, mung bean, alfalfa, and bean sprouts;
- Unprocessed vegetables; and
- Unprocessed leafy salad vegetables.

They noted in their analysis that most of the studies collected reported only the presence–absence of *L. monocytogenes* in fresh vegetable samples. Their analysis showed that while prevalence-only data can be used to help estimate mean concentration in food, this is not ideal. They note that analyses which combine prevalence and concentration data are more complex to undertake but can result in a much better estimation of the statistical distribution of *L. monocytogenes* in fresh vegetables, which is essential for good risk assessment.

Three studies focusing on *L. monocytogenes* risk if fresh vegetables were published in 2010. Carrasco et al. (2010) presented a probabilistic quantitative microbial risk assessment (QMRA) for *L. monocytogenes* in RTE lettuce salads in Spain, which also included risk management recommendations. These authors evaluated 11 specific risk management strategies falling into four specific categories:

1 Use of modified atmosphere packaging;
2 Reduced shelf life (limited to between 4 and 9 days);
3 Exclusion of high-risk individuals as salad consumers; and
4 Utilization of three different microbiological sampling plans during primary production.

The top three interventions were: the use of modified atmosphere packaging to slow or stop the growth of *L. monocytogenes*; reducing the allowed shelf life to only 4 days; and preventing high-risk individuals from consuming salads.

Franz et al. (2010) conducted a QMRA of *L. monocytogenes* and two other pathogens for leafy green vegetables consumed at salad bars in The Netherlands. Results from the other two pathogens will be discussed in the corresponding sections below. The authors modeled *L. monocytogenes* growth in the supply chain during distribution using data from 19 shipments from one processing plant to one restaurant. Four different datasets (each collected over 24 h during the holding of samples at the restaurant) were used to model *L. monocytogenes* growth in the restaurant.

The authors used Monte Carlo risk modeling to predict expected illnesses in one year from both baseline conditions and from a breakdown in restaurant salad bar cooling where the product was held at 18°C. The predicted growth of *L. monocytogenes* was minimal for both baseline (0.5 log increase) and cooling breakdown (0.6 log increase), with less than 1 case expected per year for both scenarios, assuming about half a million daily visitors to the restaurant. These same authors expanded on this work in an article published later that same year (Tromp et al., 2010). The original article used fixed storage times, while the subsequent work considered variable storage times taking into account supply chain logistics. This change resulted in an approximately fourfold increase in the probability of infection with *L. monocytogenes*, and the corresponding increase in the mean number of estimated cases per year.

Rijgersberg et al. (2010) show the application of discrete-event modeling in a QMRA for *L. monocytogenes* in fresh-cut iceberg lettuce. The authors performed discrete-event modeling using Aladin (Agro Logistic Analysis and De-sign INstrument), which is a discrete-event simulation software environment in order to study baseline conditions and six other scenarios. Those scenarios included:

1 batches with a higher initial level of the pathogen;
2 shortening the use-by date of the product;
3 encouraging the sale of products close to the end of shelf-life; and
4 more flexible 'just in time' ordering options for the retailer.

This risk assessment did not predict the probability of illness but instead focused on the probability that *L. monocytogenes* would exceed 100 CFU per gram at the time of consumption. The risk model also addressed the issue of 'shrinkage' (products that exceed their use by date while at production or retail), as well as out-of-stock conditions (no product to sell). The investigators concluded that a higher potential risk could be offset by reducing the use-by date of the product, but that this might increase the probability of shrinkage and products being out of stock. These could generally be offset by utilizing techniques that encouraged the purchase of products close to the end of shelf life as well as more flexible 'just in time' ordering for the retailer.

Gholipour et al. (2020) surveyed 126 samples of wastewater effluent, sewage sludge, or livestock manure as potential sources of *Listeria* spp. and *L. monocytogenes*. The authors also performed QMRA to assess the risk faced by farmers and those handling solid sledge. While the authors did not detect *L. monocytogenes* in wastewater or wastewater-irrigated crops they did detect the organism in half (13 of 26) of all solid sludge samples and 8% (2/25) of manure samples. The authors concluded that (depending upon the dose-response model used) the risk to solid sludge handlers might be quite high with a 44% chance of infection per year which is well above the proposed US Environmental Protection Agency (EPA) limit of one in 10 000 per year.

Despite this potentially high risk to solid sludge handlers, the authors also concluded that the risk to consumers of wastewater-irrigated crops was very low.

Despite evidence in the literature of links to listeriosis and fresh fruit including cantaloupe (Centers for Disease Control and Prevention, 2012), berries (Palumbo et al., 2013), stone fruits (Chen et al., 2016), and caramel apples (Centers for Disease Control and Prevention, 2015), no published QMRAs on *L. monocytogenes* in fruit were located.

3 Microbial risk assessment of *Salmonella*

Salmonella probably causes more foodborne disease cases than any other bacterial pathogen (Scallan et al., 2011). The organism is well known for its ability to survive under dry conditions for long periods of time. It has been historically associated with foods of animal origin like meat and eggs but increasingly has also been associated with a variety of plant-based foods. Risk assessments for this organism have focused on nuts and seeds as well as fresh produce, and these are discussed below.

3.1 Nuts and seeds

Low-moisture foods emerged as a significant salmonellosis risk in the early 2000s, and risk assessments started around this time for foods in this category. In one of the earliest examples, Danyluk et al. (2006) estimated the risk associated with the consumption of raw almonds. They also evaluated the risk reduction associated with a theoretical 5-log reduction process and a commercial propylene oxide process. The authors estimated a high probability (78%) that one or more cases of salmonellosis per year would arise from the consumption of raw almonds if no interventions were used. The commercial propylene oxide treatment reduced this risk significantly to 0.01%. When a hypothetical 5-log reduction was evaluated with different assumed standard deviations, this also reduced the predicted yearly risk of salmonellosis significantly to less than one percent.

The risk assessment cited above (Danyluk et al., 2006) informed the decision by the Almond Board of California to require all California-grown almonds sold in North America to be given a minimum 4-log *Salmonella* treatment. The same team of researchers updated their QMRA 6 years later (Lambertini et al., 2012) to assess the risk of salmonellosis from almonds given current treatments. This analysis also evaluated the ability of the production system to control risk given an increase in prevalence or concentration of *Salmonella*, and the effect of leaving some fraction of the crop untreated. They also sought to determine those conditions that might retroactively predict a large 2001 salmonellosis outbreak linked to almonds.

The investigators estimated the current risk of salmonellosis from almonds to be very low (1 case per 125 billion servings), or one case about every 20 years. Their analysis predicted that an increase in prevalence (to 25%), concentration (25 MPN/100 g), or leaving 0.05% of the crop untreated would result in more than 1 case/year. They predicted that the 2001 outbreak could have been caused by a prevalence of 65% and an average concentration of 120 MPN/100 g in raw almonds, which was consistent with observations from the time. If these almonds had been given a 4-log reduction, the 'outbreak' would have been a single case or less. A subsequent QMRA from US FDA research considering variability and uncertainty of parameters separately reached the same conclusion (Santillana Farakos et al., 2017b).

These seminal studies have been followed by a number of other publications on salmonellosis in other nuts, including pistachios (Lambertini et al., 2017). These investigators used published data on *Salmonella* prevalence and concentration in pistachios, crop volume data from the industry, as well as times and temperatures during processing, handling, and storage based on experiments and expert opinion. This QMRA predicted that the incidence of salmonellosis in the United States would be less than 1 case every 10 years assuming a 4-log treatment and typical consumer storage.

Researchers from the US Food and Drug Administration (FDA) were the first to publish a QMRA on the risk of salmonellosis from pecans in the United States (Santillana Farakos et al., 2017a). Their model considered *Salmonella* survival, growth, and recontamination from shelling to the consumer. They modeled microbial reduction treatment from 1 to 5 log CFU. The baseline model (cold conditioning, no microbial reduction treatment, and no home cooking) predicted about 529 cases of salmonellosis per year. As expected, any microbial treatment proportionally decreased risk. Hot conditioning and microbial reduction treatment (for both shelled and in-shell nuts) had a significant impact on risk. Pecans as ingredients in cooked products (e.g. a pecan pie) were very low risk.

Academic researchers collaborated with the American Peanut Council to model the risk of salmonellosis from the consumption of peanuts in the United States (Casulli et al., 2019). Prevalence and concentration data for *Salmonella* on raw, shelled peanuts were used with probability distributions of simulated log reductions during production steps to predict illness. Relevant time-temperature and survival data were obtained from published literature, industry surveys, or expert opinion. When current peanut processing and consumption patterns in the United States were used as the baseline, the QMRA predicted a greater than 10% chance of more than 1 case of salmonellosis per year and a greater than 3% chance of more than 10 cases of salmonellosis per year. Reduction from thermal inactivation had the greatest impact on risk, followed by consumer storage time. When the variability in the assumed log reduction increased, this also increased salmonellosis risk. Conclusions indicated that risk could best be managed by reducing contamination in the field, control of storage, and monitoring of appropriate critical limits during roasting.

In the same year, FDA researchers published a QMRA for salmonellosis from the consumption of walnuts in the United States (Santillana Farakos et al., 2019). Their baseline model was developed with data from surveys of walnuts from California operations, and the authors evaluated the impact of 0- to 5-log reductions during processing. The baseline model predicted one case of salmonellosis per 100 million servings, or an estimated six cases of salmonellosis per year in the United States. A 3-log reduction during processing resulted in a mean risk of <1 case per year. When the authors estimated the risk per serving associated with the consumption of walnuts with *Salmonella* data from a 2014 to 2015 US retail survey, the risk per serving estimates was two orders of magnitude larger than those of the baseline model. The authors concluded that further research was needed to understand variability in *Salmonella* contamination across the walnut supply chain.

Industry and academic researchers collaborated to evaluate the risk of salmonellosis from dry-roasted sunflower seeds based on current industry practices (Kottapalli et al., 2020). This research was driven by the requirements of the US FDA Food Safety Modernization Act, which mandates scientific and risk-based preventive controls. The investigators inoculated *Salmonella* onto sunflower seeds which were processed to simulate commercial manufacturing. Inactivation models predicted a 5-log reduction in *Salmonella* when sunflower seeds were roasted to a water activity of ~0.61. Saleable levels of water activity (0.03 to 0.04) were only achieved when sunflower seeds were roasted to achieve a >7-log reduction in *Salmonella*. A QMRA based on values from earlier studies, expert opinion, and the above-mentioned models were developed. The QMRA predicted a very low risk per serving (less than 1 in a billion) when seeds were roasted at conditions that resulted in a saleable product. The authors concluded that sunflower seeds roasted to saleable parameters should not represent a public health risk from *Salmonella*.

3.2 Fresh produce

In one of the earliest examples of QMRA for fresh produce, investigators developed a QMRA of the sprout production process to determine the most-effective points for *Salmonella* or *E. coli* O157:H7 control (Montville and Schaffner, 2005). Factors included initial pathogen concentration and prevalence, seed disinfection, sampling of seeds prior to sprouting, sampling of irrigation water, or sampling of the finished product.

Pathogen concentration and uniformity of seed contamination had a large effect on the fraction of contaminated batches. Sprout sampling and irrigation water sampling at the end of the sprouting process were predicted to be more effective in pathogen detection than seed sampling prior to production. Seed disinfection reduced the proportion of contaminated batches but could also reduce the ability to detect the pathogen when present. The volume sampled and pathogen detection limit were important variables in determining sampling effectiveness.

As noted earlier, Dutch researchers (Franz et al., 2010) conducted a QMRA leafy green vegetables consumed at salad bars in The Netherlands for three pathogens including *Salmonella*. Initial prevalence and concentration were based on a prior recent survey of Dutch fresh produce intended for raw consumption. The simulations predicted 187 *Salmonella* cases per year from salad bars. Given the uncertainty in pathogen prevalence and the number of salad bar visitors, the number of cases might range from 81 to 281. Temperature abuse (18°C) during display resulted in 1.6 times more *Salmonella* cases. The authors concluded that the number of *Salmonella* cases predicted by the baseline model was in proportion to the number expected from epidemiological data.

Researchers in Texas developed a QMRA to evaluate risk mitigation measures during the processing of baby spinach (Puerta-Gomez et al., 2013) including temperature control during harvest, washing, and irradiation. Data on *Salmonella* growth or reduction at each processing stage was based on experimental data or information from the literature. The authors evaluated six scenarios with different assumptions regarding initial *Salmonella* concentration, degree of and distribution for cross-contamination. They found that high levels of cross-contamination during harvest or washing increased prevalence within a given lot, and that irradiation at 1 kGy reduced prevalence from 84% to 0.1%, for highly contaminated lots.

Colombian researchers estimated the risk of salmonellosis from the consumption of raw unwashed vegetables irrigated with water from the Bogotá River (Henao-Herreño et al., 2017). The investigators tested water samples from the Bogotá River at six locations since untreated water from the river is commonly used for the irrigation of lettuce, broccoli, and cabbage. They developed a QMRA to estimate the daily risk of salmonellosis resulting from the use of this water source for irrigated vegetables. They estimated the daily probability of illness to be well above 50% for lettuce, cabbage, or broccoli.

Brazilian researchers developed a QMRA to predict the risk of illness caused by *Salmonella* in leafy greens based on practices commonly used in Brazilian leafy-green processing facilities (Maffei et al., 2017). The model considered in-field practices, postharvest washing, retail storage, and home storage. Scenarios evaluated two different initial pathogen concentrations, three different initial prevalence levels, and seven different chlorine concentrations during postharvest washing. The QMRA model showed that higher chlorine concentrations resulted in a lower risk of illness. When the use of <5 mg/L of chlorine was simulated, most (> 96%) illnesses arose from cross-contamination. When chlorine concentration varied according to a triangular distribution (10, 120, and 250 mg/L), no illnesses arising from cross-contamination were predicted.

FDA researchers used risk assessment to evaluate the impact of sprout seed treatment and spent irrigation water testing to mitigate the risk of salmonellosis from the consumption of contaminated alfalfa sprouts (Chen et al., 2018). The risk model they developed considered variability and uncertainty in *Salmonella* contamination in seeds as well as *Salmonella* growth and spread during sprout production. Their model predicted that the risk reduction from a 1-log reduction seed treatment was comparable to spent irrigation water testing. Both these interventions gave about a five- to sevenfold reduction in risk compared to baseline conditions. Better (i.e. >1 log reduction) seed treatments were predicted to provide a better risk reduction than spent irrigation water testing.

4 Microbial risk assessment of *E. coli*

Pathogenic *E. coli* gained prominence as a foodborne pathogen because of its widespread association with undercooked meat and was one of the first pathogens targeted for QMRA by the US Food Safety and Inspection Service (FSIS) in ground beef (2001). Because of its subsequent association with a variety of plant foods, risk assessments for those foods have also been developed.

4.1 Fresh produce

As noted earlier, Dutch researchers (Franz et al., 2010) conducted a QMRA of leafy green vegetables consumed at salad bars in The Netherlands for three pathogens including *E. coli* O157:H7. Initial prevalence and concentration were based on a recent survey of Dutch fresh produce intended for raw consumption. The simulations predicted 166 *E. coli* O157:H7 cases per year from salad bars. Given the uncertainty in pathogen prevalence and the number of salad bar visitors, the number of cases might range from 42 to 551. Temperature abuse (18°C) during display resulted in 1.6 times more cases of *E. coli* O157:H7. The authors concluded that the number of *E. coli* O157:H7 cases predicted by the baseline model was in proportion to the number expected from epidemiological data.

A study by US researchers was published in 2011 to provide a QMRA for *E. coli* O157:H7 risk in leafy greens from farm to consumption (Danyluk and Schaffner, 2011). A secondary objective of this work was to replicate the notable 2006 US outbreak of *E. coli* linked to the consumption of spinach based on available data. The risk model developed assumed that all contamination arose from exposure in the field, and the investigators assumed prevalence (per serving) ranged from 0.01% to 1% and concentration per serving ranged from −2 log CFU/g to 0 log CFU/g.

The model predicted that cut leafy greens that are temperature abused will support the growth of *E. coli* O157:H7, and populations of the organism may increase by as much as 1 log CFU/day under optimal temperature conditions. The model predicted that levels in the field of −1 log CFU/g and 0.1% prevalence could have resulted in an outbreak approximately the size of the 2006 *E. coli* O157:H7 outbreak. The investigators also identified important data gaps including information on retail storage times and correlations between storage time and temperature. They also noted the need for models for *E. coli* O157:H7 growth in leafy greens and identified that these models should include lag time where possible. Most importantly the model also predicted that a majority of simulated cases arose from leafy green servings which became cross-contaminated during the washing process. The investigators emphasized the importance of additional data and models to validate this finding.

Swedish researchers collected laboratory data on the survival of *Escherichia coli* O157 on lettuce as a function of temperature and light intensity and used that information to perform a screening-level QMRA (Ottoson et al., 2011). The risk assessment evaluated risk management strategies including rinsing leaves, lengthening the hold time between the last irrigation and harvest, and changes to irrigation water quality. The investigators found that growth temperature and light intensity each had a significant effect on bacterial survival. The pathogen survived better on plants grown at 11°C vs. 18°C or 25°C, as well as on plants exposed to lower light intensities. Pathogen concentration decreased over seven days after inoculation with the greatest reduction (3.0 log CFU) observed after seven days. Even one day after inoculation a reduction of greater than 1.0 CFU was observed. The QMRA compared the relative risk for 20 scenarios including two water quality levels, five different holding times, and rinsing or not rinsing before consumption. The greatest risk reduction came from the highest water quality standards, the longest holding time, and rinsing prior to consumption. The model identified a best-case scenario of 3.4 illnesses per 10 000 servings and a worst-case of 1390 illnesses per 10 000 servings.

US researchers developed a QMRA model for *E. coli* O157:H7 in fresh-cut lettuce from field to consumption to evaluate the effects of different intervention strategies on the reduction in illness (Pang et al., 2017). The model considered irrigation water or soil as potential sources of initial contamination, and a baseline plus intervention strategies (chlorine, ultrasound combined with organic acid, irradiation, bacteriophage treatment, and consumer washing). The baseline level of risk was 1 illness per 10 million servings (or over 2000 illnesses per year in the United States). Any of the intervention strategies evaluated (chlorine, ultrasound and organic acid, irradiation, bacteriophage, or consumer washing) significantly reduced the estimated number of illnesses compared to the baseline (~11- to 18-fold reductions). Retail and home storage temperatures had the greatest effect on the predicted number of illnesses.

Bozkurt et al. (2021) studied the risk of illness from *E. coli* O157:H7 in Australian romaine lettuce. This risk assessment included different preharvest contamination conditions, seasonal differences in *E. coli* concentration at harvest, and microbial growth kinetics during transportation and storage. Scenarios evaluated the effect of season, solar radiation, rainfall, manure, water source, and irrigation method. The baseline model predicted the annual mean of *E. coli* O157:H7 levels ranging from 0.2 to 3.4 log CFU/g and prevalence from 2% to 6%. The model predicted that increased solar radiation, use of drip irrigation, exclusion of manure, and lack of rainfall would all reduce *E. coli* levels. The sensitivity analysis found that more sunny days and a longer growing time had the greatest effect on lowering contamination, while the transfer of soil to the crop by splashed irrigation water and the number of rainy days in the harvesting season had the greatest effect on raising levels of *E. coli* O157:H7 contamination on lettuce preharvest. Postharvest washing interventions reduced the probability of illness by 1.4–2.7-log fold.

5 Microbial risk assessment of viruses

Almost two dozen foodborne or waterborne viral risk assessments have been published over the last two decades, often focusing on commonly contaminated foods (often fresh produce) and environmental sources associated with outbreaks (Miranda and Schaffner, 2019). Early risk assessments focused on irrigation water quality and a variety of viruses, while more recently published risk assessments have focused on norovirus in a variety of food products.

5.1 Irrigation water used for produce

Five of the earliest viral risk assessments focused specifically on irrigation water and sometimes included fresh produce: rotavirus in reclaimed wastewater (van Ginneken and Oron, 2000); a screening-level risk assessment for enterovirus in irrigation water (Petterson et al., 2001); hepatitis A virus in irrigation water used to irrigate cantaloupe, iceberg lettuce, and bell peppers (Stine et al., 2005); broccoli, cabbage, cucumber, and lettuce irrigated with nondisinfected secondary effluent containing rotavirus (Hamilton et al., 2006); and rotavirus in irrigation water used for lettuce (Mara et al., 2007).

The publication of a dose–response model for norovirus (Teunis et al., 2008) opened the way for norovirus risk assessment and a deluge of articles followed. Bassett and McClure (2008) investigated risk from 14 pathogens in fresh fruit using a fit-for-purpose approach. Their analysis included risk from hepatitis A and norovirus and concluded washing as an appropriate risk management action. Fiona Barker et al. (2013) created a probabilistic norovirus model to assess risks from greywater irrigation of home-grown lettuce in Australia. Their analysis showed that bathroom water posed a lower risk than laundry water but cautioned that, given the high secondary attack rate of norovirus, other uninvestigated exposure pathways might also pose significant risks. The same research group (Barker, 2014) also investigated the risk from norovirus in vegetables irrigated with municipal wastewater using three different approaches which yielded very different risk estimates. The author concluded that current wastewater reuse behaviors in the region studied were unlikely to cause excess norovirus risk. Mok et al. (2014) developed a QMRA to assess norovirus risk from lettuce, broccoli, cabbage, Asian vegetables, and cucumber after irrigation with treated wastewater. The authors concluded that while waste stabilization pond treatment did not result in sufficient risk mitigation, the use of additional treatments did give acceptable risks for cucumber and broccoli. Antwi-Agyei et al. (2015) conducted a farm-to-fork risk assessment for the use of wastewater in agriculture in Ghana and used it to estimate norovirus risk. The authors concluded that key risk factors were irrigation water and soil at the farm level, but their connection to downstream risk was unclear. The authors suggested it may be more effective to prioritize interventions at markets and kitchens due to cost, ease of implementation, and the high degree of contamination found in street food.

5.2 Viruses in foods

From 2015 viral risk research began to increasingly focus on foods and distribution channels and less on agricultural water. Bouwknegt et al. (2015) developed a farm-to-fork QMRA for norovirus and hepatitis A virus in European leafy green vegetable and berry fruit supply chains. These authors found cross-contamination via handling posed a larger risk than irrigation water, cross-contamination from food contact surfaces, or produce wash water. Jacxsens et al. (2017) developed a QMRA exposure model for norovirus using raspberries considering preharvest, harvest, and further handling. These authors concluded that norovirus contamination was predominantly driven by fruit pickers vs contaminated pesticide water. Hand washing/disinfection reduced norovirus levels by one order of magnitude as did a mild heat treatment.

Miranda and Schaffner (2018) created a QMRA model to replicate the largest known norovirus outbreak which infected about 11 000 people over a 3-week period (Bernard et al., 2014). The input that had the greatest effect on increase in the number of illnesses was a high norovirus concentration in water used for pesticide application and this was the only variable capable of producing an outbreak like that observed in Germany in 2012.

Researchers at the US FDA developed a risk assessment of norovirus transmission in food establishments (Duret et al., 2017), focusing on sick food workers experiencing diarrhea and vomiting and potential control measures. Their results showed that none of the prevention strategies examined could prevent norovirus transmission from symptomatic employees in the restaurant. The most effective intervention was high compliance with a work policy to exclude sick workers from working. This same team followed up with more analysis of the impact of compliance with mitigation strategies and the frequency of restaurant surface cleaning and sanitizing (Fanaselle et al., 2022). They concluded that compliance with exclusion and hand hygiene policies had the greatest impact on illnesses, washing hands before gloving would reduce norovirus transfer, while surface cleaning and sanitizing would have the least impact on illnesses. The model predicted that simply restricting the ill worker from handling food was not effective without additional provisions.

6 Conclusion and future trends

QMRA for the risks posed by foodborne pathogens in plant foods is a dynamic area of research which has grown tremendously since the Codex definitions for microbial risk assessment were published (Codex Alimentarius Commission, 1999). Advances in the field have been spurred on by guidance documents from the Food and Agriculture Organization and the World Health Organization (2008, 2009, 2021). The last two decades have seen increasing interest in QMRA in a variety of foods including those from both animal and plant sources. A large number of researchers around the world have continued to contribute to the rapidly growing scientific literature, and this shows no signs of declining.

As quantitative risk-based tools are increasingly used for making food safety policy decisions, this further spurs the need for research in QMRA. Future trends will include the development of a variety of automated tools to facilitate risk modeling like FDA-iRISK (2021) as well as various statistical packages (Pouillot and Delignette-Muller, 2010, Devleesschauwer et al., 2016).

7 Where to look for further information

The most recent FAO/WHO Guidance document is a comprehensive free document with a high level of detail: Food and Agriculture Organization & World Health Organization. 2021. Microbiological Risk Assessment-Guidance for Food [Online]. Available at: https://www.who.int/publications/i/item/9789240024892.

Another useful resource is: https://www.foodrisk.org administered by the Joint Institute for Food Safety and Applied Nutrition (JIFSAN), University of Maryland.

8 References

Antwi-Agyei, P., Cairncross, S., Peasey, A., Price, V., Bruce, J., Baker, K., Moe, C., Ampofo, J., Armah, G. and Ensink, J. 2015. A farm to fork risk assessment for the use of wastewater in agriculture in Accra, Ghana. *PLoS ONE* 10(11), e0142346.

Barker, S. F. 2014. Risk of Norovirus gastroenteritis from consumption of vegetables irrigated with highly treated municipal wastewater - Evaluation of methods to estimate sewage quality. *Risk Analysis* 34(5), 803–817.

Bassett, J. and McClure, P. 2008. A risk assessment approach for fresh fruits. *Journal of Applied Microbiology* 104(4), 925–943.

Bernard, H., Faber, M., Wilking, H., Haller, S., Höhle, M., Schielke, A., Ducomble, T., Siffczyk, C., Merbecks, S. and Fricke, G. 2014. Large multistate outbreak of Norovirus gastroenteritis associated with frozen strawberries, Germany, 2012. *Eurosurveillance* 19. https://doi.org/10.2807/1560-7917.ES2014.19.8.20719.

Bouwknegt, M., Verhaelen, K., Rzezutka, A., Kozyra, I., Maunula, L., Von Bonsdorff, C. H., Vantarakis, A., Kokkinos, P., Petrovic, T., Lazic, S., Pavlik, I., Vasickova, P., Willems, K. A., Havelaar, A. H., Rutjes, S. A. and De Roda Husman, A. M. 2015. Quantitative farm-to-fork risk assessment model for Norovirus and hepatitis A virus in European leafy green vegetable and berry fruit supply chains. *International Journal of Food Microbiology* 198, 50–58.

Bozkurt, H., Bell, T., Van Ogtrop, F., Phan-Thien, K. Y. and McConchie, R. 2021. Assessment of microbial risk during Australian industrial practices for Escherichia coli O157: H7 in fresh cut-cos lettuce: A stochastic quantitative approach. *Food Microbiology* 95, 103691.

Carrasco, E., Pérez-Rodríguez, F., Valero, A., García-Gimeno, R. M. and Zurera, G. 2010. Risk assessment and management of *Listeria monocytogenes* in ready-to-eat lettuce salads. *Comprehensive Reviews in Food Science and Food Safety* 9(5), 498–512.

Casulli, K. E., Calhoun, S. and Schaffner, D. W. 2019. Modeling the risk of salmonellosis from consumption of peanuts in the United States. *Journal of Food Protection* 82(4), 579–588.

Centers for Disease Control and Prevention. 2012. Multistate outbreak of listeriosis linked to whole cantaloupes from Jensen Farms. *Colorado*. Available at: https://www.cdc.gov/listeria/outbreaks/cantaloupes-jensen-farms/index.html [Accessed June 12, 2020].

Centers for Disease Control and Prevention. 2015. Multistate outbreak of listeriosis linked to commercially produced, prepackaged caramel apples [Online]. Available at: https://www.cdc.gov/listeria/outbreaks/caramel-apples-12-14/index.html [Accessed June 12, 2020].

Chen, Y., Burall, L. S., Luo, Y., Timme, R., Melka, D., Muruvanda, T., Payne, J., Wang, C., Kastanis, G., Maounounen-Laasri, A., De Jesus, A. J., Curry, P. E., Stones, R., K'Aluoch, O., Liu, E., Salter, M., Hammack, T. S., Evans, P. S., Parish, M., Allard, M. W., Datta, A., Strain, E. A. and Brown, E. W. 2016. *Listeria monocytogenes* in stone fruits linked to a multistate outbreak: Enumeration of cells and whole-genome sequencing. *Applied and Environmental Microbiology* 82(24), 7030–7040.

Chen, Y., Pouillot, R., Santillana Farakos, S. M., Duret, S., Spungen, J., Fu, T. J., Shakir, F., Homola, P. A., Dennis, S. and Van Doren, J. M. 2018. Risk assessment of salmonellosis from consumption of alfalfa sprouts and evaluation of the public health impact of sprout seed treatment and spent irrigation water testing. *Risk Analysis* 38(8), 1738–1757.

Codex Alimentarius Commission. 1999. *Principles and Guidelines for the Conduct of Microbiological Risk Assessment*. CAC/GL-30.

Council for Agricultural Science Technology. 2006. *Using Risk Analysis to Inform Microbial Food Safety Decisions*. CAST Ames, IA.

Crépet, A., Albert, I., Dervin, C. and Carlin, F. 2007. Estimation of microbial contamination of food from prevalence and concentration data: Application to *Listeria monocytogenes* in fresh vegetables. *Applied and Environmental Microbiology* 73(1), 250–258.

Danyluk, M. D., Harris, L. J. and Schaffner, D. W. 2006. Monte Carlo simulations assessing the risk of salmonellosis from consumption of almonds. *Journal of Food Protection* 69(7), 1594–1599.

Danyluk, M. D. and Schaffner, D. W. 2011. Quantitative assessment of the microbial risk of leafy greens from farm to consumption: Preliminary framework, data, and risk estimates. *Journal of Food Protection* 74(5), 700–708.

Devleesschauwer, B., Faes, C., Havelaar, A. and Speybroeck, N. 2016. QMRA-package: Parametric models for quantitative microbial risk assessment [Online]. Available at: https://rdrr.io/github/brechtdv/QMRA/man/QMRA-package.html.

Duret, S., Pouillot, R., Fanaselle, W., Papafragkou, E., Liggans, G., Williams, L. and Van Doren, J. M. 2017. Quantitative risk assessment of Norovirus transmission in food establishments: Evaluating the impact of intervention strategies and food employee behavior on the risk associated with Norovirus in foods. *Risk Analysis* 37(11), 2080–2106.

Fanaselle, W., Pouillot, R., Papafragkou, E., Liggins, G., Williams, L. and Van Doren, J. M. V. 2022. Evaluation of the impact of compliance with mitigation strategies and frequency of restaurant surface cleaning and sanitizing on control of Norovirus transmission from ill food employees using an existing quantitative risk assessment model. *Journal of Food Protection* 85(8), 1177–1191.

Fiona Barker, S., O'Toole, J., Sinclair, M. I., Leder, K., Malawaraarachchi, M. and Hamilton, A. J. 2013. A probabilistic model of Norovirus disease burden associated with greywater irrigation of home-produced lettuce in Melbourne, Australia. *Water Research* 47(3), 1421–1432.

Food and Agriculture Organization and World Health Organization. 2008. Exposure assessment of microbiological hazards in food: Guidelines [Online]. Available at: https://www.who.int/publications/i/item/9241546891 [Accessed March 11, 2023].

Food and Agriculture Organization and World Health Organization. 2009. Risk characterization of microbiological hazards in food: Guidelines [Online]. Available at: https://www.who.int/publications/i/item/9789241547895 [Accessed March 11, 2023].

Food and Agriculture Organization and World Health Organization. 2021. Microbiological risk assessment-guidance for food [Online]. Available at: https://www.who.int/publications/i/item/9789240024892 [Accessed March 11, 2023].

Food and Drug Administration Center for Food Safety and Applied Nutrition, Joint Institute for Food Safety and Applied Nutrition & Risk Sciences International. 2021. FDA-iRISK. Available at: https://irisk.foodrisk.org [Accessed March 11, 2023].

Food Safety and Inspection Service and United Stated Department of Agriculture. 2001. Risk assessment of E. coli O157:H7 in ground beef [Online]. Available at: https://www.fsis.usda.gov/node/2003 [Accessed March 11, 2023].

Franz, E., Tromp, S. O., Rijgersberg, H. and Van Der Fels-Klerx, H. J. 2010. Quantitative microbial risk assessment for *Escherichia coli* O157: H7, *Salmonella*, and *Listeria monocytogenes* in leafy green vegetables consumed at salad bars. *Journal of Food Protection* 73(2), 274–285.

Gholipour, S., Nikaeen, M., Farhadkhani, M. and Nikmanesh, B. 2020. Survey of *Listeria monocytogenes* contamination of various environmental samples and associated health risks. *Food Control* 108, 106843.

Haas, C. N., Rose, J. B. and Gerba, C. P. 2014. *Quantitative Microbial Risk Assessment*. John Wiley & Sons.

Hamilton, A. J., Stagnitti, F., Premier, R., Boland, A. M. and Hale, G. 2006. Quantitative microbial risk assessment models for consumption of raw vegetables irrigated with reclaimed water. *Applied and Environmental Microbiology* 72(5), 3284–3290.

Henao-Herreño, L. X., López-Tamayo, A. M., Ramos-Bonilla, J. P., Haas, C. N. and Husserl, J. 2017. Risk of illness with Salmonella due to consumption of raw unwashed vegetables irrigated with water from the Bogotá River. *Risk Analysis* 37(4), 733-743.

Jacxsens, L., Stals, A., De Keuckelaere, A., Deliens, B., Rajkovic, A. and Uyttendaele, M. 2017. Quantitative farm-to-fork human Norovirus exposure assessment of individually quick frozen raspberries and raspberry puree. *International Journal of Food Microbiology* 242, 87–97.

Kottapalli, B., Nguyen, S. P. V., Dawson, K., Casulli, K., Knockenhauer, C. and Schaffner, D. W. 2020. Evaluating the risk of salmonellosis from dry roasted sunflower seeds. *Journal of Food Protection* 83(1), 17–27.

Lambertini, E., Barouei, J., Schaffner, D. W., Danyluk, M. D. and Harris, L. J. 2017. Modeling the risk of salmonellosis from consumption of pistachios produced and consumed in the United States. *Food Microbiology* 67, 85–96.

Lambertini, E., Danyluk, M. D., Schaffner, D. W., Winter, C. K. and Harris, L. J. 2012. Risk of salmonellosis from consumption of almonds in the North American market. *Food Research International* 45(2), 1166–1174.

Maffei, D. F., Sant'ana, A. S., Franco, B. D. and Schaffner, D. W. 2017. Quantitative assessment of the impact of cross-contamination during the washing step of ready-to-eat leafy greens on the risk of illness caused by Salmonella. *Food Research International* 92, 106–112.

Mara, D. D., Sleigh, P. A., Blumenthal, U. J. and Carr, R. M. 2007. Health risks in wastewater irrigation: Comparing estimates from quantitative microbial risk analyses and epidemiological studies. *Journal of Water and Health* 5(1), 39–50.

Miranda, R. C. and Schaffner, D. W. 2018. Farm to fork quantitative microbial risk assessment for Norovirus on frozen strawberries. *Microbial Risk Analysis* 10, 44–53.

Miranda, R. C. and Schaffner, D. W. 2019. Virus risk in the food supply chain. *Current Opinion in Food Science* 30, 43–48.

Mok, H. F., Barker, S. F. and Hamilton, A. J. 2014. A probabilistic quantitative microbial risk assessment model of Norovirus disease burden from wastewater irrigation of vegetables in Shepparton, Australia. *Water Research* 54, 347-362.

Montville, R. and Schaffner, D. 2005. Monte Carlo simulation of pathogen behavior during the sprout production process. *Applied and Environmental Microbiology* 71(2), 746–753.

National Research Council. 1983. *Risk Assessment in the Federal Government: Managing the Process*. National Research Council.

Ottoson, J. R., Nyberg, K., Lindqvist, R. and Albihn, A. 2011. Quantitative microbial risk assessment for Escherichia coli O157 on lettuce, based on survival data from controlled studies in a climate chamber. *Journal of Food Protection* 74(12), 2000–2007.

Palumbo, M., Harris, L. and Danyluk, M. 2013. Outbreaks of foodborne illness associated with common berries, 1983 through May 2013. Food science and human nutrition department, UF/IFAS extension [Online]. Available at: https://edis.ifas.ufl.edu/fs232 [Accessed August 8, 2020].

Pang, H., Lambertini, E., Buchanan, R. L., Schaffner, D. W. and Pradhan, A. K. 2017. Quantitative microbial risk assessment for *Escherichia coli* O157: H7 in fresh-cut lettuce. *Journal of Food Protection* 80(2), 302–311.

Petterson, S. R., Ashbolt, N. J. and Sharma, A. 2001. Microbial risks from wastewater irrigation of salad crops: A screening-level risk assessment. *Water Environment Research* 73(6), 667–672.

Pouillot, R. and Delignette-Muller, M. L. 2010. Evaluating variability and uncertainty separately in microbial quantitative risk assessment using two R packages. *International Journal of Food Microbiology* 142(3), 330–340.

Puerta-Gomez, A. F., Kim, J., Moreira, R. G., Klutke, G.-A. and Castell-Perez, M. E. 2013. Quantitative assessment of the effectiveness of intervention steps to reduce the risk of contamination of ready-to-eat baby spinach with Salmonella. *Food Control* 31(2), 410–418.

Rijgersberg, H., Tromp, S., Jacxsens, L. and Uyttendaele, M. 2010. Modeling logistic performance in quantitative microbial risk assessment. *Risk Analysis* 30(1), 20–31.

Santillana Farakos, S. M., Pouillot, R., Davidson, G. R., Johnson, R., Son, I., Anderson, N. and Van Doren, J. M. 2019. A quantitative risk assessment of human salmonellosis from consumption of walnuts in the United States. *Journal of Food Protection* 82(1), 45–57.

Santillana Farakos, S. M. S., Pouillot, R., Johnson, R., Spungen, J., Son, I., Anderson, N., Davidson, G. R. and Doren, J. M. V. 2017a. A quantitative assessment of the risk of human salmonellosis arising from the consumption of pecans in the United States. *Journal of Food Protection* 80(9), 1574–1591.

Santillana Farakos, S. M. S., Pouillot, R., Johnson, R., Spungen, J., Son, I., Anderson, N. and Doren, J. M. V. 2017b. A quantitative assessment of the risk of human salmonellosis arising from the consumption of almonds in the United States: The impact of preventive treatment levels. *Journal of Food Protection* 80(5), 863–878.

Scallan, E., Hoekstra, R. M., Angulo, F. J., Tauxe, R. V., Widdowson, M. A., Roy, S. L., Jones, J. L. and Griffin, P. M. 2011. Foodborne illness acquired in the United States–major pathogens. *Emerging Infectious Diseases* 17(1), 7–15.

Schaffner, D. W. 2008. *Microbial Risk Analysis of Foods*. ASM Press.

Stine, S. W., Song, I., Choi, C. Y. and Gerba, C. P. 2005. Application of microbial risk assessment to the development of standards for enteric pathogens in water used to irrigate fresh produce. *Journal of Food Protection* 68(5), 913–918.

Teunis, P. F., Moe, C. L., Liu, P., Miller, S. E., Lindesmith, L., Baric, R. S., Le Pendu, J. and Calderon, R. L. 2008. Norwalk virus: How infectious is it? *Journal of Medical Virology* 80(8), 1468–1476.

Tromp, S. O., Rijgersberg, H. and Franz, E. 2010. Quantitative microbial risk assessment for *Escherichia coli* O157: H7, *Salmonella enterica*, and *Listeria monocytogenes* in leafy green vegetables consumed at salad bars, based on modeling supply chain logistics. *Journal of Food Protection* 73(10), 1830–1840.

U.S. Department of Health and Human Services, Center for Food Safety and Applied Nutrition, Food and Drug Administration, and U.S. 2003. Department of agriculture Food Safety, and Inspection Service (U.S. FDA/FSIS). Quantitative assessment of the relative risk to public health from foodborne *Listeria monocytogenes* among selected categories of ready to-eat foods [Online]. Available at: https://www.fda.gov/food/cfsan-risk-safety-assessments/quantitative-assessment-relative-risk-public-health-foodborne-listeria-monocytogenes-among-selected [Accessed February 5, 2021].

Van Ginneken, M. and Oron, G. 2000. Risk assessment of consuming agricultural products irrigated with reclaimed wastewater: An exposure model. *Water Resources Research* 36(9), 2691–2699.

Vose, D. 2008. *Risk Analysis: A Quantitative Guide*. John Wiley & Sons, West Sussex, England.

Part 3

Improving safety along the value chain

Chapter 7

Advances in understanding sources of pathogenic contamination of fresh produce: soil and soil amendments

Alexis Omar, University of Delaware, USA; Manan Sharma, USDA-ARS, USA; and Kalmia E. Kniel, University of Delaware, USA

1 Introduction

Soils are often enriched with biological soil amendments of animal origin (BSAAO) to increase nutrient values, enhance water-holding capacity, and support crop growth and yield. Soil amendments can be delivered to soils in the form of raw animal manure, treated or composted manures, and compost teas. Livestock production and generation of manure are critical to sustainable food systems. According to the Food and Agriculture Organization of the United Nations (FAO), livestock contributes to 40% of the global value of agricultural output and supports the livelihoods and food and nutrition security of almost 1.3 billion people. BSAAO promote soil structure and function. Poultry litter and mammalian manures provide nitrogen, phosphorous, and secondary nutrients like calcium, magnesium, and sulphur, along with micronutrients important for crop production (Tewolde et al., 2016; Chen and Jiang, 2014; Sharma and Reynnells, 2016). Inclusion of organic material from BSAAOs enhance the density, structure, and water-holding capacity of the soils thereby increasing crop yield (Rees et al., 2014; Gurtler et al., 2018). With all the benefits to soil

http://dx.doi.org/10.19103/AS.2023.0121.11

and crops aside, raw and in some cases, even treated BSAAOs, are at risk of contamination by zoonotic bacteria, protozoa and viruses.

The following sections of this chapter will focus on key issues of food safety, including contamination by pertinent pathogens and presence of antimicrobial resistance genes, related to the use BSAAO in agricultural production. The case studies focus on recent key outbreaks associated with fresh produce where soil amendments were questioned in the outbreak investigation and environmental assessment. Future trends in research focus on novel aspects and potential for remediation and use of biocontrol.

2 Pathogen survival in soils containing biological amendments

The persistence of bacterial and viral pathogens in raw animal manure may be based on the manure type, the method of application and/or incorporation into soils, the type of soil, storage of manure before application on to soils, and the microbial diversity present and nutrient ratios in manure-amended soils (Hutchison et al., 2005; Avery et al., 2004; Franz et al., 2008). Persistence and survival of bacterial pathogens in manure-amended soils depend highly on geographical and environmental factors. Field conditions may be different than greenhouse or growth chamber conditions, and therefore the context of these results in the literature should be examined critically. Extensive descriptions of the benefits and limitations of each set of laboratory, growth chamber, greenhouse settings (model systems) which could be used to evaluate the persistence of pathogens in manure are well covered in the literature (Harris et al., 2013).

There are several studies which describe the prevalence of bacterial and parasitic pathogens in various types of manure. A survey of British livestock manures from dairy cattle, poultry, swine and sheep showed that at least 30% of manures contained at least one enteric pathogen (Hutchison et al., 2004). Specifically, *E. coli* O157, *Salmonella* spp., and *Listeria* spp. were present in 12-21%, 8-18%, and 20-30%, respectively, in fresh cattle, pig and sheep manures. Furthermore, manures in Great Britain showed *Cryptosporidium parvum* and *Giardia intestinalis* in 5-29% and in 4-21%, samples tested respectively (Hutchison et al., 2004). Seasonality affected the prevalence of *E. coli* O157 and *Listeria* spp. populations in a study examining enteric pathogens in British livestock animal wastes, with *E. coli* O157 levels elevated in May and June compared to in other months; similarly, *Listeria* spp. prevalence in animal manures increased in the warmer months as well compared to in other months (Hutchison et al., 2004). No seasonal differences in prevalence were observed for *Salmonella* populations (Hutchison et al., 2005). A study in Florida at 12 different farms found that out of 518 bovine manure samples collected, 9%

(58/518) contained either *stx1* and/or *stx2* genes, potentially indicative of the presence of shiga-toxin producing *E. coli*. Of those samples which were *stx*-positive, the geometric mean of *stx*-positive *E.coli* was 0.53 log MPN/g (Baker et al., 2019).

Examining the survival of pathogens in manure and faeces is important, but of greater interest is the survival of enteric pathogens in manure once it has been applied and incorporated into soil as fertilizer or as an organic amendment. Pathogens may survive differently in specific amendments than in soils containing that amendment (Teichmann et al., 2020). The majority of studies examining the persistence of *E. coli*, enterohemorrhagic *E. coli* (EHEC), or *Salmonella* spp. in raw (untreated) animal manure in soils have been based on the use of the manure as an organic fertilizer. Table 1 shows the guidance/regulations provided by the U.S. federal government or other trade-associations which provide food safety guidance for their membership on handling of either untreated (raw) or treated (composted, heat-treated) biological soil amendments (BSA).

Several field studies have examined the use of untreated and treated animal manure over time. Twelve separate field trials conducted in the mid-Atlantic U.S. (Maryland and Pennsylvania) over four years at three different farms evaluated the survival of multi-strain inoculum of *E. coli* (non-pathogenic and attenuated-O157), inoculated at low or high levels in soils, amended with either poultry litter, dairy cattle manure, horse manure, or left unamended, applied either on the surface or tilled into soils, and in some cases managed organically or conventionally. Spatiotemporal factors (site, year, and season) affected survival durations of *E. coli* in manure-amended soils more than agricultural factors (manure type, organic or conventional management of soils, and depth of application) (Sharma et al., 2019). The maximum survival duration observed was 219 days with levels of *E. coli* ranging from 0.14–2.32 log CFU/g soil depending on the inoculum level, amendment type and depth of application. Weather effects (rainfall, daily average temperature) also affected *E. coli* survival durations more than agricultural effects. However, among the agricultural effects, manure type influenced *E. coli* survival duration over time more than depth or organic/ conventional soil type. In 60% of the comparisons made, poultry litter frequently supported longer survival durations of *E. coli* compared to other manure types. Figure 1 shows the interdependence of agricultural factors (manure type) and temporal factors (year) in describing the percentage of plots with specific amendment types which supported *E. coli* survival (dpi100mort) for less than or greater than 90 days (the current interval which FDA currently has 'no objection' to between application of an untreated soil amendment and harvest of a crop unlikely to touch the soil). In this specific study, *E. coli* had longer survival durations in the year 2013 (regardless of site) than any other year (2011, 2012, 2014). Using data from this same study,

Table 1 Guidance and regulations in the United States include information on the handling of untreated and treated biological soil amendments (BSA).

Organization/program/ federal agency	*California Leafy Green Products Handler and Marketing Agreement*	*USDA National Organic Program*	*A Guide to Tomato Good Agricultural Practices (T-GAPS)*	*Standards for Growing, Harvesting, Packing, and Holding of Produce for Human Consumption (FDA Supplemental Proposed Rule, 2014) (FSMA Final Rule, 2015)*
Manure	Prohibit the use of untreated manure on fields intended to grow leafy greens and meet the requirement to wait 1 year before planting	Uncomposted animal manures applied at least 90 days prior to harvest of crops where edible portion does not come into contact with the soil and at least 120 days prior to harvest where edible portion of plant contacts soil	Untreated animal manure not allowed in tomato fields or greenhouses	No objection to the 90/120-day interval as stated in the USDA National Organic Program
Compost	Active compost must achieve a temperature of 131°F (55°C) for either 3 days (within vessel/ aerated static pile) or 15 days/5 turnings (windrow composting) and requires a 45-day interval between application and harvest	Compost mixed or managed so that all feedstock heats between 131°F (55°C) and 170°F (77°C) for at least 3 days; must have documentation showing suitability for soil application or 45 day curing before soil application	Records required with dates of composting, methods utilized, and application	Composted soil amendment must achieve temperature of 131°F (55°C) for either 3 consecutive days (within/ vessel/aerated static pile) followed by adequate curing or 15 days/5 turnings (windrow composting) followed by adequate curing along with a zero day interval between application

Compost and agricultural teas (CTs)	If heat treatment process used to inactivate human pathogens of significance meets the microbial acceptance criteria, then no time interval is needed between application and harvest; if not validated then a 45-day interval between application and harvest is required	CTs must be made from compost that has maintained temperature of at least 131°F (55°C) for 3 days; CT with additives must adhere to 90/120-day interval before harvest (same as raw manure)	None given	Agricultural tea (CT with additive) considered as untreated manure and must be applied as such; if CT is prepared and used in <1 h, then it can be used without restriction
Reference	California Leafy Greens Marketing Agreement. (n.d.). *Food Safety Program*. [online] Available at: https://lgma.ca.gov/food-safety-program (1)	www.ams.usda.gov. (n.d.). *Soil Building - Manures & Composts | Agricultural Marketing Service*. [online] Available at: https://www.ams.usda.gov/grades-standards/soil-building-manures-composts (2)	Tomato Best Practices Manual A Guide to Tomato Good Agricultural Practices (T-GAP) & Tomato Best Management Practices (T-BMP) Purpose. (n.d.). [online] Available at: http://fvreports.freshfromflorida.com/5G_TomBPM.pdf (3)	www.regulations.gov. (n.d.). *Regulations.gov*. [online] Available at: https://www.regulations.gov/document/FDA-2011-N-0921-18558 (4)

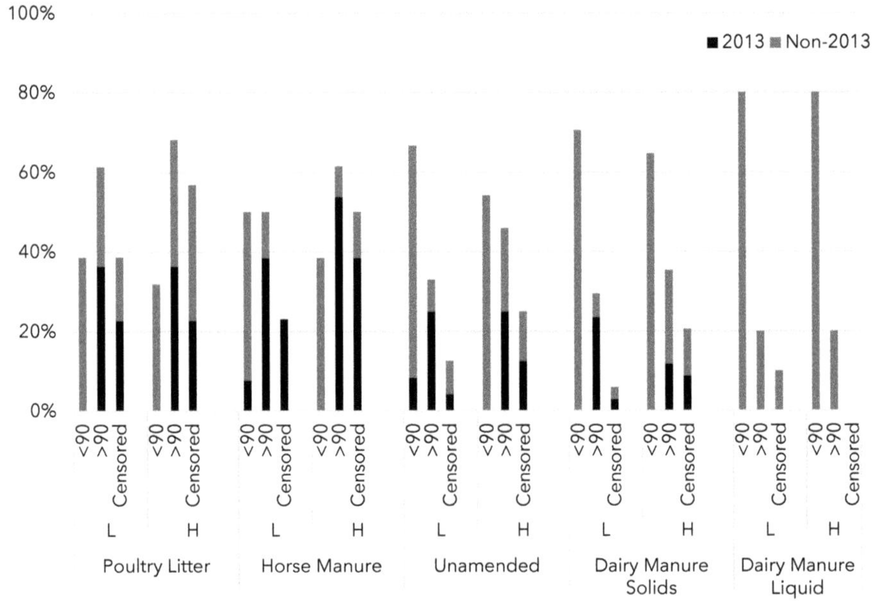

Figure 1 The percentage of plots with specific amendment types which supported *E. coli* survival for less than or greater than 90 days. Low (L) and high (H) inoculum levels were assessed. Within each amendment type and inoculum level, the percentages of profiles (across all depths and management types) which had dpi100mort values of less than 90 (<90), greater than 90 (>90), or censored (not reaching the viable threshold of 0.11 log MPN/gdw (gram dry weight) by end of the sampling range) were determined. The blue/orange sections of each bar represent 2013/non-2013 profiles. Censored indicates the percentage of survival curves where the estimated log CFU value did not reach the detection limit - 0.11 log CFU/gdw - by the maximum days post-inoculation used in the model.

other authors found that days post inoculation (application), inoculum level, amendment type, days of rain since the previous sampling day, and soil moisture content were the five highest ranked (most influential) factors in predicting the quantitative levels (not survival duration) of *E. coli* (Pang et al., 2020) and that poultry litter supported higher quantitative levels of *E. coli* compared to dairy manure, horse manure, or no amendment when added to soils. Other work, performed on the western and eastern parts of Canada at two different sites, determined that the survival duration of attenuated *E. coli* O157:H7 levels in soils amended with dairy cattle manure liquid were not affected by spatial (site) or depth differences (Bezanson et al., 2012). Survival did not exceed 22 days after an initial *E. coli* O157:H7 population was applied at 5 log CFU/g. The shorter survival durations compared to studies in the U.S. Mid-Atlantic may be due regional or seasonal differences, but also to the fact that a single strain of *E. coli* was used in Canada versus a multi-strain inoculum of *E. coli* in the mid-Atlantic U.S. study. Other studies evaluating the influence of spatial and temporal factors on persistence of non-pathogenic, indigenous populations of *E. coli* in bovine manure-amended soils have shown that different sites did not influence survival durations as much as time after application, depth, and manure application rate (Çekiç et al., 2017). Higher manure application rates, perhaps containing higher *E. coli* populations, supported longer survival durations than lower applications. This same study also showed survival durations were greater in fall seasons (252–280 days) as compared to summer seasons (56–112 days), and that soil and air temperatures were significantly lower in the fall (17–19°C) compared to the summer (26–27°C) season (Çekiç et al., 2017).

Survival of pathogens in soils containing BSAAO is critical to evaluate because the longer survival duration at higher levels, the increased likelihood of transfer of the pathogen to the edible portion of the crop. Several field and growth chamber studies have evaluated the likelihood of transfer from soils containing BSAAO to edible fruits and vegetable commodities. Using soil nutrient data or common physicochemical factors (soil moisture, nitrogen content, etc.) to predict survival and/or transfer to crops would be beneficial to many produce safety stakeholders. *E. coli* survival duration in soils containing either poultry litter, composted poultry litter, or heat-treated poultry litter pellets were evaluated over two seasons in Delaware (the mid-Atlantic U.S.) and their transfer to growing cucumbers in the field covered with plastic mulch or uncovered were evaluated. Nutrient and physicochemical measurements were evaluated for their ability to predict *E. coli* levels in BSAAO-amended soils. Nitrate levels on day 30 and moisture content on day 40 were strong predictors of *E. coli* survival in soils. However, similar to the combination of temporal agricultural factors which Sharma et al. (2019) observed, knowledge of the combination of year, amendment and mulch type was a better predictor than any single nutrient or physical measurement for survival of *E. coli* in soils planted with cucumbers (Litt et al., 2021). Transfer of *E. coli* to cucumbers from soils was best predicted by year (and associated rainfall amounts) compared to survival duration in soils. Longer survival durations of *E. coli* and more transfer of *E. coli* from unamended or amended soils to cucumbers was observed in year 1 (77 $\pm$ 30 days) to achieve a 4 log CFU/g reduction than in year 2 (56 $\pm$ 21 days). The cumulative rainfall observed in year 1 (24.94 in) was nearly twice the amount observed in year 2 (12.63 in), influencing the survival durations and transfer to cucumbers. In year 1, 98% of cucumbers harvested (regardless of which BSA treatment they were grown in) contained *E. coli*; in year 2 only 35% of cucumbers contained *E. coli*. These results showed that *E. coli* survival in soils was extended by the addition of several poultry litter-based soil amendments (heat-treated poultry pellets, composted poultry litter) commonly used in organic production of fruits and vegetables and is highly dependent on temporal variation in rainfall (Litt et al., 2021). Other multi-season studies evaluating survival of *E. coli* in soils containing treated amendments also showed that *E. coli* survived for longer durations in soils amended with poultry litter compost or heat-treated poultry pellets compared to soils amended with dairy manure composts (Limoges et al., 2021). This study also revealed *E. coli* survived for longer durations in clay soils compared to sandy loam soils, with water potential and daily temperature impacting survival durations as well. While the use treated BSAAOs limits the microbial contamination that can be introduced through application directly to soils, treated BSAAO-amended soils may provide opportunities for pathogens to survive for longer durations when introduced to soils through contaminated irrigation water (Acheamfour et al., 2021) or animal intrusion.

Several studies conducted in greenhouses support conclusions from field studies described above. Greenhouse and growth chamber conditions may provide more consistent controlled temperature conditions (and scheduled irrigation) which can then describe the effects of specific soil properties and the effects of specific BSAAO's on pathogen survival. The same multi-strain inoculum of non-pathogenic and O157-*E. coli* survived for longer durations in poultry litter-amended soils (regardless of if they were sandy or clay soils) in both small and large pots when stored in a greenhouse for up to 56 days (Sharma et al., 2016). Other studies have shown that levels of *E. coli* O157:H7 inoculated in artificial soils were greater (4.45 log CFU/g) when filtered soil rinse (containing nutrients but less competing microorganisms) was added to the mecosm, compared to pathogen levels (1.83 log CFU/g) when unfiltered soil rinse (containing competing microorganisms) when stored for 7 days at 30°C (Baker et al., 2021). This work shows that indigenous microorganisms in soils and microbial communities in soils may also play a role in affecting survival of enteric pathogens like *E. coli* O157:H7 and *Salmonella*.

Studies comparing the survival of 12 different *Salmonella* serovars in different types of soils revealed weekly irrigation (compared to daily irrigation) promoted longer survival durations for most *Salmonella* serovars in sandy or clay soils; poultry litter-amended soils extended survival of 10/12 *Salmonella* serovars compared to unamended soils, and survival was also influenced by specific soil-serovar interactions (Bardsley et al., 2021). Similar results with a single serovar (*S.* Newport) showed that soils amended with heat-treated poultry pellets (HTPP), a commonly used treated BSAAO, supported a significantly slower decline (0.10–0.12 log CFU/g/day) of *Salmonella* Newport levels than unamended soils (0.33–0.38 log CFU/g/day), and promoted more transfer of the pathogen from soils to leaves of spinach plants during the study (Shah et al., 2019a). Higher levels of soluble carbon and specific nitrogen compounds in HTPP-amended soils influenced the extended survival of *Salmonella* Newport compared to unamended soils.

Nutrient runoff from soils into waterways can cause potential environmental issues and pollute waterways. Runoff from soils containing BSAAOs may also provide nutrients for pathogens in irrigation water sources to proliferate and potentially contaminate produce. *Salmonella* Newport grew to higher populations in simulated soil runoff containing HTPP and levels increased by 3–5 log CFU/mL compared to runoff that did not contain HTPP, where levels increased by 1–3 log CFU/mL (Shah et al., 2019b). Data from this study also showed that HTPP-amended runoff alleviates some nutrient stress placed on *Salmonella* Newport as *S.* Newport strains lacking the *rpoS* gene (a global stress response gene) were not able to grow in simulated soil runoff without HTPP. Similar results were observed with non-pathogenic *E. coli* and their ability to proliferate rapidly in poultry litter-based compost soil runoff and dairy-manure based compost soil runoff, increasing their levels between 6 and 9 log CFU/mL (Neher et al., 2019). In both experiments, runoff which contained less indigenous soil bacteria promoted higher levels of both *E. coli* and *Salmonella* and more rapid growth, indicating that indigenous bacterial populations in soils may compete with enteric bacterial foodborne pathogens and limit their growth.

3 *E. coli* and *Salmonella* survival in manure dust

While contact with manure-amended soils could potentially cause contamination of fruits and vegetables, there are several reports of manure dust potentially transported by wind that may contaminate leafy greens and other fruit or vegetable commodities. As referenced earlier, manure dust is a potential source of *Salmonella* contamination on peaches which caused an outbreak of salmonellosis in 2020. Dust containing soil and manure particles containing pathogenic *E. coli* or *Salmonella* generated from the movement of cattle herds or from other animal rearing or operations may become airborne and subsequently contaminate growing vegetable crops. A multi-season study showed that spinach, mustard and turnip greens planted adjacent to a cattle-feed lot were contaminated with both non-pathogenic and pathogenic *E. coli* O157:H7 originating from the cattle feedlot (Berry et al., 2015). *E. coli* O157:H7 was recovered from 1.8% of leafy green samples planted 180 m (590 ft) away from the feedlot, which was a significantly lower prevalence than the 3.5% of leafy green samples which contained the pathogen when grown only 60 m (197 ft) away, but not significantly different from the 2.2% of samples containing *E. coli* O157:H7 at the 120 m (400 ft) distance (Berry et al., 2015). The current LGMA standards state that a 1200-ft distance be kept between a concentrated animal feeding operation (CAFO) and a field of leafy greens, but shorter distances can be used for smaller animal operations (LGMA, 2021). Poultry manure dust and the relative abundance of poultry-associated microorganisms (*Staphylococcaceae*) were in greater amounts in trees of an almond orchard 35 m from a poultry house than those trees which were 120 m away in the orchard (Theofel et al., 2020). *Salmonella* was not detected in any poultry manure dust in this study, perhaps because *Salmonella* populations in the dust were not above 10^6–10^7 CFU/g, or the moisture levels in the poultry manure dust were elevated. Recovery of *Salmonella* from poultry litter dust was shown to be dependent upon the original level of the pathogen present and the moisture level of the poultry litter (Pal et al., 2021). When levels of *Salmonella* were less than 10^6–10^7 CFU/g, *Salmonella* spp. was not recovered from poultry litter. Similarly, when the moisture level of poultry litter was increased to levels about 25%, the predicted probability of detecting *Salmonella* fell drastically (Pal et al., 2021). Other work has shown *Salmonella* spp. survival in turkey manure dust was inversely related to the moisture content and the size of the manure particles (Oni et al., 2015). A multi-strain inoculum of *Salmonella* spp. survived for >290 days in manure dust with a particle size of >125 micrometres and with a 5% moisture content, longer than in larger particle sizes (212, 355, and 500 micrometres) and at higher moisture contents (10%, 15%). *Salmonella* spp. in manure dust on the leaves of growing spinach plants declined much less rapidly (1-2 log CFU) over a period of 14 days compared to *Salmonella* populations applied as a water spray (3.5–4.5 log CFU) to the leaves of spinach plants (Oni et al., 2015). Manure dust can not only transport pathogens for distances greater than 400 ft but can also support the prolonged survival of bacterial pathogens and facilitate enhanced persistence on leafy green foliar surfaces. Environmental factors (wind speed, dry conditions) along with adjacent land use issues (application of animal manure to fields adjacent to those growing fruit/vegetable crops) may influence the transfer of dust containing foodborne pathogens to contaminate produce.

4 Prevalence of antibiotic resistance genes in manure-amended soils

Bacteria present in dairy manure have been known to be a reservoir of antibiotic-resistant genes associated with several enteric pathogens (Wichmann et al., 2014). Dairy manure from this study was found to contain 80 distinct antibiotic-resistant genes (ARGs). Soils which received manure as organic fertilizer (dairy manure applied at a rate of 6900 gal/acre, and swine manure at 3600 gal/acre) in Ontario, Canada, contained more antibiotic-resistant bacteria than soils which did not receive manure (Wichmann et al., 2014). It should be noted that the application of manure did not affect the presence of the antibiotic resistance genes, but a different report stated a greater number of antibiotic-resistant coliform bacteria were recovered from manured soils compared to unmanured soils (Marti et al., 2013). The presence of antibiotic-resistant bacteria on vegetables grown in manured soils were not considerably different from those grown in non-manured soils; however, there were unique antibiotic resistance markers (macrolide-lincosamide-streptogramin type A resistance genes (*erm*) and incompatibility (*Inc*) groups) recovered from vegetables grown in manured soils which were not present in non-manured soils. These studies suggest that manure-amended soils may increase the diversity of antibiotic-resistant coliforms in soils. Other types of BSA, like poultry litter, have also altered the levels and types of antibiotic-resistant bacteria (ARB) present in soils (Chaves-Ulate et al., 2021). Soils amended with poultry litter contained increased levels of bacteria resistant to ciprofloxacin and tylosin compared to unamended soils, and levels of these resistant sub-populations remained high 128 days after introduction of poultry litter (Chaves-Ulate et al., 2021). Other studies have shown that in general, the relative abundances of 12 ARGs were lower in rhizosperic and bulk soil planted with either endive or lettuce compared to control soils. Bacterial communities influenced the prevalence of ARGs in soils, specifically the class I integron-integrase gene (*int1*) and several tetracycline resistance genes (*tetA, tetBP,tetC, tetM,* and *tetX*) (Wang et al., 2022). The continuous application of swine manure to soils elevated the relative abundance of ARGs in soil compared to soils which received no swine manure application. These authors found that the presence and relative abundance of ARGs was more affected by the bacterial community in the soils compared to soil physicochemical factors, and the most frequently detected ARGs were: multidrug (MDR – multiple ARGs), macrolide-lincosasme-streptogramin B (MLSB), and B-lactamase (Han et al., 2022). The wheat phyllosphere examined from swine manure-amended soils also contained higher levels of ARGs compared to the control (no swine manure) wheat phyllosphere. Results from this study also showed the continued application of swine manure depressed the relative abundance of ARGs over time, perhaps due to lack of physiological fitness of enteric or intestinal bacteria harbouring these ARGs in soils for a variety of reasons. And while there was no statistical correlation between the presence of the ARGs in soils and the levels transposases or integrases in soils, there was a significant correlation between ARGs and transposases and integrases in the wheat phyllosphere (Han et al., 2022).

The use of poultry litter as biological soil amendment showed similar results to the use of swine manure with respect to the bacterial community present affecting the prevalence of ARGs in soils (Peng et al., 2021). This work also attributed a vastly greater percentage of the ARGs in soil to indigenous bacterial communities in the soil or to unknown sources rather than to the introduction of poultry litter, heat-treated poultry litter or mushroom residues. Broadly, poultry litter contributed 1.41 times the total abundance of ARGs to amended soils compared to heat-treated poultry litter or mushroom residues, indicating that heat treatment of poultry litter can reduce the ARG burden poultry litter-based amendments (Peng et al., 2021). Another study has reported that soils amended with dairy cattle manure from animals which did not receive any antibiotics may encourage the growth of bacteria expressing beta-lactamase genes originally present in the soil before the addition of the manure (Udikovic-Kolic et al., 2014). At the outset of the study, soils amended with inorganic fertilizers contained higher populations of beta-lactamase containing bacteria then those amended with dairy cattle manure that received no antibiotic treatments. However, from 4 days to 130 days after the application of the manure, applied at a rate of 20 kg manure/bed (unknown size), the relative abundance of specific alleles of the beta-lactamase gene ($bla_{CEP\text{-}04}$) in manure-amended soils was higher than the abundance of $bla_{CEP\text{-}04}$ in the soils receiving inorganic fertilizers. The application of manure was determined to encourage the growth of *Pseudomonas* spp., *Pscychrobacter pulmonis* and *Janthinobacterium* spp, all of which are known to contain beta-lactamases (Udikovic-Kolic et al., 2014). The use of manure may alter the overall population of the soil microbiome by increasing the proportion of ARB in agricultural fields.

5 Case studies

Soil amendments may have played a role in crop contamination associated with past outbreaks. When considering adjacent land use, which may be defined as that within five hundred feet of agricultural land actively engaged in properly conducted agricultural operations (according to LawInsider.com), the proximity of a CAFO to produce fields may increase the likelihood of contamination. Manure contamination can occur in the field, along with potential windborne transmission of dust particles carrying pathogens could come into contact with adjacent fields, produce, or agricultural water. There are three outbreaks which provide potential evidence for transmission of pathogens across commodities and the risks to food safety and public health associated with soil amendments. These outbreaks which involved potential windborne transmission of contaminated dust particles include the 2018 Romaine lettuce outbreak in Yuma, Arizona, the 2020 outbreak associated with contaminated red onions in Holtville and Bakersfield, California, and the 2020 outbreak associated with contaminated peaches in Cutler, Kerman, and Sanger, California.

5.1 2018 Romaine lettuce outbreak

5.1.1 Epidemiological summary

In early April 2018, FDA and the Centers for Disease Control and Prevention (CDC) began investigating a multi-state outbreak of *E. coli* O157:H7 infections linked to Romaine lettuce. When this outbreak was declared, it was considered the largest outbreak of *E. coli* O157:H7 infections since 2006 (FDA Memorandum,

2018). 210 people were infected across 36 states (Anon, 2018). 96 people were hospitalized, including 27 people who developed a type of kidney failure associated with hemolytic uremic syndrome (Anon, 2018). Five deaths were reported from Arkansas, California, Minnesota (2), and New York (Anon, 2018). CDC laboratory testing identified the outbreak strain of *E. coli* O157:H7 in water samples taken from an irrigation canal in Wellton area, near the Yuma, AZ produce growing region, and which was genetically similar to environmental isolates of *E. coli* O157:H7 found by an adjacent cattle ranch, also referred to as a CAFO (Anon, 2018). Traceback of the Romaine lettuce that was consumed by infected individuals determined that the origin of the outbreak was from the Yuma produce growing region. This region consists of farms located in Imperial County, California, and Yuma County, Arizona. The traceback identified a total of 36 fields in 23 farms in the Yuma growing region as potentially supplying the contaminated Romaine lettuce which was consumed and linked to the outbreak (Memorandum, 2018). Whole genome sequencing (WGS) showed that the *E. coli* O157:H7 found in the canal water was closely related genetically to the *E. coli* O157:H7 from ill people (Anon, 2019). Windborne transmission of contaminated dust, from the adjacent cattle farm onto Romaine lettuce crops or canal water is speculated to be the source of contamination (Memorandum, 2018).

5.1.2 FDA environmental assessment investigation

An Environmental Assessment (EA) was conducted by the FDA in the Yuma growing region soon after confirmation of traceback. The EA team relied on the FDA Food Modernization Act (FSMA) Produce Safety Rule and Good Agricultural Practices (GAPs) as the basis for assessing factors that may have been the source of contamination of Romaine lettuce on farms that were identified in the traceback investigation (Nutrition, 2021a). Areas of focus included: agricultural water, soil amendments, growing and harvesting practices, animal intrusion, adjacent land use, and employee health and hygiene practices (Nutrition, 2021a). The areas of focus for the manufacturing and processing operations included: facility and equipment sanitary design, cleaning and sanitizing practices, washing, and drying of fresh-cut Romaine lettuce and storage of Romaine lettuce (Nutrition, 2021a).

The outbreak strain of *E. coli* O157:H7 was identified in the water from three separate locations along the 3.5 mile stretch of an irrigation canal near Wellton in Yuma County, Arizona (Nutrition, 2021a). This outbreak strain was not found in any other samples collected during the EA testing in the Yuma growing region or in Imperial County, CA (Nutrition, 2021a). As a direct result the FDA concluded that the water from the irrigation canal where the outbreak strain was identified was most likely the cause of contamination of the Romaine lettuce that was consumed during this outbreak (Nutrition, 2021a). However, the FDA will not rule out other potential sources of *E. coli* O157:H7 contamination of Romaine lettuce that could have occurred (Nutrition, 2021a). There are several potential ways in which the contaminated canal water could have contaminated Romaine lettuce: through direct application to the crop (overhead irrigation), or through the use of the irrigation canal water to dilute the crop protection chemicals that were then applied to the Romaine lettuce crop by aerial or ground-based spray (Nutrition, 2021a). The EA does not address when the irrigation canal may have become contaminated with the outbreak isolate.

During their initial site visit to the Yuma growing region, the EA team assessed the animal feeding operation that was adjacent to the 3.5 mile stretch of Wellton irrigation canal water where the outbreak strain of *E. coli* O157:H7 was found (FDA Memorandum, 2018). This operation mainly consisted of Holstein steers. The steer pen flooring had not been changed in over a decade and was made up of native sand and organic material that had been compacted over time (Memorandum, 2018). Facilities managers informed the EA team that potential nutrient permeability as well as microbiological pathogen transport down through the pen floor and into the groundwater was very low (Memorandum, 2018). CAFO managers provided soil samples from the cattle pen for analysis to address this claim. The EA team observed operations moving the manure from the CAFO to the north composting facility (Memorandum, 2018). During the on-site activities, the EA team noticed the trucks used a different route when hauling manure to the west composting facility than that of the north composting facility. The EA team saw no potential for cross-contamination between raw manure and finished compost, as trucks used to transport finished compost were specifically dedicated to this purpose (Memorandum, 2018). Trucks used to haul raw manure to the composting facilities were different and specifically used for this purpose.

This CAFO also had several wells, an irrigation canal, and two cattle pen drainage retention ponds to use as water sources (Memorandum, 2018). During rain events the two retention ponds are utilized to contain the runoff from the feedlots, where the excess water is directed to channels that run along the pens and flow into the retention ponds in order to mitigate risk of contaminating the ground water or irrigation canal. The CAFO is allowed to use this water from the retention ponds for dust abatement purposes.

5.2 2020 Red onions outbreak

5.2.1 Epidemiological summary

On 19 June 2020, the FDA, CDC and Canadian state and local partners began an investigation on a multi-state outbreak of *Salmonella* Newport infections that was linked to red onions. There were 1,127 confirmed cases across 46 states (CDC, 2020). There were no deaths reported but 167 people were hospitalized (CDC, 2020). Epidemiological and traceback investigations conducted by the Public Health Agency of Canada and The Canadian Food Inspection Agency (CFIA) identified red onions from Thomson International Inc. were the source of this outbreak (CDC, 2020). It was also found that other types of onions (such as white, yellow, and sweet yellow) were also contaminated with *Salmonella* Newport as they were also grown, harvested, and processed alongside the originally contaminated red onions (CDC, 2020). On 1 August 2020, Thomson International Inc. recalled all red, yellow, white, and sweet yellow onions because there was potential for contamination with *Salmonella* (CDC, 2020). WGS analysis of 732 *Salmonella* isolates from infected individuals revealed that only two isolates contained antibiotic resistance genes. One isolate had predicted resistance to ampicillin, and the other had predicted resistance to tetracycline. FDA's traceback investigation identified a packing facility and multiple farms that supplied red onions that could have been potential sources of contamination (Nutrition, 2021b). Field-level investigations were conducted after identifying the suspect vehicle and firm within Thomson International Inc (Nutrition, 2021b); however, the field investigations were limited since most of the onions had already been harvested and distributed. The FDA has completed over 2000 product and environmental analyses from multiple Thomson International Inc. locations and surrounding areas which include water, soil, and scat samples (Nutrition, 2021b). Although a conclusive root cause could not be identified, several potential contributing factors to the 2020 red onion outbreak were identified, including the hypothesis that contaminated irrigation water used in a growing field in Holtville, California may have led to contamination of the onions (Nutrition, 2021c). This is further described in the following section.

5.2.2 FDA environmental assessment investigation

The FDA and CDC began an Environmental Assessment investigation on suspected locations (farms, processing facilities) where the outbreak could have originated from. First, their investigation took them to a farm in Holtville, California. A myriad of birds, squirrels and bird scat were observed around the produce fields and equipment at the onion operation (USDA, 2021a). Adjacent farmland housed an alfalfa production with hundreds of sheep observed grazing and sheep tracks and scat were found directly adjacent to an onion field that was identified by the traceback investigation (FDA, 2021a). The sheep located on the adjacent land, along with the amount of bird activity observed, are potential sources and vectors of *Salmonella* contamination. Several water samples from irrigation canals, drainage, and water seepage networks were collected and contained *Salmonella*. Two samples collected from the drains and drainage networks contained multiple serovars of *Salmonella* but did not match the outbreak isolate (FDA, 2021a). The cross-contamination of irrigation canals and field drainage or seepage networks could be a potential route of contamination between irrigation water to onions, especially if these areas were habitats for wildlife (FDA, 2021a).

The FDA and CDC investigation next led them to a farm in Bakersfield, California. They observed birds, bird nests and cats along with animal scat in and around the buildings including evidence of pests on food contact surfaces (FDA, 2021a). Samples of the irrigation water and sediment from an on-farm tail water collection basin, and environmental swabs of the tail pump water were collected and several contained *Salmonella*. It was noted by investigators that frogs were observed in the water used to irrigate the onion fields and bird activity in and around the water could have been a source of contamination (FDA, 2021a). A CAFO was also identified less than half a mile from the onion fields and immediately adjacent to an irrigation canal which supplied the water to the onion crop (FDA, 2021a). From this location, Bakersfield California, five samples contained *Salmonella* although none were identified as the *Salmonella* Newport outbreak isolate (FDA, 2021a).

Even though a conclusive root cause could not be identified, there were several potential factors that could have caused the *Salmonella* outbreak of infections associated with onions. Contaminated irrigation water used in the fields may have been a source of contamination (FDA, 2021a). Another potential source of *Salmonella* is adjacent and nearby land use to the onion fields, which may have generated soil runoff into the growing fields or water sources (FDA, 2021a). Visual observations were made by the investigators, and they observed several potential sources of contamination: sources of irrigation water, sheep grazing on adjacent land, and signs of animal intrusions. However, it is important to note that *Salmonella* isolates were detected in two sediment subsamples and two water subsamples collected during this investigation and they closely matched clinical isolates from a sprout-associated outbreak from 2016 and 2018 (FDA, 2021a).

5.3 2020 Peaches

5.3.1 Epidemiological summary

In 2020, there was another multi-state outbreak involving potential dust contamination. An outbreak of *Salmonella* Enteritidis infections was linked to the consumption of peaches. In total this outbreak infected 101 individuals, across 17 states, including 28 hospitalizations (Quality Assurance & Food Safety, n.d.).

This was the first case where *Salmonella* was linked to peaches. The FDA and CDC along with state partners and Canadian public health officials began their investigation between August 2020 and October 2020. The traceback evidence prioritized two subsequent investigations of peach packing/holding operations and peach orchards located in Cutler, Kerman, and Sanger, California (Nutrition, 2021d). Over 700 tests on environmental, peach, and peach tree leaf samples were conducted (Nutrition, 2021d). An orchard where peaches contaminated with *Salmonella* Alachua (but not *S. enteritidis*) originated was adjacent to a poultry operation. The outbreak isolate was further linked to other isolates by WGS to 2019 and 2020 chicken isolates (Nutrition, 2021d). During a follow-up investigation, orchards adjacent to a cattle feedlot contained *Salmonella* Montevideo and was genetically similar to the 2018–2020 beef and cattle isolates (Nutrition, 2021d).

5.3.2 FDA environmental assessment investigation

The traceback investigation identified several locations as sources of contamination however a single source was unable to be determined. A tandem investigation was soon led by the CFIA to address cases in Canada. Traceback evidence obtained during these investigations identified several peach packing/holding operations (including packing lines within packing operations) and peach orchards in Cutler, Kerman, and Sanger California as potential outbreak sources (FDA, 2021b). On-site and environmental assessment investigations were conducted by the FDA (FDA, 2021b). Environmental samples were taken at suspected packing houses and product samples from the packing houses and orchards. All environmental samples were negative for the *S. Alachua* outbreak strain (FDA, 2021b).

A follow-up investigation expanded sampling into the orchards, determined by traceback, and began sampling of peach products, orchard tree leaves, and soil drag-swabs (FDA, 2021b). *Salmonella* Montevideo was recovered from two of the orchard leaves collected nearby the adjacent cattle operations (FDA, 2021b). These *Salmonella* Montevideo isolates were genetically similar to 2018–2020 beef and cattle isolates (FDA, 2021b), and peach and peach tree leaves contained *Salmonella* Alachua. These samples were collected near the adjacent poultry farm which reported genetically similar isolates to 2019 and 2020 chicken isolates (FDA, 2021b). While none of the environmental *Salmonella* isolates recovered were the outbreak *S.* Enteritidis isolate, these results show that fugitive dust from adjacent agricultural operations can impact produce farms and can provide a potential route of contamination of produce grown on adjacent or nearby lands (FDA, 2021b).

6 Recommendations

As a result of these investigations into adjacent land use and produce outbreaks, several recommendations regarding growing and harvesting should be considered:

- Be cognizant of and assess risks that could be affected by adjacent and nearby land use, especially if it relates to fugitive dust exposure from adjacent or nearby livestock and poultry.
- Make sure that all agricultural water (water that directly contacts the harvestable portion of the crop) is safe and adequate for its intended use (including agricultural water used for crop production chemicals).
- Assess and mitigate risks related to land uses near or adjacent to growing fields that may contaminate agricultural water or leafy greens crops directly (e.g. nearby cattle operations or dairy farms, manure, or composting facilities).
- Consider additional tools such as pre-harvest and/or post-harvest sampling and testing of products to help inform the risk assessment and clarify the need for specific prevention measures.
- Verify that food safety procedures, policies, and practices, including supplier controls for fresh-cut processors, are developed, and consistently implemented on farms (both domestic and foreign) and in fresh-cut produce manufacturing/ processing food facilities to limit the potential for contamination and/ or spread of human pathogens.
- Local knowledge and actions are critical in resolving potential routes of contamination of crops and in the development and implementation of short- and long-term remediation measures to reduce the potential for another outbreak associated with leafy greens or other fresh produce commodities (Nutrition, 2021a).
- When pathogens are identified through microbiological surveys, pre-harvest testing of produce, or post-harvest testing of produce implement industry-led root cause analyses to determine how the contamination likely occurred and then implement appropriate prevention and verification measures and lastly.
- Improve traceability through increased digitization, interoperability, and standardization of traceability records which would expedite traceback and help remove contaminated product from the marketplace more quickly, thereby preventing further illnesses. This is not only important for growers, but also critical for shippers, manufactures and retailers as well, to improve overall traceability throughout the supply chain (FDA, 2021b).

7 Conclusion

Scientific research must contribute to improving productivity and supporting sustainable storage and use of soil amendments. Given that the use of soil amendments remains critical to crop production and enhancing crop yield, and that livestock production continues to grow, thereby increasing the amount of BSAAO being produced; the use of BSAAO's in fruit and vegetable production must be balanced with preventing foodborne contamination scenarios. Across the globe animal agriculture supports production of soil amendments useful for equally enriching soils and crop yields. The survival of bacterial pathogens, like *Salmonella*, pathogenic *E. coli* and *Listeria*, can be extended in soils containing BSAAO and therefore, there is a potential for transmission of pathogens to raw agricultural commodities (Litt et al., 2021). The role of bacterial community analysis in the soil bacterial microbiome in production fields is a growing area of research and may also impact crop yield as well as bacterial survival (Shu et al., 2022; Peng et al., 2021). Bacterial communities in soils may affect the presence and abundance of antibiotic resistance genes in soils, and microbial community analysis should also be utilized to determine the effect on enteric foodborne bacterial pathogens, that can also be transferred to manure-amended soil grown root vegetables in association with bacterial communities dominant in root and plant surfaces (Wang et al., 2022). Other research studies have shown that microbial population concentration and diversity in BSAAO, like poultry litter, can be affected by straightforward parameters like storage time and location, as well as temperature and moisture (Chen and Jiang, 2014). Overall, the growing body of global literature suggests that soil fertility and microbial diversity are both positively connected to the use of BSAAO, and these benefits must be coupled with mitigation strategies to reduce the risk of pathogen transfer from soils to produce.

8 Future trends in research

8.1 Soil amendment remediation

Soil amendments have become more widely involved with remediation purposes. With soil contamination by potentially toxic elements (PTEs) leading to adverse environmental impacts. Using soil amendments to help reduce this waste could be an efficient and effective control method. As livestock production and the amount of poultry waste products (poultry litter) is continuously increasing, this waste could be applied to improve the quality of contaminated soils (Palansooriya et al., 2020). Soil amendments such as manure, compost, biochar, clay minerals, phosphate compounds, coal fly ash, and liming materials can be widely used to immobilize PTEs (Palansooriya et al., 2020). For example, the use of biochar, a solid material obtained from the thermochemical conversion of biomass in an oxygen-limited environment, can be utilized in carbon sequestration, carbon farming, climate change mitigation and soil pollution remediation (Wu et al., 2017). Biochar can increase soil nutrient availability, microbial activity, soil organic matter, water retention and crop yields in soils (Wu et al., 2017). The interaction of biochar with compost and other soil amendments led to the high adsorption of several pollutants (Wu et al., 2017). Further research is needed to confirm efficacy of adsorption rates of pollutants from biochar but provides exciting possibilities for the future of remediating soil amendments.

8.2 Fungal biocontrol

The term 'bioremediation' may conjure up images of removal of heavy metals from contaminated soils and chemical breakdown of pesticides associated with industrial use; however, through the novel use of biocontrol agents, bioremediation of manure is an emerging field of research. Of the microbes used to remediate contaminated sites, ligninolytic white-rot fungi (WRF) can degrade a wide variety of toxic or persistent environmental contaminants (Dorbert, 2022). WRF are capable of detoxifying and degrading pollutants in the environment. WRF are part of the basidiomycetes family also known as, 'higher fungi' and have the unique capability of degrading lignin. While most fungi have the capability to break down cellulose and hemicellulose, lignin is very recalcitrant with a complex chemical structure (Pointing, 2001). The chemical structural similarities between lignin and many recalcitrant contaminants such as pesticides, PCBs, explosives, heavy metals, etc., make WRF a promising candidate for the remediation of these pollutants (Rodriguez-Couto, 2017).

Two species of WRF that can breakdown these components include *Phanerochaete chrysosporium* and *Pleurotus ostreatus*. They also produce ligninolytic enzymes. These fungi species secrete a system of enzymes comprising lignin peroxidase (LiP), manganese dependent peroxidases (MnP), phenol oxidase or laccase and glycol oxidases (Wu et al., 1996). The secretion of these enzymes allows the fungi to increase their ligninolytic activity and allow the fungi to degrade a wide range of materials such as bacteria and nematodes (Barron and Thorn, 1987). Ligninolytic activity is triggered by nutrient limiting conditions or by 'starving' the fungi. When key nutrients such as nitrogen and phosphorus are limited then the predatory behaviours of the fungi can be activated (Carroll and Wicklow, 1992). To immobilize its prey the fungi will secrete a mycotoxin, then its hyphae will infiltrate its prey and absorb its nutrients (Barron and Thorn, 1987). Studies have shown that the WRF, *P. ostreatus*, is able to immobilize and destroy nematodes and bacteria (Barron and Thorn, 1987).

Using a filamentous fungus as a system to inactivate pathogenic bacteria as a pre-harvest preventative method to reduce risk of crop contamination may be an option to reduce microbial pathogens. Fungal bioreactor-based systems have been predominately used in biotechnology industries for enzyme production and have been used for decades in environmental engineering, for bioremediation purposes to degrade persistent and toxic chemicals and contaminants (Couto and Toca-Herrera, 2007; Linko, 1988). Studies have shown a significant reduction in *E. coli* levels in wastewater and stormwater runoff through the use of filamentous fungi (Chirnside et al., 2013). Use of fungal bioreactor systems to inactivate pathogens could increase access to treating wastewater, stormwater, and agricultural runoff as these important mixtures and watershed areas may contain pathogens like those present in animal manures.

9 Where to look for further information

The following articles provide a good overview of the subject:

1. Sharma, M., Millner, P. D., Hashem, F., Vinyard, B. T., East, C. L., Handy, E. T., White, K., Stonebraker, R. and Cotton, C. P. (2019). Survival of Escherichia coli in Manure-Amended Soils Is Affected by Spatiotemporal, Agricultural, and Weather Factors in the Mid-Atlantic United States. *Appl Environ Microbiol.* 2019 Feb 20;85(5):e02392-18. doi: 10.1128/AEM.02392-18. Print 2019 Mar 1.PMID: 30552193.
2. Sharma, M. and Reynnells, R. (2016). Importance of Soil Amendments: Survival of Bacterial Pathogens in Manure and Compost Used as Organic Fertilizers. *Microbiol Spectr.* 2016 Aug;4(4). doi: 10.1128/microbiolspec.PFS-0010-2015. PMID: 277 26763.

Key research in this area can be found at the following organizations:

1. Soil Health Institute https://soilhealthinstitite.org.
2. FDA Final Rule on Produce Safety https://www.fda.gov/food/food-safety-modernization-act-fsma/fsma-final-rule-produce-safety.
3. Food Safety Resource Clearinghouse information on soil amendments https://foodsafetyclearinghouse.org/resources/fda-fact-sheet-biological-soil-amendments-animal-origin-bsaao.

10 References

Acheamfour, C. L., Parveen, S., Hashem, F., Sharma, M., Gerdes, M. E., May, E. B., Rogers, K., Haymaker, J., Duncan, R., Foust, D., Taabodi, M., Handy, E. T., East, C., Bradshaw,

R., Kim, S., Icallef, S. A., Callahan, M. T., Allard, S., Anderson-Coughlin, B., Craighead, S., Gartley, S., Vanroe, A., Kniel, K. E., Solaiman, S., Bui, A., Murray, R., Craddock, H. A., Kulkarni, P., Golstein, R. E. R., Sapkota, A. R. and Gralnick, J. A. 2021. Levels of Salmonella enterica and Listeria monocytogenes in alternative irrigation water vary based on water source on the eastern shore of Maryland. *Microbiology Spectrum* 9, e00669-21.

AMS and USDA n.d. Soil building - manures & composts. *Agricultural Marketing Service* [online]. Available at: https://www.ams.usda.gov/grades-standards/soil-building-manures-composts [Accessed 31 Mar. 2022].

Anon. 2018. *Outbreak of E. coli Infections Linked to Romaine Lettuce* [online]. Available at: https://www.cdc.gov/ecoli/2018/o157h7-11-18/index.html.

Anon. 2019. *Multistate Outbreak of E. coli O157:H7 Infections Linked to Romaine Lettuce (Final Update)* [online]. Available at: https://www.cdc.gov/ecoli/2018/o157h7-04-18/index.html.

Avery, L. M., Hill, P., Killham, K. and Jones, D. L. 2004. Escherichia coli O157 survival following the surface and sub-surface application of human pathogen contaminated organic waste to soil. *Soil Biology and Biochemistry* 36(12), 2101–2103.

Baker, C. A., De, J., Bertoldi, B., Dunn, L., Chapin, T., Jay-Russell, M., Danyluk, M. D. and Schneider, K. R. 2019. Prevalence and concentration of stx+ E. coli and E. coli O157 in bovine manure from Florida farms. *PLoS ONE* 14(5), e0217445.

Baker, C. A., De, J. and Schneider, K. R. 2021. Influence of soil microbes on Escherichia coli O157:H7 survival in soil rinse and artificial soil. *Journal of Applied Microbiology* 131(3), 1531–1538.

Bardsley, C. A., Weller, D. L., Ingram, D. T., Chen, Y., Oryang, D., Rideout, S. L. and Strawn, L. K. 2021. Strain, soil-type, irrigation regimen, and poultry litter influence salmonella survival and die-off in agricultural soils. *Frontiers in Microbiology* 12, 590303.

Barron, G. L. and Thorn, R. G. 1987. Destruction of nematodes by species of Pleurotus. *Canadian Journal of Botany* 65(4), 774–778.

Berry, E. D., Wells, J. E., Bono, J. L., Woodbury, B. L., Kalchayanand, N., Norman, K. N., Suslow, T. V., López-Velasco, G., Millner, P. D. and Schaffner, D. W. 2015. Effect of proximity to a cattle feedlot on Escherichia coli O157:H7 contamination of leafy greens and evaluation of the potential for airborne transmission. *Applied and Environmental Microbiology* 81(3), 1101–1110.

Bezanson, G., Delaquis, P., Bach, S., Mckellar, R., Topp, E., Gill, A., Blais, B. and Gilmour, M. 2012. Comparative examination of Escherichia coli O157:H7 survival on romaine lettuce and in soil at two independent experimental sites. *Journal of Food Protection* 75(3), 480–487.

CALGMA Food Safety Standards: https://lgma.ca.gov/food-safety-program.

California Leafy Greens Marketing Agreement. n.d. *Food Safety Program* [online]. Available at: https://lgma.ca.gov/food-safety-program [Accessed 31 Mar. 2022].

California Leafy Greens Handler Marketing Agreement. 2013. Commodity specific food safety guidelines for the production and harvest of lettuce and leafy greens. *Food Safety Program* [online]. Available at: https://lgma.ca.gov/food-safety-program [Accessed 31 Mar. 2022].

California Leafy Greens Handler Marketing Agreement. 2021. *Commodity Specific Food Safety Guidelines for the Production and Harvest of Lettuce and Leafy Greens.* Available at: https://lgma-assets.sfo2.digitaloceanspaces.com/downloads/August-2021-CA-LGMA-Metrics_FINAL-v20211208_A11Y.pdf.

Carroll, G. C. and Wicklow, D. T. 1992. *The Fungal Community. Its Organization and Role in the Ecosystem*. The CRC Press: Boca Raton, pp. 535–539.
CDC. 2020. *Outbreak of Salmonella Newport Infections Linked to Onions*. Centers for Disease Control and Prevention (U.S.) [online]. Available at: https://www.cdc.gov/salmonella/newport-07-20/index.html.
Çekiç, S. K., De, J., Jubair, M. and Schneider, K. R. 2017. Persistence of indigenous Escherichia coli in raw bovine manure-amended soil. *Journal of Food Protection* 80(9), 1562–1573.
Chaves-Ulate, C., Granados-Chinchilla, F. and Rodríguez, C. 2021. Fertilization with poultry litter increases the abundance of antibiotic-resistant bacteria in tropical soil: a microcosm study. *Water, Air, and Soil Pollution* 232(10).
Chen, Z. and Jiang, X. 2014. Microbiological safety of chicken litter or chicken litter-based organic fertilizers: a review. *Agriculture* 4(1), 1–29. doi: 10.3390/agriculture4010001.
Chirnside, A. E. M., Harris, J. P. and Gregory, N. 2013. Degradation of harmful bacteria in simulated wastewater by the White rot fungus Pleurotus ostreatus. *World Environmental and Water Resources Congress*. Available at: http://udspace.udel.edu/handle/19716/11573.
Couto, S. R. and Toca-Herrera, J. L. 2007. Laccase production at reactor scale by filamentous fungi. *Biotechnology Advances* 25(6), 558–569. doi: 10.1016/j.biotechadv.2007.07.002.
Dorbert, A. 2022. *Quantifying Fungal Growth in Nutrient Limited Systems for Bioremediation Purposes*. University of Delaware, Newark, DE.
FDA. 2021a. *Investigation Report: Factors Potentially Contributing to the Contamination of Red Onions Implicated in the Summer 2020 Outbreak of Salmonella Newport*. FDA. Available at: https://www.fda.gov/food/cfsan-constituent-updates/fda-releases-investigation-findings-following-summer-2020-outbreak-linked-peaches.
FDA. 2021b. *Investigation Report: Factors Potentially Contributing to the Contamination of Peaches Implicated in the Summer 2020 Outbreak of Salmonella Enteritidis*. FDA; USDA.
Food and Drug Administration 2014. Standards for growing, harvesting, packing, and holding of produce for human consumption. *FSMA Proposed Rule for Produce Safety*. Available at: http://www.fda.gov/Food/GuidanceRegulation/FSMA/ucm334114.htm.
Food and Drug Administration 2015. Standards for growing, harvesting, packing, and holding of produce for human consumption. *FSMA Final Rule for Produce Safety*. Available at: https://www.regulations.gov/document/FDA-2011-N-0921-18558.
Franz, E., Semenov, A. V., Termorshuizen, A. J., De Vos, O. J., Bokhorst, J. G. and Van Bruggen, A. H. C. 2008. Manure-amended soil characteristics affecting the survival of E. coli O157:H7 in 36 Dutch soils. *Environmental Microbiology* 10(2), 313–327.
FSMA Final Rule on Produce Safety: https://www.regulations.gov/document/FDA-2011-N-0921-18558.
Gurtler, J. B., Doyle, M. P., Erickson, M. C., Jiang, X., Millner, P. and Sharma, M. 2018. Composting to inactivate foodborne pathogens for crop soil application: A review. *Journal of Food Protection* 81(11), 1821–1837. doi: 10.4315/0362-028X.JFP-18-217.
Han, X.-M., Hu, H.-W., Li, J.-Y., Li, H.-L., He, F., Sang, W.-C., Liu, Y.-H. and Ma, Y.-B. 2022. Long-term application of swine manure and sewage sludge differently impacts antibiotic resistance genes in soil and phyllosphere. *Geoderma* 411, 115698.

Harris, L. J., Berry, E. D., Blessington, T., Erickson, M., Jay-Russell, M., Jiang, X., Killinger, K., Michel, F. C., Millner, P., Schneider, K., Sharma, M., Suslow, T. V., Wang, L. and Worobo, R. W. 2013. A framework for developing research protocols for evaluation of microbial hazards and controls during production that pertain to the application of untreated soil amendments of animal origin on land used to grow produce that may be consumed raw. *Journal of Food Protection* 76(6), 1062–1084.

https://www.ams.usda.gov/grades-standards/soil-building-manures-composts.

Hutchison, M. L., Walters, L. D., Avery, S. M., Munro, F. and Moore, A. 2005. Analyses of livestock production, waste storage, and pathogen levels and prevalences in farm manures. *Applied and Environmental Microbiology* 71(3), 1231–1236.

Hutchison, M. L., Walters, L. D., Avery, S. M., Synge, B. A. and Moore, A. 2004. Levels of zoonotic agents in British livestock manures. *Letters in Applied Microbiology* 39(2), 207–214.

Land Use n.d. [online] Available at: https://www.fcs.uga.edu/docs/3_Land_Use.pdf [Accessed 4 Apr. 2022].

Law Insider. n.d. *Adjacent to Agriculture Definition*. Available at: https://www.lawinsider.com/dictionary/adjacent-to-agriculture [Accessed 4 Apr. 2022].

Limoges, M., Neher, D. A., Weicht, T. R., Millner, P. D., Sharma, M. and Donnelly, C. 2021. Differential survival of generic E. coli and Listeria spp. in Northeastern U.S. soils amended with dairy manure compost, poultry litter compost, and heat-treated poultry pellets and fate in raw edible radish crops. *Journal of Food Protection* 85(12), 1708–1715.

Linko, S. 1988. Production and Characterization of extracellular Lignin peroxidase from Immobilized Phanerochaete chrysosporium in a 10-l Bioreactor. *Enzyme and Microbial Technology* 10(7), 410–417. doi: 10.1016/0141-0229(88)90035-X.

Litt, P. K., Kelly, A., Omar, A., Johnson, G., Vinyard, B. T., Kniel, K. E. and Sharma, M. 2021. Temporal and agricultural factors influence Escherichia coli survival in soil and transfer to cucumbers. *Applied and Environmental Microbiology* 87(7), e02418-20.

Marti, R., Scott, A., Tien, Y. C., Murray, R., Sabourin, L., Zhang, Y. and Topp, E. 2013. Impact of manure fertilization on the abundance of antibiotic-resistant bacteria and frequency of detection of antibiotic resistance genes in soil and on vegetables at harvest. *Applied and Environmental Microbiology* 79(18), 5701–5709.

Memorandum. 2018. *Memorandum to the File on the Environmental Assessment; Yuma 2018 E. coli O157:H7 Outbreak Associated with Romaine Lettuce*. FDA [online]. Available at: https://www.fda.gov/media/120690/download.

National Organic Program. *Soil Fertility and Crop Nutrient Management Practice Standard. 7 C.F.R. 205.203*.

Neher, D. A., Cutler, A. J., Weicht, T. R., Sharma, M. and Millner, P. D. 2019. Composts of poultry litter or dairy manure differentially affect survival of enteric bacteria in fields with spinach. *Journal of Applied Microbiology* 126(6), 1910–1922.

NOP Compost Teas: https://www.ams.usda.gov/sites/default/files/media/Compost_FINAL.pdf.

NOP Manure: https://www.ams.usda.gov/sites/default/files/media/5006.pdf.

NOP Manure & Compost: https://www.ams.usda.gov/sites/default/files/media/Manure%20in%20Organic%20Production%20Systems_FINAL.pdf.

Nutrition, C. for F. S. and A. 2021a. *Environmental Assessment of Factors Potentially Contributing to the Contamination of Romaine Lettuce Implicated in a Multi-State Outbreak of E. coli O157:H7*. FDA [online]. Available at: https://www.fda.gov/

food/outbreaks-foodborne-illness/environmental-assessment-factors-potentially-contributing-contamination-romaine-lettuce-implicated.
Nutrition, C. for F. S. and A. 2021b. *Outbreak Investigation of Salmonella Newport: Red Onions (July 2020).* FDA [online]. Available at: https://www.fda.gov/food/outbreaks-foodborne-illness/outbreak-investigation-salmonella-newport-red-onions-july-2020#:~:text=On%20August%201%2C%202020%2C%20Thomson [Accessed 31 May 2022].
Nutrition, C. for F. S. and A. 2021c. *Factors Potentially Contributing to the Contamination of Red Onions Implicated in the Summer 2020 Outbreak of Salmonella.* FDA, Newport [online]. Available at: https://www.fda.gov/food/outbreaks-foodborne-illness/factors-potentially-contributing-contamination-red-onions-implicated-summer-2020-outbreak-salmonella#:~:text=Executive%20Summary [Accessed 31 May 2022].
Nutrition, C. for F. S. and A. 2021d. *FDA Releases Investigation Findings Following Summer 2020 Outbreak Linked to Peaches.* FDA [online]. Available at: https://www.fda.gov/food/cfsan-constituent-updates/fda-releases-investigation-findings-following-summer-2020-outbreak-linked-peaches [Accessed 31 Mar. 2022].
Oni, R. A., Sharma, M. and Buchanan, R. L. 2015. Survival of Salmonella enterica in dried turkey manure and persistence on spinach leaves. *Journal of Food Protection* 78(10), 1791–1799.
Pal, A., Bailey, M. A., Talorico, A. A., Krehling, J. T., Macklin, K. S., Price, S. B., Buhr, R. J. and Bourassa, D. V. 2021. Impact of poultry litter Salmonella levels and moisture on transfer of Salmonella through associated in vitro generated dust. *Poultry Science* 100(8), 101236.
Palansooriya, K. N., Shaheen, S. M., Chen, S. S., Tsang, D. C. W., Hashimoto, Y., Hou, D., Bolan, N. S., Rinklebe, J. and Ok, Y. S. 2020. Soil amendments for immobilization of potentially toxic elements in contaminated soils: A critical review. *Environment International* 134, 105046. doi: 10.1016/j.envint.2019.105046.
Pang, H., Mokhtari, A., Chen, Y., Oryang, D., Ingram, D. T., Sharma, M., Millner, P. D. and Van Doren, J. M. 2020. A predictive model for survival of Escherichia coli O157:H7 and generic E. coli in soil amended with untreated animal manure. *Risk Analysis* 40(7), 1367–1382.
PDF: https://lgma-assets.sfo2.digitaloceanspaces.com/downloads/August-2021-CA-LGMA-Metrics_FINAL-v20211208_A11Y.pdf.
Peng, S., Wang, Y., Chen, R. and Lin, X. 2021. Chicken manure and mushroom residues affect soil bacterial community structure but not the bacterial resistome when applied at the same rate of nitrogen for 3 years. *Frontiers in Microbiology* 12, 618693.
Pointing, S. B. 2001. Feasibility of bioremediation by white-rot fungi. *Applied Microbiology and Biotechnology* 57(1–2), 20–33. doi: 10.1007/s002530100745.
Quality Assurance & Food Safety. n.d. *FDA, Releases Report on Summer 2020 Salmonella Outbreak Linked to Peaches* [online]. Available at: https://www.qualityassurancemag.com/article/fda-releases-report-on-summer-2020-salmonella-outbreak/#:~:text=FDA- [Accessed 31 Mar. 2022].
Rees, H. W., Chow, T. L., Zebarth, B., Xing, Z., Toner, P., Lavoie, J. and Daigle, J.-L. 2014. Impact of supplemental poultry manure application on potato yield and soil properties on a loam soil in north-western New Brunswick. *Canadian Journal of Soil Science* 94(1), 49–65. doi: 10.4141/cjss2013-009.

Rodriguez-Couto, S. 2017. Industrial and environmental applications of white-rot fungi. *Mycosphere* 8(3), 456–466. doi: 10.5943/mycosphere/8/3/7.

Shah, M. K., Bradshaw, R., Nyarko, E., Handy, E. T., East, C., Millner, P. D., Bergholz, T. M. and Sharma, M. 2019a. Salmonella enterica in Soils Amended with Heat-Treated Poultry Pellets Survived Longer than Bacteria in Unamended Soils and More Readily Transferred to and Persisted on Spinach. *Applied and Environmental Microbiology* 85(10), e0034-19.

Shah, M. K., Bradshaw, R., Nyarko, E., Millner, P. D., Neher, D., Weicht, T., Bergholz, T. M. and Sharma, M. 2019b. Survival and growth of wild-type and rpoS-deficient salmonella Newport strains in soil extracts prepared with heat-treated poultry pellets. *Journal of Food Protection* 82(3), 501–506.

Sharma, M., Millner, P. D., Hashem, F., Camp, M., Whyte, C., Graham, L. and Cotton, C. P. 2016. Survival and persistence of nonpathogenic Escherichia coli and attenuated Escherichia coli O157:H7 in soils amended with animal manure in a greenhouse environment. *Journal of Food Protection* 79(6), 913–921.

Sharma, M., Millner, P. D., Hashem, F., Vinyard, B. T., East, C. L., Handy, E. T., White, K., Stonebraker, R. and Cotton, C. P. 2019. Survival of Escherichia coli in manure-amended soils is affected by spatiotemporal, agricultural, and weather factors in the mid-Atlantic United States. *Applied and Environmental Microbiology* 85(5), e02392-18.

Sharma, M. and Reynnells, R. 2016. Importance of soil amendments: survival of bacterial pathogens in manure and compost used as organic fertilizers. *Microbiology Spectrum* 4(4), PFS-0010. doi: 10.1128/microbiolspec.PFS-0010-2015.

Shu, X., He, J., Zhou, Z., Xia, L., Hu, Y., Zhang, Y., Zhang, Y., Luo, Y., Chu, H., Liu, W., Yuan, S., Gao, X. and Wang, C. 2022. Organic amendments enhance soil microbial diversity, microbial functionality and crop yields: A meta-analysis. *Science of the Total Environment* 829, 154627. doi: 10.1016/j.scitotenv.2022.154627.

Teichmann, J., Litt, P. K., Sharma, M., Nyarko, E. and Kniel, K. E. 2020. Influence of poultry litter amendment type and irrigation events on survival and persistence of salmonella Newport. *Journal of Food Protection* 83(5), 821–828.

Tewolde, H., McLaughlin, M. R., Way, T. R. and Jenkins, J. N. 2016. Optimum poultry litter rates for maximum profit versus yield in cotton production. *Crop Science* 56(6), 3307–3317. doi: 10.2135/cropsci2016.04.0257.

Theofel, C. G., Williams, T. R., Gutierrez, E., Davidson, G. R., Jay-Russell, M., Harris, L. J. and Schaffner, D. W. 2020. Microorganisms move a short distance into an almond orchard from an adjacent upwind poultry operation. *Applied and Environmental Microbiology* 86(15), e00573-20.

The Counter. *Last Year's Salmonella Outbreak Linked to Peaches May Have Been Caused by 'fugitive dust'* [online] Available at: https://thecounter.org/salmonella-outbreak-peaches-fugitive-dust-fda/ [Accessed 31 Mar. 2022].

T-GAP: http://fvreports.freshfromflorida.com/5G_TomBPM.pdf.

Tomato best practices manual: A guide to tomato good agricultural practices (T-GAP) & tomato best management practices (T-BMP) purpose. n.d. [online]. Available at: http://fvreports.freshfromflorida.com/5G_TomBPM.pdf [Accessed 31 Mar. 2022].

Tomato Best Practices Manual. 2007. *A Guide to Tomato Good Agricultural Practices (T-GAP) and Tomato Best Management Practices (T-BMP) 2007*. Available at: http://fvreports.freshfromflorida.com/5G_TomBPM.pdf.

Tomato Best Practices Manual. 2018. *A Guide to Tomato Good Agricultural Practices (T-GAP) and Tomato Best Management Practices (T-BMP) 2007*. Available at: http://fvreports.freshfromflorida.com/5G_TomBPM.pdf.

Udikovic-Kolic, N., Wichmann, F., Broderick, N. A. and Handelsman, J. 2014. Bloom of resident antibiotic-resistant bacteria in soil following manure fertilization. *Proceedings of the National Academy of Sciences of the United States of America* 111(42), 15202–15207.

Wang, F., Sun, R., Hu, H., Duan, G., Meng, L. and Qiao, M. 2022. The overlap of soil and vegetable microbes drives the transfer of antibiotic resistance genes from manure-amended soil to vegetables. *Science of the Total Environment* 828, 154463. doi: 10.1016/j.scitotenv.2022.154463.

Wichmann, F., Udikovic-Kolic, N., Andrew, S., Handelsman, J. and Zhou, J. 2014. Diverse antibiotic resistance genes in dairy cow manure. *mBio* 5, e01017-13.

Wu, F., Ozaki, H., Terashima, Y., Imada, T. and Ohkouchi, Y. 1996. Activities of ligninolytic enzymes of the White rot fungus, Phanerochaete chrysosporium and its recalcitrant substance degradability. *Water Science and Technology* 34(7–8), 69–78. doi: 10.2166/wst.1996.0603.

Wu, H., Lai, C., Zeng, G., Liang, J., Chen, J., Xu, J., Dai, J., Li, X., Liu, J., Chen, M., Lu, L., Hu, L. and Wan, J. 2017. The interactions of composting and biochar and their implications for soil amendment and pollution remediation: A review. *Critical Reviews in Biotechnology* 37(6), 754–764. doi: 10.1080/07388551.2016.1232696.

www.ams.usda.gov. n.d. *Soil Building - Manures & Composts | Agricultural Marketing Service* [online]. Available at: https://www.ams.usda.gov/grades-standards/soil-building-manures-composts [Accessed 31 Mar. 2022].

www.regulations.gov n.d. *Regulations.gov* [online]. Available at: https://www.regulations.gov/document/FDA-2011-N-0921-18558 [Accessed 31 Mar. 2022].

Chapter 8

The role of Good Agricultural Practices (GAPs) in preventing pathogenic microbial contamination of fresh produce

Thomas P. Saunders and Elizabeth A. Bihn, Produce Safety Alliance – Cornell University, USA

1 Introduction

The implementation of GAPs on farms and in packinghouses has grown over the past 24 years resulting in an increase in written farm food safety plans and targeted practices aimed at reducing the risks of microbial contamination. Growers and packers have expanded their mindset and management practices to include food safety in addition to other production practices such as pest control, crop production, market access, and quality concerns. Proving that GAPs have prevented microbial contamination is difficult, but

http://dx.doi.org/10.19103/AS.2023.0121.14
Published by Burleigh Dodds Science Publishing Limited, 2023.

there are clear indications that the produce industry has made significant strides forward with implementing food safety practices such as the presence of portable toilet facilities in fields, the increase in worker training programs, and the establishment of groups focused on produce safety extension, education, and research. The National Good Agricultural Practices (GAPs) Program, the Center for Produce Safety (CPS), the California Leafy Green Products Handler Marketing Agreement (LGMA), and the Produce Safety Alliance (PSA) are all groups that have been established since 1999 and are focused on produce safety (*Center For Produce Safety*, 2022). Focus on food safety is not limited to US programs, the World Health Organization (WHO), Food and Agriculture Organization of the United Nations (FAO), and the European Union have also developed produce safety educational materials and training programs that have been widely adopted by growers (Joint FAO/WHO, 2003).

Unfortunately, despite the strides forward in the adoption of GAPs and other food safety practices, produce-associated foodborne illness outbreaks continue to occur (Wadamori et al., 2017; Carstens et al., 2019; FDA, 2020; *Reported foodborne outbreaks due to fresh produce in the United States and European Union: trends and causes*). This does not necessarily mean that GAPs have been ineffective, as there are many reasons why produce-associated outbreaks continue to happen. Among these reasons are that not all farms and packinghouses have adopted GAPs, surveillance technologies are better at identifying outbreaks, and pathogens continue to evolve (Lynch et al., 2009; Wadamori et al., 2017). In addition, agricultural resources are changing including the availability of water, changing land characteristics and use impacted by human development and animal agriculture, and a scarcity of personnel to complete production tasks (National Research Council et al., 2010; Cosgrove and Loucks, 2015; LEI International Policy 2018; Rosa et al., 2020). Research indicates GAPs are effective at reducing microbial risks during production, but implementing GAPs happens amid many other activities on farms and in packinghouses that are dynamic and ever-changing (Hamilton et al., 2015; Shaw et al., 2015; Sinkel et al., 2018).

The focus of this chapter is to review the role of GAPs in preventing microbial contamination of fresh produce over time. This will include summarizing how GAPs have evolved over time, the impact they have had on the produce industry, current GAPs and how they are implemented, and projecting their relevance into the future. There is always a hope that some technology will be developed that can remove all microbial risks on produce while maintaining the raw and fresh characteristics, but at this time that technology is not available or widely adopted so the focus remains on preventing contamination. Thus, our focus remains on GAPs.

2 Evolution of Good Agricultural Practices from voluntary to regulation

GAPs, as a concept, were introduced in the 1998 *Guide to Minimize Microbial Food Safety Hazards for Fresh Fruits and Vegetables* (FDA, 1998) to address the concern over continued foodborne illness outbreaks associated with fresh produce. Encompassing practices from pre-planting through postharvest activities, GAPs were outlined as voluntary guidelines to encourage the produce industry to adopt practices to reduce microbial risks associated with fresh produce (Bihn and Gravani, 2005). Though the 1998 Guide did not define the term 'Good Agricultural Practices', it was subsequently defined as any agricultural management practice or operational procedure that reduces microbial risks or prevents contamination of fruits and vegetables on the farm or in packing areas. This essentially means that any grower or packer, regardless of farm size or commodities grown or packed, can implement GAPs to reduce risks. Growers and packers first need to understand microbial risks to assess their operations to identify where risks exist and persist. They also need knowledge about what practices would reduce these risks as well as the resources (e.g. time, money, and personnel) needed to implement the practices. GAPs have now been included in third-party audits, marketing agreements, and both state and federal regulations (*Good Agricultural Practices (GAP) Audits | Agricultural Marketing Service;* FDACS, 2022; LGMA, 2007; Federal Register, 2015) (See Fig. 1).

In 1999, the National GAPs Program was established with 15 state land-grant universities and their extension personnel collaborating (CALS, 2022a). Around this same time, there was also a group in the Southeast US and personnel at the University of California, Davis focused on GAPs implementation. Around 2000, growers began receiving letters from buyers asking about their farm-specific food safety practices (CALS, 2022a). The requirement by buyers to implement food safety practices remains the primary driver of GAPs implementation on farms and in packinghouses. Growers and packers found that by implementing GAPs, they were able to maintain markets and access new markets, though there was not always a price premium. Sometimes the reason for implementing GAPs was to get access to better buyers, buyers that paid on time, and consistently bought products. By 2003, the National GAPs Program included 25 state land-grant universities and their extension personnel. This highlights the funding and attention, in just five short years, across academia, government, and industry in the US to respond to food safety concerns and begin the implementation of GAPs. Despite the focus, there were still many growers who were unaware of GAPs.

In 2006, the US Spinach outbreak pushed produce safety to the top of the news feed and impacted consumer awareness and buying habits (www.ers.usda.gov, 2010). This led to more commodity-specific GAPs. The Leafy Greens

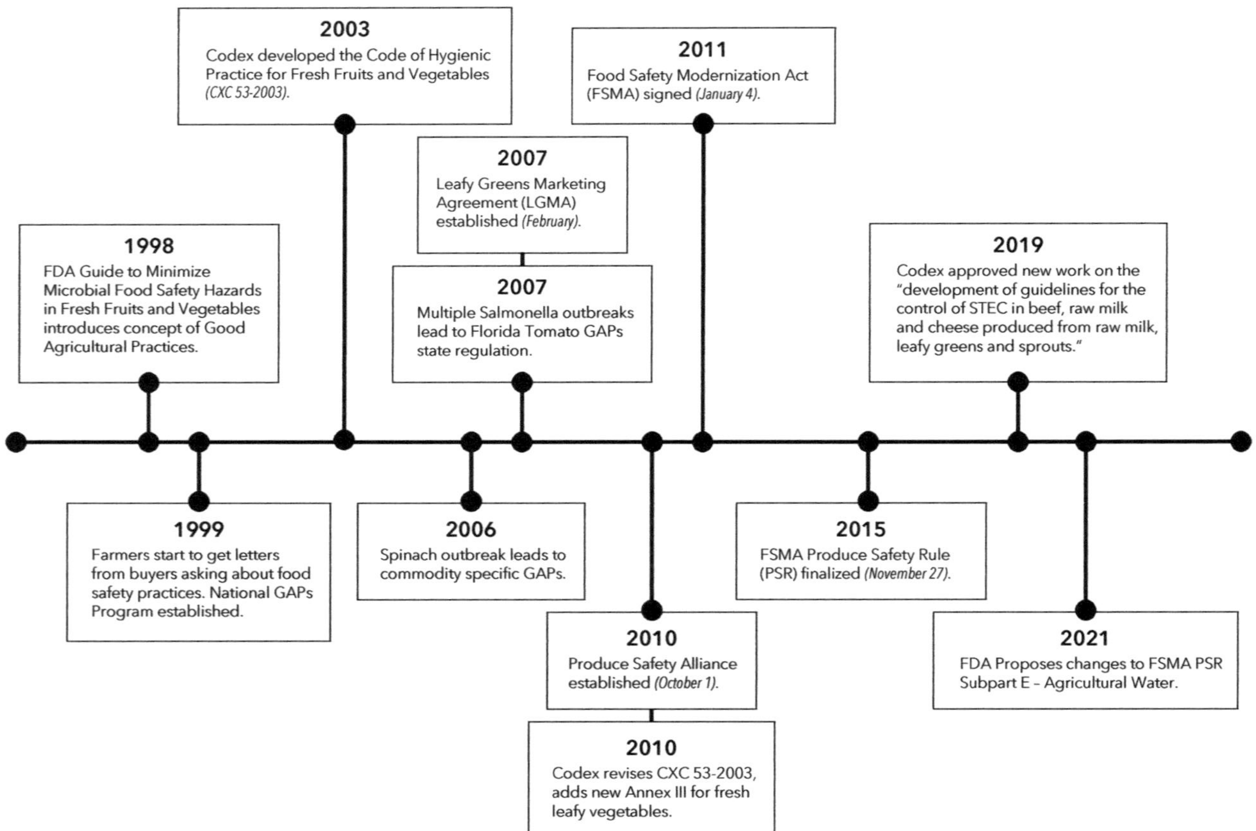

Figure 1 GAPs, produce safety requirements and training programs have evolved over time. This timeline includes key events in this evolution to highlight both US and international efforts. *Source*: Design credit: Elizabeth Bihn, Thomas Saunders, and Nicholas Luongo.

Marketing Agreement (LGMA) was formed in 2007, designed to help ensure that leafy greens growers implement food safety practices to produce the safest product they could and rebuild trust with consumers (LGMA, 2007). Marketing agreements are initiated by industry but governed by a government entity. Although voluntary to join, once a grower signs on, they are required to follow the practices. The LGMA shows a shift from voluntary compliance to mandated requirements. It was not just leafy greens growers that were experiencing this shift. Due to multiple *Salmonella* outbreaks traced back to the tomato industry, Florida Tomato GAPs were started in 2007 and were adopted as a statewide regulation in July 2008 (FDACS, 2022). This was the first statewide regulation of produce safety practices. While tomato growers could always apply GAPs to their operation, TGAPs encompassed regulatory inspections and audits by the Florida Department of Agriculture and Consumer Services (FDACS), an annual registration of all tomato packers in Florida, and a training component to help educate workers at all levels of the tomato industry about food safety practices (FDACS, 2022).

In December 2009, FDA announced in the public docket that work would begin on a regulation to govern fruit and vegetable production. Congress also began working on a law to update the food safety system in the US. On January 4, 2011, President Obama signed the Food Safety Modernization Act (FSMA) into law. The FSMA *Standards for the Growing, Harvesting, Packing, and Holding of Produce for Human Consumption* (a.k.a. Produce Safety Rule) was finalized on November 27, 2015. The intent of the Produce Safety Rule (PSR) is to 'establish science-based minimum standards for the safe growing, harvesting, packing, and holding of produce' (Federal Register, 2015). The PSR is the first federal regulation imposed on fruit and vegetable farms and packinghouses subject to the rule. Many of the practices required in the PSR are GAPs, though there are requirements beyond GAPs as well. The PSR does not cover all fruit and vegetable operations.

3 Reasons for adopting Good Agricultural Practices

Growers and packers legally subject to the FSMA PSR are motivated to implement food safety practices such as GAPs to meet these regulatory requirements and avoid negative consequences that may be associated with not complying but not all growers are legally subject to the FSMA PSR. Some growers are required to implement GAPs by their buyers to secure markets for the fruits and vegetables they grow. This provides a clear economic incentive to implement GAPs, but there are many growers that sell into markets that do not demand or require any food safety practices, so what inspires these growers to implement GAPs? One survey found some growers do it because they think it is the right thing to do (Schmit et al., 2020). The survey did not assess why the growers thought it was the right thing to do, but survey results indicated that even when there was no pressure to complete a farm food safety plan after being GAPs trained, many participating growers completed their farm food safety plans despite having no market or regulatory pressure. Whether this motivation comes from concern about produce-related foodborne illnesses or a drive to open new market channels, it is clear that once growers understand GAPs they move forward with implementing practices. This study also showed clear economic benefits to doing so, beyond reducing microbial risks and preventing microbial contamination of fruits and vegetables.

4 Why Good Agricultural Practices remain the foundation for fresh produce safety

Many people, including growers and packers, have described GAPs as 'common sense' because in the simplest of terms, keeping fecal contamination off of produce makes sense to most people. Implementing this common-sense idea is more complicated in practice because no two operations are alike, there are differences in production commodity to commodity as well as region to region, and no two human beings tasked with implementing practices are alike. As defined earlier, the flexibility of GAPs implementation allows them to be adapted to different farms, packinghouses, and commodities but this does not mean there are no challenges. There are key attributes that make the successful implementation of GAPs more likely and many of these begin with the commitment of growers and packers (Bihn and Gravani, 2005; Amekawa, 2009).

5 Commitment and training

Commitment and training are likely the most critical components of GAPs as you need to both understand the task to complete it properly and be motivated to do it when it needs to be done. From a training perspective, adult learners need to feel vested in the idea and GAPs usually make sense to growers (Knowles, 1984). It is also something that resonates with farm employees, especially those that hand-harvest fresh fruits and vegetables. Many of the most effective GAPs are concepts that improve other parts of people's lives, not just the safety of the fresh produce they grow, such as their overall health and well-being. Things like washing hands after going to the bathroom, not working when sick, and taking care of injuries so they heal properly are basic practices that benefit everyone. They also happen to be critical to food safety.

Training as a critical component of GAPs is also something that most growers understand as vital to their operations, both for their own professional growth and for their employees. Traditionally, GAPs training for growers have been done in person and accessibility was determined by the grower's location.

Many states have state extension educators, departments of agriculture personnel, and land-grant university faculty that offer educational opportunities and assistance, but there are areas where these services are not available or not prioritized (Singleton and Val Hillers, 2006). This ability to find and attend is not only a challenge domestically but internationally as well (Tobin et al., 2013). The COVID-19 pandemic provided for training adaptation as it forced most training online and in remote formats. It turns out, GAPs concepts can be effectively communicated whether in person, remote, or online (Bugingo et al., 2022). What matters is the quality of the trainers and the engagement with the learners. Remote and online formats of GAPs training have provided the opportunity for instructors to interact with adult learners in a new capacity that may never replace traditional class room format, but it met training needs during the pandemic. A topic that once was easily discussed in person, because it would engage multiple growers at one time, has been adapted to a poll question or some form of anonymous engagement to provide an equivalent level of participation. Instructors are able to promote a healthy learning environment where participants are still having their needs as adult learners met, while successfully conveying the core content of GAPs. An unexpected benefit of remote trainings has been the ability to reach populations and participants that could not afford, either monetarily or the time required, to attend an in-person course. Regardless of the format, motivation of the grower still remains a significant factor for how much they learn, but it is also impacted by the instructor's knowledge and ability to convey information. While any training still takes an investment of time, the ability to learn remotely has allowed growers to attend single and multi-day training programs while staying on-the farm where they can continue day-to-day activities and management requirements.

Growers and packers do not need to be microbiologists, but they do need to understand how microorganisms exist and persist on their farms and in their packinghouses. They need basic knowledge of the different microorganisms that are human pathogens and are likely to contaminate fresh fruits and vegetables. There has been discussion about whether or not it is fair to expect this of growers and packers who may have no formal training in microbiology or have taken classes. Food companies and processing plants usually have specially trained personnel who are solely dedicated to implementing food safety practices, yet growers and packers are expected to carry out many of the same activities. It is true that many farms and packinghouses have hired food safety personnel to implement their GAPs, but all farms, even those with one grower need to be able to prevent microbial contamination of the crops they grow. Whether fair or not, it is a necessity. The good news is that most growers know their operation very well, so once they gather the knowledge about microbial risks, they can begin applying the information to assess risks and implement practices.

As mentioned above, growers and packers must be committed to this process and believe this is necessary, otherwise, they will not invest the required time and resources. When growers and packers are trained but either misunderstand the information or decide to ignore the information, risks will persist in their operations. An example of this was in recent news, where a routine inspection found that a Michigan produce grower was using untreated human waste on fields that were growing produce that was then sold directly to consumers and local grocery stores (Desk, 2022). This grower was under a routine inspection to meet compliance with the FSMA PSR, and it was not their first inspection. Whether this grower made a conscious effort to ignore GAPs and regulatory requirements or simply misunderstood expectations, it highlights that training is no guarantee of implementation.

It is important that training be clear and includes why practices are necessary. It is also important that trainers remember that growers and packers are balancing many other goals beyond food safety, where food safety practices apply to other farm goals those practices can be highlighted. Helping growers incorporate food safety practices into overall farm management increases the likelihood of long-term implementation as it becomes part of the farm culture. Training that a grower or farm food safety specialist receives must be extended to all farm employees. There are a wide variety of resources and options for worker training programs to effectively and efficiently promote policies and habits that will reduce risk on the farm. Investing the time in training can benefit growers and farm workers alike, creating confidence in the knowledge that can ensure food safety practices are used throughout the operation every day, to help prevent contamination of fresh produce. Establishing a farm food safety culture is the ultimate goal.

6 Scalability of Good Agricultural Practices

Perhaps one of the most important aspects of GAPs is that they are intended for implementation no matter the size of the farm or packinghouse. A key part is the grower's knowledge of microbial risks and how their operation functions. As an example, it is critical that everyone involved in harvesting fresh produce or touching food contact surfaces wash their hands prior to doing either of these tasks. This requires access to soap, clean water, and a sanitary way to dry your hands. It also requires 20 s of scrubbing hands, so training workers is really important. On a small farm, growers might rely on one hand-washing station or sink in a centrally located hand-washing station where workers must wash their hands before handling produce or coming in contact with food contact surfaces. That grower might have a greater concern with waste disposal or the expense that single-use paper towels can carry and instead invest in single-use washable towels that can be used to sanitarily dry hands as well as cut down

Published by Burleigh Dodds Science Publishing Limited, 2023.

on cost and waste. Since there are fewer employees using the hand-washing facilities than in a larger operation, this grower is able to provide enough reusable towels and launder those towels, which allows them to be used over and over again. While on a large farm, a grower must maintain multiple hand-washing stations around their operation, and laundering hand towels may not be a feasible option. The larger grower provides single-use paper towels at each station and monitors when supplies must be replaced. Both growers are able to meet the requirement of providing a sanitary way for workers to dry their hands but are able to adapt their practices to not only meet food safety standards, but also personal preferences, sustainability goals, and cost-management requirements.

Scalability of GAPs could also be reflected in how growers and packers use water in their operations. Some farm operations may have resources available to electronically monitor and add sanitizer to postharvest water that they use to wash their produce to reduce cross-contamination risks. They may also have a dedicated food safety professional on staff to ensure all water-related food safety practices are completed and documented. Considering the cost of chemicals, time required, and management complexity to maintain and monitor postharvest water to ensure its sanitary quality, some farm operations may decide to not wash produce before selling it. While washing is a common practice for many fresh produce commodities, there is no requirement for produce to be washed prior to selling it. Farm operations that do not have the means to properly maintain and monitor postharvest water may adapt practices to the field or dry pack, where water is not introduced at or after harvest. This is a good choice since postharvest water that is not properly maintained can increase microbial risks and turn a small contamination problem into a big contamination issue, especially in batch and dump tank water. GAPs help provide growers with the tools to maintain food safety on the farm but do not limit growers to one specific means of achieving the food safety goal. GAPs provide adaptability for each farm to reduce microbial risks in ways that fit within their operation, budget, and personnel limits so they can successfully implement food safety practices.

6.1 Commodity considerations

Just as farms are diverse in size, they are also diverse in what they grow. Some farms will focus on one or two crops, while other farms will grow 30 or more commodities. For those farms that grow only one or two crops, they may be interested in following commodity-specific guidance such as for melons (FDA, 2022a) or mushrooms (LaBorde, 2018). Commodity-specific guidance can be very useful because they address common food safety concerns and can provide very specific guidance that is relevant to producers and their operations. They

Published by Burleigh Dodds Science Publishing Limited, 2023.

often are highly driven by industry and very crop specific. They provide printed or online publications to help growers understand practices and because these examples are all focused on a specific commodity, these instructions are often easier to follow and understand.

Farms that grow many different commodities usually group commodities together based on practices and likely risks of contamination. For instance, field-packed commodities will have different risks and management practices than commodities that are bulk picked, then washed, and packed. Practices like top icing or hydrocooling are going to have very specific management practices that are likely only relevant to some of the crops a farm may be growing. There are resources and templates available for growers that produce diverse commodities, though the guidance is more general and growers may need to consider how the guidance can be applied in their operation. For many growers, it takes time and assistance from other growers or educators to help develop a program that best fits their operation. This task can be tedious and frustrating, so it is important that growers have access to technical assistance when they need it. Later in the chapter, we will discuss effective ways to help growers adapt food safety practices and ensure that they are sustainable for their operation.

7 Feasibility of implementing Good Agricultural Practices

When an operation investigates the feasibility of implementing GAPs on the farm, there are many factors to consider. The cost of implementing a food safety program, such as GAPs or any similar food safety program, is usually the first consideration. Costs associated with implementing food safety practices have been identified as the primary barrier and directly correlate with the long-term feasibility of maintaining food safety programs on the farm (Schmit et al., 2020). In the study, costs were broken into the following categories: training costs, implementation costs associated with labor, testing costs, disposable supplies, and modifications such as annual improvements and other additional safety costs. This study also considered additional estimated food safety program costs including labor costs and capital investments, which other estimates seemed to exclude (Schmit et al., 2020). These results indicated food safety improvement costs averaged US$312 per acre across all survey participants' farms, varying across farm size with a demonstrated economy of scale, with costs dropping per acre from small farms to large (Schmit et al., 2020). This study concluded that farms adhering to GAPs, including participation in third-party audits, economically benefit from their food safety efforts.

In the ideal world, every grower and packer would have the resources available to be trained and implement GAPs properly, benefit economically from these efforts, and be able to maintain these food safety practices. Unfortunately, that is not always the case. Even though studies indicated it does not always require a significant amount of money to implement GAPs, there are other resource limitations that may prohibit some growers from participating in GAPs such as the availability of time (Schmit et al., 2020). This time commitment encompasses tasks previously mentioned like training for the grower or food safety person on the farm, establishing Standard Operating Procedures (SOP), farm food safety plans, and implementing practices. Many small farm operations have only one or two people on staff, so finding additional time in addition to financial resources can limit the ability to implement GAPs.

8 Assessing risks

The key to effective GAPs implementation is understanding the hazards at each stage of production and packing, identifying which hazards are likely to be risks, and implementing practices that effectively reduce these risks. This process of assessing risks needs to be done in all operations regardless of the size of the farm, the commodities grown, or the markets served. This requires growers and packers to learn enough about microbial risks to assess their operation. Taking a step-by-step approach can help growers focus on risks at every step of the operation, while not getting overwhelmed by the task. Growers and packers know the most about their own operation, but if they have not been given the tools to understand why food safety is important and how to assess microbial risks, they may not make it a priority or they may not do an effective job. This comes back to the idea of GAPs being common sense. Beginning with a very basic assessment of keeping feces from humans and animals off of fruits and vegetables during production and packing, then moving to ways other microbial, chemical, and physical hazards can lead to contamination of farm inputs, food contact surfaces, and eventually fresh produce.

The terms 'hazard' and 'risk' are often used interchangeably when discussing microbial contamination. It is important to make a clear distinction between the two terms. Hazard is defined by the FDA as any biological, chemical, or physical agent that has the potential to cause injury (*CFR,* 2022). While risk has been defined as the combination of the probability of occurrence of harm and the severity of that harm ('Guidance for Industry Q9 Quality Risk Management', 2006). Those definitions often do not clear up the confusion and clearing up the confusion is made more difficult when documents such as the FSMA PSR focus on hazards. Fruit and vegetable farms have many hazards. Fresh produce is often grown in open fields, subject to contamination by wildlife and humans, soils amended with animal manures, and surface water that has the potential to spread contamination. Fresh produce is usually harvested by hand, so worker health and hygiene can impact the microbial contamination potential. These

Figure 2 This figure highlights the distinction between hazards and risks. Risk occurs when the hazard is present, comes into contact with fresh produce, and persists long enough to be consumed. *Source*: Design credit: Don Stoeckel, Rob Way, Thomas Saunders, and Elizabeth Bihn.

are all hazards, but they may not all be risks depending on practices and the situation. For example, if soil amendments made from animal manures are composted or treated to reduce potential pathogen contamination, they may not be a risk. Growers will never have enough time or money to address all of the hazards that exist on a farm, which is why growers need to assess which of these hazards is most likely to result in risks on their farms. Through GAPs training, growers learn how to assess their operations for potential hazards and the likelihood that the hazard may become a risk. Focusing time and effort on addressing risks is the best use of resources to reduce food safety risks to the fresh produce they are growing and packing. Figure 2 highlights the distinction between hazards and risks. Risk occurs when the hazard is present, comes into contact with fresh produce, and persists long enough to be consumed.

8.1 Practices: worker training

In assessing risks, a good place to begin is to identify every time fresh produce is handled by workers. Microbial contamination can be easily introduced to fresh produce during handling by workers, so having an awareness of every time produce is handled on the farm is critical to assessing risks posed by workers (Osborne et al., 2006).

Controlling risks introduced by workers is focused on three specific areas: training, providing proper facilities, and effectively implementing practices. Though that may sound simple in some ways, it is quite complex. Beginning

Published by Burleigh Dodds Science Publishing Limited, 2023.

with training, all employees on the farm should be provided with food safety training even if they are not directly handling fresh produce or food contact surfaces. The primary reason for the training is to make it clear to everyone that the microbial safety of fresh produce is a company-wide expectation and goal. Training is the key to establishing a food safety culture and a well-established food safety culture ensures everyone is engaged in food safety. Functionally, this means everyone is working toward the same goal, which is very important because if some employees are not trained, they could be making it harder for other employees to implement food safety practices because they are unaware of how their personal actions are increasing food safety risks. For instance, workers who do not wash their hands after using the toilet can contaminate door handles and other surfaces, allowing microbial pathogens to spread to those that do handle fresh produce. Beyond basic produce safety training for all employees, specific training programs should be developed by assessing each task for those that handle fresh produce or food contact surfaces and identifying key food safety concepts they should know. For instance, those that harvest should be trained to make sure harvest containers are clean to avoid harvesting produce with feces on it or produce that is in an area that has been significantly impacted by animal intrusion and to handle fresh produce in a way that prevents damage, including never handling fresh produce that has dropped to the ground. They should also be trained to notify their supervisors if they see some other issue that is a food safety concern. This process should be done for all farm areas that impact fresh produce safety including packing, sorting, storing, transporting, water management, and chemical applications.

Training should be done in a language that workers understand. It should be conducted before workers begin their work on the farm. Update training should be conducted on a regular basis and whenever problems or issues are identified. The most effective training strategy will include different delivery modalities and opportunities for workers to practice what they have been taught (Fleming and Mills, 1992). The training should include expectations that include why the expectations are relevant to food safety and to the task workers are asked to complete.

Once workers are trained, providing the necessary facilities and resources is critical to ensuring workers can implement food safety practices. This requires both an awareness of what is needed and organization to ensure it is available. For instance, toilet and hand-washing facilities often need to be portable so they can be moved from field to field. Facilities must be continuously stocked and cleaned as necessary. Hygienic facilities are usually pretty obvious, but other resources can be more difficult to find and to train workers to use, such as test strips and titration kits to monitor postharvest water quality. In this case, expiration dates have to be monitored and training becomes much more extensive for those completing these tasks. This increases expenses and the

Published by Burleigh Dodds Science Publishing Limited, 2023.

level of organization and coordination needed to ensure tasks are completed properly. This is why a step-by-step assessment of the operation is so important.

Lastly, growers and packers need to ensure the effective implementation of practices is happening. This can be done in many ways, but it usually begins with monitoring. This can include monitoring facilities to make sure they are being used, monitoring recordkeeping logs, and paying attention to workers' actions as they go about their tasks. This is also where establishing a food safety culture will be most valuable. No single grower has the time to monitor everything on their farm, even if they are the sole employee on the farm. Empowering others to take an active role in implementing food safety practices means everyone is working to reach food safety goals. Investing in training, facilities, and resources so that all employees know how to implement food safety practices and have access to the necessary resources while they work makes implementing food safety practices possible. This remains an area of struggle for many fresh produce operations. Data collected from the On-Farm Readiness Review program identifies worker training as an area where farms need to improve practices (Melendez et al., 2022).

8.2 Practices: preharvest water

Water has the potential to lead to pathogenic contamination of fresh produce during growing, harvesting, and postharvest handling practices ('USDA GAP Program Water FAQs', 2022). Many factors impact potential contamination, such as source of the water, how the water is applied, and timing of the water application. Water availability is also an issue as farms are impacted by changing climate and the impact of increasing water demand from people and industries including agriculture.

In assessing risks associated with water, reviewing the source of water used on the farm and in the packinghouse is a good place to begin. Surface water sources, exposed to the environment and thus external sources of contamination, will usually have the highest risks and take the most effort to assess and manage. Surface water sources include rivers, streams, lakes, ponds, reservoirs, and well water pumped into open holding tanks or ponds as well as other water sources open to the environment. Running water sources, such as streams and rivers, tend to have more variable water quality over time when compared to non-running sources such as ponds, but the only way to know the quality of a surface water source is to test it and assess environmental factors that may be impacting its water quality such as run-off, wildlife, industrial contamination, and other factors. How often the water should be tested is determined by its use and the assessment of environmental factors impacting its quality. Right now, if the surface water is used in a way that contacts the harvestable portion of the crop during production, it is recommended that the

Published by Burleigh Dodds Science Publishing Limited, 2023.

water is tested for quantified generic *E. coli*. Even though generic *E. coli* is not a human pathogen, it is an indication of fecal contamination from humans and other warm-blooded animals. It is rare that water is tested for human pathogens, but if there is a specific known risk some farms may test for specific human pathogens. Untreated surface water should never be used at or after harvest since the likelihood of contamination is higher in surface water.

As with surface water, if a protected ground water source is used, it should be tested and the environment around the sources should also be assessed. If the source is a well, the casing, cap, area around the well-head, and recharge area should all be assessed to make sure there is nothing cracked or present that could lead to contamination. A similar process should be followed if it is a captured spring or other ground water sources. The protected ground water source should be tested for generic *E. coli* but a presence/absence test is acceptable as long as no *E. coli* is present. If there is *E. coli* present, then a quantified test should be performed. If the protected ground water source is also being used as a drinking water source, it should be tested to ensure it meets drinking water standards for chemical contaminants as well as microbial contaminants. Protected ground water that is free from detectable generic *E. coli* can be used at and after harvest in direct contact with fresh produce.

In the United States, water supplied by municipalities, towns, or cities is usually treated and tested so does not need to be tested by the farm. There are some circumstances where the pipes between the municipal treatment facility and the farm may be in poor repair or have other concerns that may warrant the farm testing the water to confirm its quality.

After the source of the water has been assessed, it is important to assess how the water is delivered and used throughout the farm as well as when the water is applied to the crop. Any water that contacts the harvestable portion of the crop increases the microbial risks. Overhead applications of water are also a less efficient use of water and can lead to plant pathogen and postharvest quality concerns. In some instances, such as water mixed with protective sprays, the water must be applied directly to the crop to get the desired effect. In this case, growers should be mixing the chemicals with water that has no detectable generic *E. coli* to reduce risks. They should also be aware of spray equipment sanitation to ensure the equipment is not introducing microbial risks.

For growers with large fields, converting to drip irrigation can be very expensive so they may continue to use overhead irrigation. In these instances, reducing risks can be done by using a water source that has no detectable or low generic *E. coli* or one that is protected from external contamination. The other option to reduce risks is to apply the water either before the harvestable portion is present or as far from the harvest date as possible. Studies have shown that there is a reduction in pathogens from contaminated water applied overhead over time in field environments but the reduction over time does not mean zero (Cabral, 2010; Alegbeleye and Sant'Ana, 2023). In addition, it has been shown that pathogens can begin to multiply if the environmental conditions are right, so although risks are reduced they are not eliminated.

As growers assess the risks associated with water use, it is good to focus on the source of the water, application, and timing. By assessing these three things, growers will be able to decide if they have risks and the best way to modify their water management to reduce the risks.

8.3 Practices: harvest and postharvest water

For water used at or after harvest, the quality of the water is very important because it has the biggest potential to contaminate fresh produce and persist through consumer consumption (Estrada-Acosta et al., 2014; Allende and Monaghan, 2015; Alegbeleye and Sant'Ana, 2023; Sandoval, 2015). The lowest risk approach is to use a high-quality source of water that has no detectable generic *E. coli*. If growers are using dump tanks or flumes, the potential for spreading contamination is significant unless they are using a sanitizer to reduce the risks of cross-contamination. Setting up effective water sanitation programs can be very complex and requires monitoring to ensure it is working properly. Fortunately, there are many options to consider and in the United States, these options are regulated predominantly by the Environmental Protection Agency. Chemicals that are labeled for this use will outline exactly how they are to be used in order to be effective. Monitoring them can still be a challenge, but the research to ensure their effectiveness at levels specified on the label has been done. Different sanitizers require different parameters, so it is critically important that growers know what they are using and how to use it effectively. Traditionally, farms have used chlorine-based products but peroxyacetic acid products are becoming more prevalent on small farms because they are easier to manage. Larger produce operations may use ozone, chlorine gas, or other systems. Farms also can invest in oxidation–reduction potential monitoring as well as inline injection systems. The scope of this chapter does not allow for an in-depth discussion of sanitizers, but the key point is to recognize the value sanitizers bring to managing microbial risks. In addition to human pathogens, sanitizers often also reduce plant pathogens which improves postharvest quality and shelf life. Any food safety practice that also benefits crop quality is easier for growers to justify the effort and expenses associated with the implementation.

8.4 Practices: soil amendments

Soil composition and amendments impact the overall productivity of fields, and they also can impact the safety of the fruits and vegetables grown (Ramos et al., 2021). Soils also have an impact on plant and human pathogen survival

Image 1 This image shows a food safety workshop participant monitoring the compost pile temperature during the composting process. This is a key step in some validated treatment processes for biological soil amendments of animal origin. Photo credit: Rob Way, copyright of Cornell University.

as well as overall human health (Steffan et al., 2018; Samaddar et al., 2021). Some human microbial pathogens, such as *Listeria monocytogenes*, and heavy metal contaminants, such as selenium, are naturally occurring in some soils, while some contamination results from the addition of soil amendments (Vivant et al., 2013; Li et al., 2019; Pepper et al., 2021).

In assessing food safety risks associated with soils and soil amendments, it is important to consider how contamination of the harvestable part of the crop occurs such as through direct contact with the soil and soil amendments or through other mechanisms such as splash. This requires identifying hazards within the soil and the soil amendments to assess the likelihood, they are risks for the crops that are being grown and harvested. For example, not all soil contaminants can be translocated by all crops and even if the crop can move the contamination through the roots, it may not deposit the contamination into the harvestable part of the crop. Some crops are grown off the ground, so unlikely to be directly contaminated by the soil and soil amendments. Although some soils and soil amendments have biological and chemical hazards, they are not likely to be food safety risks. The important point is that an informed assessment be done and if there are risks, that the risks be mitigated.

Published by Burleigh Dodds Science Publishing Limited, 2023.

Mitigating heavy metal soil risks through remediation can be very difficult, so growers should assess the soils before buying or renting new land. Reducing microbial risks associated with soil amendments can be done in many ways. Some soil amendments are treated to reduce microbial risks, such as heat-treated poultry pellets or composted cow manure. If soil amendments are being sourced from off-farm sources, suppliers should be asked for records to verify the processes they are using are effective at reducing microbial risks. Many growers will choose to compost manure on their farms. As long as they are following a validated composting process, this is an effective way to reduce microbial risks. With all treatment processes, records should be kept to ensure the processes were done properly and completely. Image 1 The grower is seen taking temperature to make sure they maintain the required conditions to reach a treated compost to use on the farm. It is also important to consider chemical risks as well, especially if using biosolids. The presence of per and polyfluoroalkyl substances (PFAS) in biosolids and their impact to the soil as well as human health when land is applied is well-documented (Pepper et al., 2021). For these reasons, growers may want to avoid biosolids or other soil amendments that present chemical hazards.

8.5 Practices: wildlife and domesticated animals

Animals are a very natural part of the growing environment. They can be present as domesticated animals that purposefully share the farm environment with the fruits and vegetables produced on the land. In some cases, domestic animals are used in the production of fresh produce such as draft horses pulling plows or harvest wagons. Animals can also be a fixture of the natural environment as wildlife. The primary concern is that animals can carry human pathogens in their feces and spread that feces as they move through and around produce fields (CALS, 2022a). This contamination can also be moved by people and equipment used during production and harvesting of fresh produce crops. When assessing how the hazards introduced by animals may impact risks in produce operations, a place to begin would be to assess animal access to the field. Usually, growers are very aware of wildlife pressure, especially if the wildlife is likely to consume the crop. Local jurisdictions often allow growers to use nuisance permits to eliminate the risk, but some animals are protected or may not be directly consuming the crop. In this case, growers will need to consider other ways to exclude the animals from the field such as exclusion structures such as fences or deterrence such as sounds or predator decoys. For growers that are using animals during production, risk reduction strategies need to be focused not only on limiting risks associated with the animal such as its feces but also on the actions of the workers handling the animals. Ensuring they wash their hands, change their boots, and change their clothes if they have been contaminated by interacting with the animals. Domesticated animals such as pets are also a concern. Some pets, such as dogs, are used to control pests such as ground hogs and this means they may have access to fields. Ensuring fields are monitored before harvest so that fresh produce that is contaminated with feces or in areas with animal intrusion that may have resulted in crop contamination is not harvested is one way to reduce and manage associated with both domestic animals and wildlife risks.

8.6 Practices: sanitation

Cleaning and sanitizing are key steps in preventing the contamination of fresh produce. Unclean surfaces such as tables, harvest bins, packing equipment, harvest aids such as knives, and any other surfaces that come in contact with fresh produce can lead to cross-contamination. These surfaces that are in direct contact with fresh produce are the most important consideration when developing an effective sanitation program, but it is also important for farms to consider how the environment around these surfaces impacts their cleanliness and the likelihood they could get contaminated. When developing a sanitation program, it is often just as valuable to consider how things get dirty before you consider how to get them clean as there may be steps that can be taken to minimize contamination in the first place.

A good place to begin is to consider the entire flow of produce in an operation. Considering everything from the point where the harvestable portion of the crop is present and anything that is likely to contact it. During production, this may include crop thinning but for most, this will begin at harvest. Packing bins, harvest aids, and hands including those with and without gloves are likely to come into contact with produce. Developing SOPs and being sure expectations are clear by effectively training workers is critical. These could include standards for harvesting, when to clean and sanitize harvest tools, proper glove use if gloves are required, not harvesting fruits and vegetables with fecal materials on them and not harvesting into visibly contaminated containers. Maintaining a schedule to clean and sanitize harvest containers and maintaining cleanliness of containers before they are used to harvest produce are additional steps that may be relevant for those harvesting into reusable containers. For those using single-use containers, storage, and transportation to the field may be important steps to review to ensure the containers are not contaminated before use.

Moving into sorting and packing, it is important to assess surfaces such as conveyor belts and washing lines. Many fresh produce operations are using old equipment. Though there is nothing wrong with older equipment, it can pose additional challenges. First, surfaces such as wood can be cleaned but not sanitized. Older equipment is often difficult to take apart. If you cannot see or access a surface, you cannot clean it. Cleaning always comes before sanitizing, as you cannot sanitize a dirty surface. For this reason, there has been a continued push to promote the use of surfaces and equipment in fresh produce operations that are not only easily cleaned but that have surfaces appropriate for the use of a sanitizer that can effectively reduce the microbial contamination on the food contact surfaces (National Good Agricultural Practices Program, 2023). Changing from wood surfaces for packing bins, tables, and sorting equipment to surfaces such as food-grade plastics and stainless steel allows better sanitation but can come at a cost that not all growers and packers can afford. Costs though are not just based on purchasing new equipment. Growers and packers need to evaluate the time it takes to clean equipment and surfaces as well as the amount of chemicals (e.g. detergents and sanitizers) that are used. An example of this is illustrated here in Image 2, where we can see food safety workshop participants assessing the cleanability and sanitary design of a leafy greens spinner.

Sanitary design is a concept and topic that all farms and packing operations should be aware of, especially if they are buying new equipment or building new facilities. It is also something equipment manufacturers should be using when they design new equipment. Facilities and equipment that are designed for being cleaned and sanitized are made of non-porous surfaces and can be easily taken apart and reassembled. They do not have hollow areas or inaccessible niches where pathogens can hide and persist. Sanitarily designed equipment and facilities increase sanitation efficiency and effectiveness (Graham, 2006;

Image 2 This image shows food safety workshop participants assessing the cleanability and sanitary design of a leafy greens spinner. Photo credit: Rob Way, copyright of Cornell University.

Published by Burleigh Dodds Science Publishing Limited, 2023.

Schmidt and Erickson, 2017a; Schmidt and Erickson, 2017b). This can lead to a significant reduction in sanitation costs and should be a consideration as it may make buying new equipment an economically good decision as it improves safety and efficiency.

Lastly, the area of sanitary design is still evolving when it comes to produce equipment. Many fruits and vegetables are easily bruised and damaged so designing equipment needs to take these challenges into consideration. Not all new equipment is easy to clean or made of non-porous surfaces. Making some of these design changes results in more expensive equipment that may not be widely available or within reach of every grower and packer. Despite these limitations, it is very important that growers and packers be able to develop an effective sanitation program for all surfaces that can impact the safety of the fruits and vegetables that contact them. Understanding sanitation challenges and options for overcoming these challenges is a critical part of implementing practices that reduce microbial risks.

8.7 Practices: storage and transportation

Storage and transportation of fresh produce is a major concern when preventing contamination of produce before it reaches the consumer (Sanitary Transportation of Human and Animal Food, 2016). Since storage and transportation are often the last steps for a grower maintaining possession of their produce, being able to maintain the same standards as they did during growing, harvesting, and packing is a pivotal step. There are many ways to store and transport produce, but no matter how produce is being stored or transported, it should always be done in a way that prevents contamination.

Maintaining a clean storage area for packed produce prevents contamination from pests and environmental factors in the storage area. Cleaning storage areas regularly and making sure that the removal of visible dirt and debris along with the removal of any unnecessary items from storage areas are steps that can reduce risks. Keeping boxes off the floor and away from the walls, as well as having an effective pest control program, are GAPS that will reduce risks. Limiting access to storage areas to only employees that need to work in those areas will reduce the likelihood of foreign materials being brought into storage areas and will prevent dirt from being tracked in on boots and shoes.

Cold storage is not required, but if a grower is using cold storage, it must be properly maintained. Making sure that the temperature is maintained and cooling equipment is functioning correctly helps to prevent condenzation and dripping which can lead to contamination by pathogens such *L. monocytogenes*. If top icing is used, ensuring top-iced boxes are not stored over boxes that are not top iced will reduce the risks of dripping water contaminating other products. For diversified farms that may be using a cooler for multiple agricultural products such as fresh produce, eggs, and meat, maintaining produce segregation and cooler sanitation becomes even more important to prevent cross-contamination.

Transportation can either be provided by the farm operation or contracted. If transportation is provided by the farm or packinghouse, an SOP and appropriate employee training should be in place to ensure all transportation vehicles are cleaned and sanitized on a regular schedule and inspected before loading. For small operations that use transportation vehicles for not only transporting packed produce but to transport soil amendments, agricultural chemicals, or other farm necessities, the sanitation program is even more important. For those operations that contract transportation services, they should consider placing sanitation expectations in the contract. Inspecting the truck for potential issues such as improper temperature control, transportation of previous loads that could introduce risks such as livestock, and any other cargo that could introduce food safety risks should be evaluated through inspection prior to loading any fresh produce. Making sure that the transportation is not introducing new food safety risks right before fresh produce reaches the consumer is the ultimate goal of implementing GAPs during storage and transportation.

9 Good Agricultural Practices: always room for improvement

Are all growers and packers aware of GAPs? The answer is no, and that this may never be an achievable goal, but it will not be for a lack of effort by educators and regulators in the produce safety field. We have discussed many factors that include why growers and packers work to implement GAPs in their operation and how they involve aspects of personal motivation. Implementation of GAPs requires time, effort, money, training, outreach, and technical assistance. Sometimes produce safety is not a priority for growers as they may already be overwhelmed by the stress of farming or still not convinced food safety is important to their operation.

10 Produce safety adds to farming stress

Food safety is not the only worry on growers and packers minds when they attend a GAPs or other food safety training. A recent survey by the American Farm Bureau Federation (AFBF) found that farmers and farm workers (61% of the 2000 polled) are experiencing more stress and mental health challenges compared to 2020. Stressors farmers listed as having the most impact on their mental health included financial issues (80%), weather or other factors beyond their control (82%), and the state of the farm economy (80%) (American Farm Bureau Federation, 2021). The importance of the statistics from the AFBF survey is that food safety likely has an impact on a few of those top three stressors.

Buyers often require food safety practices to be in place on farms that are their suppliers, resulting in food safety impacting market access (www.ers.usda.gov, 2022). Food safety practices can also impact price and type of market that is available. In addition, for growers and packers subject to FSMA PSR, their food safety practices also impact their compliance with federal regulation. All of these could add more stress to the lives of growers and packers. Of course, foodborne illnesses associated with commodities grown on the farm also have economic impacts such as lost markets due to reduction in consumption by consumers or broad recalls. As has been outlined, the goal of GAPs implementation is to reduce food safety risks to fresh produce but in order for this to happen, growers need to have the mental bandwidth to add these actions to their operation.

Considering the mental health of growers and packers should be a key consideration when developing education and extension programs. If a grower or packer leaves a training seeing only the time and financial burden, without that proof of return on investment they are unlikely to spend effort on food safety. Conveying information in a positive way, providing examples of practice implementation that is not dependent on new or expensive equipment, and promoting food safety practices that overlap with other farm goals are ways to help growers incorporate food safety into their everyday farm practices without increasing stress. It is also important to highlight that growers may already be doing many GAPs. For example, growers may already be excluding pests from their fields to reduce crop loss, but perhaps they are not thinking about it as food safety or not keeping the records needed to pass a GAPs audit they would like to complete. Building on things growers are already doing and providing management strategies to help them reach a goal will allow them to leave training realizing they have already made progress and that they do not need to start everything from the beginning.

There are real risks to pushing fruit and vegetable producers too far. First, we must value growers as human beings and be concerned for their physical and mental health. What happens to fresh produce production when the additional work, stress, and regulations cause growers to give up growing fruits and vegetables? Many growers and packers have expressed that produce safety costs too much money and takes too much time without sufficient economic gain so they intend to stop growing fruits and vegetables. Farm land has value and when growing becomes too arduous, growers have options. In the United States, lots of fertile farm land has been sold for housing developments. In a country with lots of tillable land, this may not immediately raise alarm but land is not the only issue. In the western United States, some horticultural crops are being abandoned because there is not enough water to grow them. There are many production challenges on the horizon, so it is important that food safety goals do not add to the challenges but provide opportunities for growers and packers.

11 Good Agricultural Practices relevance for the future

GAPs themselves do not guarantee the production of safe produce, as noted by the fact that many recent outbreaks in fresh produce come from farms that passed third-party GAPs audits with very high scores. Ensuring the safety of fresh produce is ultimately the result of effective implementation of food safety practices that reduce risks and prevent contamination, but nothing is ever going to be 100% when working with fresh produce. Even with advancements in fresh produce safety and the knowledge gain of growers, fresh produce continues to be linked to many foodborne outbreaks (Wadamori *et al.*, 2017). There have been 85 confirmed multistate outbreaks linked to fresh produce between the years of 2010 and 2017 alone. Many of these multistate outbreaks are directly associated with contamination of fresh produce by pathogenic bacteria: *E. coli, L. monocytogenes,* and *Salmonella enterica* (Carstens et al., 2019). Pathogenic bacteria pose the primary concern when discussing contamination of fresh produce but contamination can occur from many other foodborne pathogens such as viruses, protozoa, and fungi (Federal Register, 2015).

It is also important to recognize risks are different for different populations. Consumers with a weakened immune system, including young children, older adults (people 65 and older), people who are pregnant, people with cancer, people with diabetes, organ transplant recipients, people living with HIV/Aids, and those who have various autoimmune diseases are more susceptible to foodborne illnesses (FDA, 2022b). Recent studies have estimated that approximately 2.7–3% of the adult population in the United States would be considered immunosuppressed, living with HIV/AIDS or other autoimmune diseases (Harpaz et al., 2016). Though it is estimated that at least 25% of the population is considered high risk for foodborne illness, taking all the previous subgroups as a collective (Smith, 1997).

While protecting consumer health is the primary goal of GAPs implementation, GAPs are also important to protecting fruit and vegetable production because the overall economic impact of foodborne illness outbreaks can be devastating to the produce industry. For example, there was an estimated short-term farm-level loss of US$25 million in the tomato industry during a previous outbreak in the United States (Ribera et al., 2012). This impact was not only felt by the farms that caused the outbreak but by large and small farm operations not involved in the outbreak. When outbreaks and recalls occur, many consumers stop buying the specific commodity, even if the farm producing the commodity was not involved in the recall. The impact trickles down to employees of farms and can result in loss of work and livelihood for many, increasing the cost that is paid during a foodborne illness outbreak.

Despite continued outbreaks and concerns about changing populations, there is reason to believe GAPs are working. Foodborne illness trends over the last 20 years in the United States have remained steady (CDC, 2017).

With improved surveillance and detection, this may mean that foodborne outbreaks have actually decreased overall. There are still lots of opportunities for education and outreach since we know not all growers have implemented GAPs and new research continues to help us better understand risks and how to minimize them.

12 Conclusion

GAPs have provided a structure for identifying and addressing microbial, chemical, and physical risks to fresh produce for more than 20 years. Because of how GAPs were conceptualized and eventually defined, all growers, regardless of farm size, commodities grown, farm location, or markets, can implement GAPs to reduce risks to the fresh produce they grow, pack, and market. The biggest challenge to GAPs implementation is ensuring that growers and packers have the knowledge to effectively identify risks and the resources to implement practices to reduce risks. It is important to recognize that currently there is no way to achieve zero risks without a 'kill step' to reduce all risks. Many consumers enjoy consuming raw produce and in its raw state, there will always be some risks, but with training and knowledge growers and packers can implement GAPs to reduce risks during the production and packing of fresh produce.

There have been advances in technology in terms of identifying pathogens, such as whole genome sequencing, but it is still often very difficult to identify the origin of the pathogens and how the pathogens ended up on the produce when produce-associated foodborne illness outbreaks occur. For this reason, growers and packers need to be proactive in assessing their operations to identify risks and implement practices. There are many growers and packers who have yet to embrace GAPs so there is always work to be done, but GAPs, when effectively implemented, have been an effective approach to reducing food safety risks in fresh produce.

13 Where to look for further information

Key research and information in this area can be found at the following organizations:

- National GAPs program: https://cals.cornell.edu/national-good-agricultural-practices-program This website provides GAPs information and resources for growers, packers, farm workers, extension educators, scientists, state and federal government personnel, and other interested stakeholders.
- The Produce Safety Alliance is dedicated to providing fundamental, science-based, on-farm food safety knowledge to fresh fruit and vegetable farmers, packers, workers, regulatory personnel, and others interested in the safety of fresh produce. This website has a vast amount of fact sheets,

Published by Burleigh Dodds Science Publishing Limited, 2023.

webinar recordings, and resources available. Those looking for additional FSMA PSR information can also easily reach out the the PSA. : https://cals.cornell.edu/produce-safety-alliance.

- When looking for food safety resources that apply specifically to leafy green crops, California Leafy Green Marketing Agreement can be a helpful place to start. This website provides food safety resources for growers, shippers, and harvesters. LGMA: https://lgma.ca.gov/.
- If an understanding of regulatory requirements included in the FSMA PSR is of interest, the FDA website is a helpful resource. FDA Produce Safety Rule: https://www.fda.gov/food/food-safety-modernization-act-fsma/fsma-final-rule-produce-safety.
- The Food Safety Clearinghouse is a collection of PSR and Preventive Controls for Human Foods resouces. This is a public site were anyone can search and view resources linked to the clearinghouse: https://www.foodsafetyclearinghouse.org/.

14 Acknowledgement

We would like to acknowledge the United States Department of Agriculture (USDA)-Agricultural Marketing Service Cooperative Agreement 18-SCIDX-NY-0001 for funding.

15 References

Agricultural Marketing Service. Good agricultural practices (GAP) audits. Available at: https://www.ams.usda.gov/services/auditing/gap-ghp (Accessed: 23 June 2022).

Alegbeleye, O. and Sant'Ana, A. S. (2023). Microbiological quality of irrigation water for cultivation of fruits and vegetables: an overview of available guidelines, water testing strategies and some factors that influence compliance, *Environmental Research* 220, 114771. https://doi.org/10.1016/j.envres.2022.114771.

Allende, A. and Monaghan, J. (2015). Irrigation water quality for leafy crops: A perspective of risks and potential solutions, *International Journal of Environmental Research and Public Health* 12(7), 7457–7477. https://doi.org/10.3390/ijerph120707457.

Amekawa, Y. (2009). Reflections on the growing influence of good agricultural practices in the global south, *Journal of Agricultural and Environmental Ethics* 22(6), 531–557. https://doi.org/10.1007/s10806-009-9171-8.

American Farm Bureau Federation. (2021). Farmer_and_Rural_Mental_Health_AFBF.pdf. Available at: https://www.fb.org/files/Farmer_and_Rural_Mental_Health_AFBF.pdf (Accessed: 2 December 2022).

Bihn, E. A. and Gravani, R. B. (2005). Role of good agricultural practices in fruit and vegetable safety. In: *Microbiology of Fresh Produce*. John Wiley & Sons, Ltd, pp. 21–53. https://doi.org/10.1128/9781555817527.ch2.

Bihn, E. A., Saunders, T. P. and Luongo, N. D. (2023). *Evolution of Produce Safety Timeline*. Cornell Agritech, Geneva, NY.

Bugingo, C., Blasini, D. E., Stoeckel, D., Clements, D., Villarreal-Silva, M., Humiston, M. C., Acuna-Maldonado, L., Pineda-Bermudez, L., George, L. J., Way, R., Orellana, R. A., Saunders, T. P., Melville, T. and Bihn, E. A. (2022). Evaluating benefits and limitations of PSA grower training remote delivery. Poster Presented at: NASDA Produce Safety Consortium (14–16 December 2022). Phoenix, AZ.

Cabral, J. P. S. (2010). Water microbiology: Bacterial pathogens and water, *International Journal of Environmental Research and Public Health* 7(10), 3657–3703. https://doi.org/10.3390/ijerph7103657.

California Leafy Greens Marketing Agreement (LGMA). (2022). Food safety program. Available at: http://lgma.ca.gov/food-safety-program (Accessed: 5 October 2022).

CALS. (2022a). National good agricultural practices program. Available at: https://cals.cornell.edu/national-good-agricultural-practices-program (Accessed: 1 December 2022).

CALS. (2022b). Wildlife & animal management. Available at: https://cals.cornell.edu/national-good-agricultural-practices-program/resources/educational-materials/decision-trees/wildlife-animal-management (Accessed: 29 November 2022).

Carstens, C. K., Salazar, J. K. and Darkoh, C. (2019). Multistate outbreaks of foodborne illness in the United States associated with fresh produce from 2010 to 2017, *Frontiers in Microbiology* 10, 2667. https://doi.org/10.3389/fmicb.2019.02667.

CDC. (2017). *National Outbreak Reporting System (NORS) Dashboard*. Centers for Disease Control and Prevention (U.S.). Available at: https://wwwn.cdc.gov/norsdashboard/ (Accessed: 6 June 2022).

Center for Produce Safety. (2022). Available at: https://www.centerforproducesafety.org/about.php (Accessed: 5 October 2022).

Code of Federal Regulations – CFR. (2022). *Title 21: Food and Drugs*. Available at: https://www.accessdata.fda.gov/scripts/cdrh/cfdocs/cfcfr/cfrsearch.cfm?fr=117.3 (Accessed: 1 December 2022).

Cosgrove, W. J. and Loucks, D. P. (2015). Water management: Current and future challenges and research directions, *Water Resources Research* 51(6), 4823–4839. https://doi.org/10.1002/2014WR016869.

Desk, N. (2022). Michigan consumers warned of produce contaminated with human waste, *Food Safety News*. Available at: https://www.foodsafetynews.com/2022/10/michigan-consumers-warned-of-produce-contaminated-with-human waste/ (Accessed: 10 November 2022).

Estrada-Acosta, M., Jiménez, M., Chaidez, C., León-Félix, J. and Castro-Del Campo, N. (2014). Irrigation water quality and the benefits of implementing good agricultural practices during tomato (*Lycopersicum esculentum*) production, *Environmental Monitoring and Assessment* 186(7), 4323–4330. https://doi.org/10.1007/s10661-014-3701-1.

FDACS. (2022). *Tomato Good Agricultural Practices (T-GAP)/Fruit and Vegetables/Agriculture Industry/Home*. Florida Department of Agriculture and Consumer Services. Available at: https://www.fdacs.gov/Agriculture-Industry/Fruit-and-Vegetables/Tomato-Good-Agricultural-Practices-T-GAP (Accessed: 29 November 2022).

Federal Register. (2015). Standards for the growing, harvesting, packing, and holding of produce for human consumption. Available at: https://www.federalregister.gov/documents/2015/11/27/2015-28159/standards-for-the-growing-harvesting-packing-and-holding-of-produce-for-human-consumption (Accessed: 8 March 2022).

Fleming, N. D. and Mills, C. (1992). Not another inventory, rather a catalyst for reflection, *To Improve the Academy* 11(1), 137–155.

Food and Drug Administration. (1998). *Guidance for Industry: Guide to Minimize Microbial Food Safety Hazards for Fresh Fruits and Vegetables*. U.S. Food and Drug Administration. Available at: https://www.fda.gov/regulatory-information/search-fda-guidance-documents/guidance-industry-guide-minimize-microbial-food-safety-hazards-fresh-fruits-and-vegetables (Accessed: 6 June 2022).

Food and Drug Administration. (2020). Factors potentially contributing to the contamination of romaine lettuce implicated in the three outbreaks of *E. coli* O157:H7 during the fall of 2019. Available at: https://www.fda.gov/food/outbreaks-foodborne-illness/factors-potentially-contributing-contamination-romaine-lettuce-implicated-three-outbreaks-e-coli (Accessed: 24 June 2022).

Food and Drug Administration. (2022a). Commodity specific food safety guidelines for the melon supply chain - 1st edition. Available at: https://www.fda.gov/food/produce-plant-products-guidance-documents-regulatory-information/commodity-specific-food-safety-guidelines-melon-supply-chain-1st-edition (Accessed: 3 February 2023).

Food and Drug Administration. (2022b). Food safety for older adults and people with cancer, diabetes, HIV/AIDS, organ transplants, and autoimmune diseases. Available at: https://www.fda.gov/food/people-risk-foodborne-illness/food-safety-older-adults-and-people-cancer-diabetes-hivaids-organ-transplants-and-autoimmune (Accessed: 7 June 2022).

Graham, D. J. (2006). Snapshots in sanitary equipment: Developing an eye for hygiene, *Food Safety Magazine*. Available at: http://67.199.119.73/articlePF.asp?id=926&sub=sub1.

Guidance for industry Q9 quality risk management. (2006). Available at: https://www.fda.gov/media/71543/download (Accessed: 1 December 2022).

Hamilton, K. E., Umber, J., Hultberg, A., Tong, C., Schermann, M., Diez-Gonzalez, F. and Bender, J. B. (2015). Validation of good agricultural practices (GAP) on Minnesota vegetable farms, *Foodborne Pathogens and Disease* 12(2), 145–150. https://doi.org/10.1089/fpd.2014.1817.

Harpaz, R., Dahl, R. M. and Dooling, K. L. (2016). Prevalence of immunosuppression among US adults, 2013, *JAMA* 316(23), 2547–2548. https://doi.org/10.1001/jama.2016.16477.

Joint FAO/WHO Codex Alimentarius Commission. (2003). Codex alimentarius: Hygienic practice for fresh fruits and vegetables. Available at: https://www.fao.org/fao-who-codexalimentarius/sh-proxy/en/?lnk=1&url=https%253A%252F%252Fworkspace.fao.org%252Fsites%252Fcodex%252FStandards%252FCXC%2B53-2003%252FCXC_053e.pdf (Accessed: 20 February 2023).

Knowles, M. S. (1984). The adult learner: A neglected species. Available at: http://files.eric.ed.gov/fulltext/ED084368.pdf.

LaBorde, L. (2018). Mushroom GAPs audit and FSMA resources, mushroom GAPs audit and FSMA resources. Available at: https://extension.psu.edu/mushroom-gaps-audit-and-fsma-resources (Accessed: 3 February 2023).

LEI International Policy. (2018) *The Food Systems Approach: Sustainable Solutions for a Sufficient Supply of Healthy Food*. The Hague: Wageningen Economic Research. https://doi.org/10.18174/451505.

Li, C., Zhou, K., Qin, W., Tian, C., Qi, M., Yan, X. and Han, W. (2019). A review on heavy metals contamination in soil: Effects, sources, and remediation techniques, *Soil and Sediment Contamination: An International Journal* 28(4), 380–394. https://doi.org/10.1080/15320383.2019.1592108.

Lynch, M. F., Tauxe, R. V. and Hedberg, C. W. (2009). The growing burden of foodborne outbreaks due to contaminated fresh produce: Risks and opportunities, *Epidemiology and Infection* 137(3), 307–315. https://doi.org/10.1017/S0950268808001969.

Melendez, M., Ehart, B., Bihn, E. A., Gunter, C. C., Kline, W., Danyluk, M. D. and Tocco, P. (2022). On-farm readiness review tool and training curriculum to help farmers assess their readiness to comply with the FSMA produce safety rule. *Food Protection Trends* 42(6), 449–459. https://doi.org/10.4315/FPT-22-009.

National Research Council. (2010). *Toward Sustainable Agricultural Systems in the 21st Century*. National Academies Press.

Osborne, D. J., Sanders, D. C., Ward, D. R. and Rushing, J. W. (2006). Gaining producer buy-in for food safety and security in the United States, *hortTechnology* 16(4), 563–565. https://doi.org/10.21273/HORTTECH.16.4.0563.

Pepper, I. L., Brusseau, M. L., Prevatt, F. J. and Escobar, B. A. (2021). Incidence of PFAS in soil following long-term application of class B biosolids, *Science of the Total Environment* 793, 148449. https://doi.org/10.1016/j.scitotenv.2021.148449.

Ramos, T. D. M., Jay-Russell, M. T., Millner, P. D., Baron, J. N., Stover, J., Pagliari, P., Hutchinson, M., Lilley, J., Rowley, N., Haghani, V., Aminabadi, P., Kenney, A., Hashem, F., Martínez-López, B., Bihn, E. A., Clements, D. P., Shade, J. B., Sciligo, A. R. and Pires, A. F. A. (2021). Survival and persistence of foodborne pathogens in manure-amended soils and prevalence on fresh produce in certified organic farms: A multi-regional baseline analysis, *Frontiers in Sustainable Food Systems* 5. Available at: https://www.frontiersin.org/articles/10.3389/fsufs.2021.674767 (Accessed: 1 December 2022).

Ribera, L. A., Palma, M. A., Paggi, M., Knutson, R., Masabni, J. G. and Anciso, J. (2012). Economic analysis of food safety compliance costs and foodborne illness outbreaks in the United States, *hortTechnology* 22(2), 150–156. https://doi.org/10.21273/HORTTECH.22.2.150.

Rosa, L., Chiarelli, D. D., Rulli, M. C., Dell'Angelo, J. and D'Odorico, P. (2020). Global agricultural economic water scarcity, *Science Advances* 6(18), eaaz6031. https://doi.org/10.1126/sciadv.aaz6031.

Samaddar, S., Karp, D. S., Schmidt, R., Devarajan, N., McGarvey, J. A., Pires, A. F. A. and Scow, K. (2021). Role of soil in the regulation of human and plant pathogens: Soils' contributions to people, *Philosophical Transactions of the Royal Society of London. Series B, Biological Sciences* 376(1834), 20200179. https://doi.org/10.1098/rstb.2020.0179.

Sandoval, K. M. (2015). Assessment of bacterial water quality in an irrigation drainage and delivery system and adoption of good agricultural practices among Pacific Northwest produce growers. *Washington State University*. Available at: https://rex.libraries.wsu.edu/esploro/outputs/graduate/Assessment-of-bacterial-water-quality-in/99900525123401842 (Accessed: 14 November 2022).

Sanitary transportation of human and animal food (2016). Available at: https://www.govinfo.gov/content/pkg/FR-2016-04-06/pdf/2016-07330.pdf (Accessed: 1 March 2023).

Schmidt, R. H. and Erickson, D. J. (2017a). Sanitary design and construction of food equipment, *EDIS*. University of Florida - IFAS Extension. Available at: https://edis.ifas.ufl.edu/pdffiles/FS/FS11900.pdf.

Schmidt, R. H. and Erickson, D. J. (2017b). Sanitary design and construction of food processing and handling equipment. Available at: https://edis.ifas.ufl.edu/pdffiles/FS/FS12000.pdf. University of Florida - IFAS Extension.

Schmit, T. M., Wall, G. L., Newbold, E. J. and Bihn, E. A. (2020). Assessing the costs and returns of on-farm food safety improvements: A survey of good agricultural practices (GAPs) training participants, *PLoS ONE* 15(7), e0235507. https://doi.org/10.1371/journal.pone.0235507.

Shaw, A. M., Strohbehn, C. H., Naeve, L. L., Domoto, P. A. and Wilson, L. A. (2015). Knowledge gained from good agricultural practices courses for Iowa growers, *Journal of Extension* 53(5), 24.

Singleton, J. and Val Hillers, V. (2006). The extension programme and food safety education in the USA, *British Food Journal* 108(9), 771–774. https://doi.org/10.1108/00070700610688395.

Sinkel, D., Khouryieh, H., Daday, J. K., Stone, M. and Shen, C. (2018). Knowledge and implementation of good agricultural practices among Kentucky fresh produce farmers, *Food Protection Trends* 38, 111–121.

Smith, J. L. (1997). Long-term consequences of foodborne toxoplasmosis: Effects on the unborn, the immunocompromised, the elderly, and the immunocompetent, *Journal of Food Protection* 60(12), 1595–1611. https://doi.org/10.4315/0362-028X-60.12.1595.

Steffan, J. J., Brevik, E. C., Burgess, L. C. and Cerdà, A. (2018). The effect of soil on human health: An overview, *European Journal of Soil Science* 69(1), 159–171. https://doi.org/10.1111/ejss.12451.

Tobin, D., Thomson, J., LaBorde, L. and Radhakrishna, R. (2013). 'Factors affecting growers' on-farm food safety practices: Evaluation findings from Penn State Extension programming, *Food Control* 33(1), 73–80. https://doi.org/10.1016/j.foodcont.2013.02.015.

USDA. (2010). *Consumers' Response to the 2006 Foodborne Illness Outbreak Linked to Spinach*. ERS. Available at: https://www.ers.usda.gov/amber-waves/2010/march/consumers-response-to-the-2006-foodborne-illness-outbreak-linked-to-spinach/ (Accessed: 1 Mar. 2023).

USDA. (2022). *ERS, Markets, Regulation, and Policy*. Available at: https://www.ers.usda.gov/topics/food-safety/markets-regulation-and-policy/ (Accessed: 1 March 2023).

USDA GAP Program. (2022). Water FAQs, p. 20. Available at: https://www.ams.usda.gov/sites/default/files/media/GAPWATERFAQS.pdf.

Vivant, A. L., Garmyn, D. and Piveteau, P. (2013). Listeria monocytogenes, a down-to-earth pathogen, *Frontiers in Cellular and Infection Microbiology* 3, 87. Available at: https://www.frontiersin.org/articles/10.3389/fcimb.2013.00087 (Accessed: 1 December 2022).

Wadamori, Y., Gooneratne, R. and Hussain, M. A. (2017). Outbreaks and factors influencing microbiological contamination of fresh produce, *Journal of the Science of Food and Agriculture* 97(5), 1396–1403. https://doi.org/10.1002/jsfa.8125.

Zotero. Reported foodborne outbreaks due to fresh produce in the United States and European Union: Trends and causes, *GAPS Public*. Available at: https://www.zotero.org/groups/425300/gaps_public/items/NWTVFB3X/item-list (Accessed: 5 June 2022).

Chapter 9

Advances in sanitising techniques and their assessment for assuring the safety of fresh produce

Silvia Vanessa Camacho Martinez, Mahdiyeh Hasani, Lara Jane Warriner and Keith Warriner, University of Guelph, Canada*

1 Introduction

The fruit and vegetable market has undergone significant changes over the last two decades. Production practices, processing, market demand, complex global distribution chains and diversification have all undergone substantive changes. As with other sectors, there has been consolidation of the industry, although there is still room for the relatively small farms given that sub-contracting remains the principal market structure (Gokarn and Kuthambalayan, 2019). In parallel with market growth, the fruit and vegetable sector continue to be linked to the majority of foodborne illness outbreaks accounting for more than any other commodity, such as meat (Murray et al., 2017; Turner et al., 2019). Ongoing efforts to enhance the food safety of fresh produce have focused on pre-harvest initiatives that are largely based on testing agricultural inputs (irrigation water and organic amendments) along with other good agricultural practice (GAP) (Parker et al., 2012). Yet, pathogens such as *Escherichia coli* O157:H7 and *Salmonella* have almost become

* Corresponding Author: kwarrine@uoguelph.ca T: +1519 8244120.

http://dx.doi.org/10.19103/AS.2023.0121.16

endemic within growing regions due to intensive animal production, pressure on water sources and climate change (Rock et al., 2019; Gerdes et al., 2022). As GAP can reduce, but not prevent, the risk of pathogens entering the fresh produce chain, there is a renewed focus on post-harvest treatments to remove field-acquired contamination. It is well known that post-harvest washing has limited efficacy in removing pathogens from fresh produce and can potentially result in cross-contamination between batches if not controlled (Murray et al., 2017). This, in turn, has driven the search for alternatives to post-harvest washing with the potential to support pathogen reduction. To further justify the cost of implementing alternative technologies, it is also desirable for treatments to extend product shelf life. This not only reduces food waste but is also compatible with extended distribution chains.

In this chapter, the latest innovations in post-harvest decontamination approaches will be described. Less focus will be placed on alternative sanitisers to chlorine such as electrolysed water, chlorine dioxide, essential oils, bacteriophages, bacteriocins and others, as they have been previously reviewed (Ali et al., 2018; Meireles et al., 2016; Feliziani et al., 2016; Tomas-Callejas et al., 2012; Ma et al., 2017; Miller et al., 2013). Instead, the focus will be on emerging technologies that are mostly in the advanced research or early commercial adoption stage. In addition, the need for a standardised protocol for validating novel processes will be discussed.

2 Importance of post-harvest decontamination

Outbreak investigations have established that fruit and vegetables can become contaminated via multiple sources from farm to fork (Warriner and Hasani, 2020). At the farm level, pathogens can enter irrigation water through sewage spills or via run-off from animal production centres (Jongman and Korsten, 2017). An additional source of contamination is the use of organic amendments that can directly introduce pathogens to soil or run-off into irrigation water (Goss et al., 2013). Animal intrusion into growing fields along with farm workers and unsanitary harvesting equipment have also been suggested to contribute to the introduction of pathogens (Soon and Baines, 2012).

Once pathogens become established within growing regions, they can persist over prolonged periods in the soil or sediments, thereby representing a continuous source of contamination (Sharma and Reynnells, 2016). This was illustrated by the observation that the strain of *E. coli* O157:H7 was recovered from the same romaine lettuce growing area over successive years (Hoff et al., 2021). There is also a link between the extended persistence of enteric pathogens in the environment and climate change, although this has yet to be conclusively established (Tirado et al., 2010).

To date, pathogen control has been focused at the farm level with reliance on testing of agricultural inputs (soil and irrigation water) in combination with GAP (Gutierrez-Rodriguez and Adhikari, 2018). The effectiveness of such approaches to prevent pathogen ingress is debatable, especially given the limitation of testing. This is underlined by the wide distribution of human pathogens within the environment along with on-going foodborne illness outbreaks/recalls linked to contaminated fruit & vegetables (Gerdes et al., 2022).

2.1 Limitation of post-harvest washing

Processed fruit and vegetables undergo a post-harvest wash to remove field-acquired contamination, and for many years, this was considered a critical control point for pathogen control. However, it was subsequently realised that post-harvest washing had limited efficacy in removing pathogens and reducing endogenous microflora (Murray et al., 2017). It was proposed that the limitation of post-harvest washing is attributed to the internalisation of pathogens into the inner tissue, in addition to being incorporated into biofilms (Olaimat and Holley, 2012). Although it has been shown that pathogens could internalise within the vascular system of growing plants, this was under artificial conditions of high cell density and young plants (Solomon and Matthews, 2006). However, it has been demonstrated that pathogens can be internalised within inner tissues via hydrocooling or via cut edges and temperature differential of water versus produce (Li et al., 2008). It was also subsequently identified that sanitisers such as hypochlorite became sequestered during commercial washing by reacting with organic material to create disinfection by-products (Murray et al., 2018a). The depletion of free chlorine was shown to lead to cross-contamination between produce batches, ultimately resulting in foodborne illness outbreaks (Murray et al., 2017).

The limitation of post-harvest washing to remove pathogens was one of the factors that led to focusing on pre-harvest operations to enhance food safety. Indeed, it has been proposed that the best that can be achieved to prevent cross-contamination by post-harvest washing is to ensure a sufficient concentration of active sanitiser, for example hypochlorite, is maintained (Gombas et al., 2017). Mitigation strategies to ensure sufficient free chlorine include increasing the rate at which fresh water is introduced (recharge rate) or to add high chlorine levels (hyperchlorination) to compensate for the free chlorine sequestered by organics (Wang et al., 2022). Neither solution is ideal as using large volumes of water negatively affects water management programmes designed to reduce usage along with wastewater generated (Trajer et al., 2021). In turn, hyperchlorination leads to the formation of disinfection by-products that leave sensory taints on produce, in addition to being potential carcinogens (Wang et al., 2022).

As hypochlorite can be sequestered by organic and inorganic constituents in wash water, alternative sanitisers are needed (Banach et al., 2015). The main focus has been on peracetic acid as it is one of the few sanitisers that is compatible with organic certification programmes (Petri et al., 2021). Although peracetic acid is less sensitive to organic material in wash water, the log count reductions achieved on produce are similar to those of chlorine within commercial processing (Petri et al., 2021). The degree to which peracetic acid prevents cross-contamination within commercial processing has yet to be demonstrated, although it has been shown under laboratory conditions (Banach et al., 2015). In a similar manner, sanitisers based on quaternary ammonium salts, chlorine dioxide or a combination of organic acids have been shown to support pathogen reduction and cross-contamination under laboratory conditions, but data under commercial conditions are lacking (Banach et al., 2015).

3 Advances in post-harvest washing processes

For certain products, such as shredded lettuce, contact with water is essential to facilitate transportation within flumes and to remove debris such as soil and insects. Consequently, operating a water-free process is not possible in every post-harvest processing facility, but innovations have improved the performance and consistency of the washing process. Specifically, focus has been placed on determining the critical concentration of free chlorine required to prevent cross-contamination and how levels can be monitored under commercial processing conditions (Murray et al., 2017). In addition, effort has been focused on reducing the chlorine demand (pre-oxidation process) of wash water to maintain free chlorine stabilisation. There has also been considerable effort to find alternative sanitisers to chlorine that are effective without negatively changing the quality of the produce, which will be elaborated upon in subsequent sections.

3.1 Monitoring chlorine

There have been various reports on the minimum chlorine concentration required to prevent cross-contamination between batches of produce (Luo et al., 2018; Mokhtari et al., 2021; Bertoldi et al., 2022). The general procedure to determine the minimum chlorine concentration is to challenge a suspension of the test pathogen with increasing concentrations of the sanitiser until a specific log reduction is attained. Such studies have been performed in simple buffered solutions that are not representative of commercial processing water. For example, the traditional approach was to apply the dilution test, which establishes the concentration of sanitiser required to reduce a population of a designated pathogen by 5-log CFU (colony-forming units) within 30 s (Gombas et al., 2017). Although the dilution test provides an estimate of the free chlorine concentration, it does not account for the conditions within wash tanks, for example, the type and loading of organic and inorganic constituents, in addition to the intrinsic tolerance of pathogens that are commonly pre-stressed with enhanced resistance. In this regard, databases have been constructed to provide characteristics of wash waters from different produce-processing facilities to aid researchers in formulating simulated waters (Mundi et al., 2017, 2021).

An approach by Teng et al. (2018) used a microfluidic system that introduced chlorine and different organic loadings. Following a 4-h stabilisation period, the test pathogen, *E. coli* O157:H7, was introduced and survivors enumerated following a defined contact time of 1–5 s (Teng et al., 2018). It was reported that by using water with no organic loading, the critical chlorine concentration to support a 5-log reduction of *E. coli* O157:H7 was 0.5 mg/L (0.5 ppm), which increased to 7.5 mg/L in the presence of an organic loading of 900 mg/L chemical oxygen demand (COD) (Teng et al., 2018). A further study using a laboratory-scale wash tank reported that a minimum chlorine concentration of 10 mg/L was required to prevent cross-contamination of pathogens during lettuce washing (Luo et al., 2011). Trials performed on a commercial scale using total aerobic count of the water to assess cross-contamination between produce batches concluded that the critical chlorine concentration was 20 mg/L (Lopez-Galvez et al., 2019).

The broad range of critical free chlorine levels to prevent cross-contamination between produce batches reflects the dynamics of the wash process. The accumulation of organic loading during the wash process is one aspect, as is the reactivity of constituents with chlorine. Indeed, it has been demonstrated that the critical chlorine concentration varies between fruit and vegetable type, along with shifting of pH that modulates the equilibrium between free chlorine and hypochlorite anions (Tudela et al., 2019). In practical terms, the critical chlorine concentration is arbitrarily selected in a commercial wash process to maintain sanitiser levels up to 100 ppm to provide an appropriate safety margin/degree of redundancy if there is an increase in the chlorine demand of the water.

In addition to selecting a critical chlorine concentration, levels in wash tanks during a continuous or semi-continuous mode require monitoring to maintain free chlorine concentrations. An early attempt to monitor the free chlorine concentration was through measuring the oxidation reduction potential (ORP) within a feedback-loop system (Davidson et al., 2017). The ORP non-selectively measures the oxidation potential of the water, which increases with chlorine content. Typically, the ORP is maintained at >600 mV by dosing tanks with chlorine along with adjusting the pH to 5.5–7.0 with an organic or inorganic acid. The ORP became established in the industry as a means of real-time monitoring of the free chlorine concentration, although limitations have been noted in commercial applications (Murray et al., 2017); for example, the free chlorine concentration oscillates over time, whereas the ORP remains constant. This discrepancy between ORP and free chlorine can be attributed to several factors, for example, ORP probes have a 5- to 10-min reaction time to reach steady-state measurement, which is not the case in the dynamic environment of commercial wash tanks. Furthermore, the linear analytical range for ORP versus free chlorine is relatively narrow (0–600 mV) and then plateaus, which negatively affects the accuracy of the readings (Van Haute et al., 2019). The placement of the ORP probe also influences the ability to control free chlorine within wash tanks. In most cases, the ORP probe measures the ORP of water at the end of the tank or during return and then dispenses hypochlorite which raises the free chlorine to the set point. Under this regime, it is assumed that the free chlorine persists in the wash tank, whereas in reality, the sanitiser reacts with organics along with the pH being displaced to the alkaline region favouring the less antimicrobial hypochlorite anion. In summation, the ORP is a convenient means of monitoring wash tanks coupled with a feedback loop to facilitate automation and data logging; however, the lack of correlation with free chlorine concentration within wash tanks is a significant limitation (Van Haute et al., 2019).

Given the limitation of ORP measurements, an alternative approach is to withdraw samples and measure the chlorine off-line using the *N*-diethyl-*p*-phenylenediamine (DPD) assay, and the hypochlorite concentration is then manually adjusted as needed. This method has the same disadvantages as the ORP method with an additional lack of continuous monitoring. There are semi-continuous chlorine analysers that withdraw samples to perform an automated DPD assay. A derivation of this approach includes Prussian blue amperometric sensors that can be used to monitor free chlorine concentrations more selectively than ORP probes (Kongkaew et al., 2021). It is also possible to measure chlorine via ion-selective electrodes, although such sensors do not discriminate the different forms of the ion, i.e. total chlorine might not be determined.

3.2 Predicting the chlorine demand of wash waters

An alternative approach to maintain the free chlorine concentration is to measure the chlorine demand and then introduce fresh water to stay below a designated threshold; this is the same principle as increasing the recharge rate but in a controlled manner. The chlorine demand of water is performed using the jar test, which involves the water sample being placed within a container and a chlorine dilution series applied. The samples are then left for 30 min prior to measuring the residual and free chlorine concentration. The residual chlorine concentration versus chlorine added is plotted to produce a breakpoint curve (Fig. 1).

The classic breakpoint curve of municipal wastewater is characterised by a hump that represents the formation and subsequent neutralisation of chloroamines via the interaction of chlorine with ammonia (Fig. 1). Once the chlorine demand has been satisfied, the free chlorine increases to exert antimicrobial action. Determining the chlorine demand enables dosing the water with sufficient hypochlorite to exceed it. The chlorine demand of spent wash water varies with produce type, along with loading, and typical values for spent lettuce wash water are 100–220 ppm (Murray et al., 2018a; Mundi et al., 2017, 2021).

Although the Jar test is an accurate and established method to determine chlorine demand, the method is laborious and time consuming, as well as being unsuitable for the real-time monitoring of wash tanks. This has led to alternative approaches to measure the chlorine demand of wash water. Through comparative studies, the water pH, turbidity, conductivity, biological oxygen demand, total amino acid concentration and soluble solids have been shown to poorly correlate with chlorine demand compared with Chemical Oxygen Demand (COD) (Abnavi et al., 2021; Van Haute et al., 2018; Zhou et al., 2014; Chen and Hung, 2017). The main limitation of COD is that the titration assay is

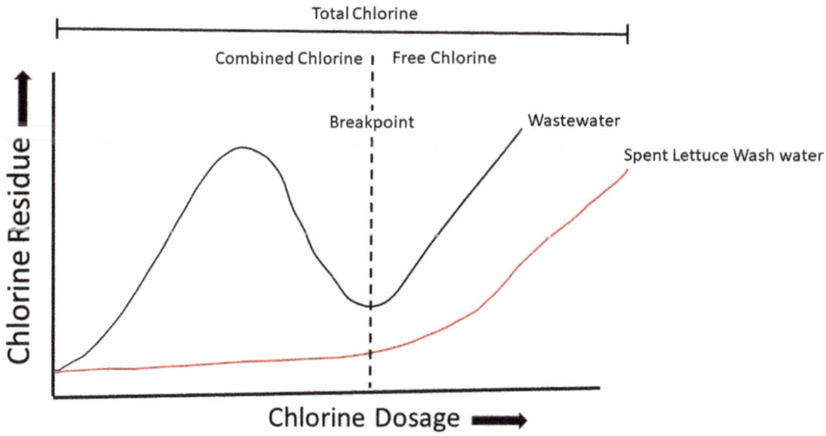

Figure 1 Examples of breakpoint curves typically obtained for municipal wastewater or spent lettuce wash water. Breakpoint curves are generated from a Jar test that doses different concentrations of hypochlorite into the water sample and then measures the residual chlorine (chlorine species present after the oxidant demand has been met). Combined chlorine represents chlorinated species, such as chloroamine, that forms a 'hump'. Free chlorine measures chlorine in gas, hypochlorous acid or hypochlorite ions, with total chlorine representing the sum of combined and free chlorine. The transition point between combined and free chlorine is referred to as the breakpoint, which defines the chlorine demand of the water sample.

measured off-line, which limits its use as a continuous monitoring method. A correlation has been found between chlorine demand and absorbance at 254 nm, along with spectral analysis studies, although neither approach has been verified with different produce types under commercial processing conditions (Van Haute et al., 2018).

3.3 Pre-oxidation of wash water

An alternative approach to maintain free chlorine levels in water, or at least reduce the amount of hypochlorite required, is to perform a pre-oxidation step. By oxidizing compounds that potentially could react with chlorine preserves levels of the sanitizer by reducing the chlorine-demand (Rouge et al., 2020). The approach has been employed in wastewater treatment to minimise the amount of hypochlorite needed to treat water, which in turn minimises the formation of polluting disinfection by-products. Pre-oxidation of water to reduce the chlorine-demand can be performed using a range of approaches that include the addition of permanganate and ozone or through germicidal ultraviolet light (UV-C) photo-oxidation (Tak and Vellanki, 2020; Rouge et al., 2020). An advanced oxidation process (AOP) has also been applied in which water is supplemented with hydrogen peroxide and then passed through a UV-C reactor (Wang et al., 2022). Hydroxyl-radicals are formed from the UV-C degradation of hydrogen peroxide, which can readily oxidise water constituents that can be degraded to carbon dioxide with prolonged treatments (Giannakis, 2019). The Photo-Fenton AOP process is based on the UV-C (254 nm)-mediated degradation of hydrogen peroxide via a ferric catalyst to generate hydroxyl-radicals (Giannakis, 2019). For example, hydroxyl-radicals generated from an AOP–hydrogen peroxide–ferric treatment applied at pH 4.5 resulted in a 48% decrease of disinfection by-products when the treated water was subsequently chlorinated (dos Santos et al., 2022). A further study probed the mechanistic action of the breakdown products of a solution of humic and fulvic acids (the main constituents in leafy greens wash water) treated with UV–hydrogen peroxide (Varanasi et al., 2018). The researchers demonstrated that the aromatic structure of the humic and fulvic acids was degraded to carboxylic acids, aldehydes and ketones (Varanasi et al., 2018). Although chlorine reacted with the degradation products, the disinfection by-products formed were volatile and biodegradable unlike persistent species such as chlorinated aromatic compounds (Xiang et al., 2020; Tak and Vellanki, 2020).

A continuous pre-oxidation treatment of spent lettuce wash water has also been described (Wang et al., 2022). The process was optimised through varying the hydrogen peroxide and ferric catalyst concentration along with the UV-C dose applied. The optimal treatment of 320 mJ/cm^2 UV-C dose, 9.6 mg/L

H_2O_2 and 9 mg/L ferric catalyst reduced the chlorine demand of the simulated wash water and generated volatile disinfection by-products that prevented cross-contamination between lettuce batches (Wang et al., 2022).

3.4 Advances in alternative sanitisers

The instability and reactivity of hypochlorite have led to the evaluation of different sanitisers that are insensitive to organic matter in water. Although peracetic acid is insensitive to organic loading, this has not been observed under commercial processing conditions (Petri et al., 2021). Organic acids are more stable with organic matter and have the added advantage of being Generally Regarded As Safe status. The most widely applied acids are lactic, citric and acetic acids, in addition to combinations that include malic, tartaric and succinic acids. In general, the initial log reduction achieved with fresh produce is negligible, given the slow action of organic acids. However, during post-treatment storage, the levels of microbes associated with fresh produce such as lettuce decline, especially in the case of acetic acid (Wang et al., 2021). In comparison, the microflora of lettuce washed in lactic acid remains relatively stable, although it has been found that *E. coli* and *Shigella* increase in abundance within the microflora, suggesting an increased food safety risk (Wang et al., 2019).

The ester surfactant, lauroyl-L-arginine (LEA), is a relatively new class of sanitiser that exhibits a broad range of antimicrobial activity with the additional benefit of breaking down to ethanol and arginine over time (Ma et al., 2020). LEA supplemented into wash water could decrease *E. coli* levels by 4-log CFU, although it had a negative effect on the sensory characteristics of shredded lettuce (Nubling et al., 2017).

Activated sodium persulphate, generated by the inclusion of ferrous sulphate and sodium hydroxide, has been evaluated as a sanitiser for washing romaine lettuce (Qi and Hung, 2019). It was found that 70 mmol/L activated persulphate could decrease *Listeria monocytogenes* and *E. coli* O157:H7 by 3.5-log CFU on lettuce, which was significantly higher than chlorine (100 mg/L) or peracetic acid (78 mg/L). The activated persulphate could also eliminate *E. coli* O157:H7 in wash water, although residual *L. monocytogenes* was recovered. The treatment did not have a negative effect on the quality of lettuce, with the advantage of producing no disinfection by-products associated with chlorine (Qi and Hung, 2019). However, the same sensitivity of chlorine to organic loading within wash water was noted (Qi et al., 2020).

To overcome the issues related to post-harvest washing, there has been interest in spraying as an alternative to immersion tanks (Beuchat et al., 1998). However, the spraying approach can also act as a cross-contamination point if the water is recycled without treatment to reduce the organic loading along with maintaining sanitiser levels. A derivation of spraying is electro-spraying (also referred to as electrostatic spraying), which enhances the coverage of the produce being treated with lower volumes of the sanitising solution (Lyons et al., 2011). Electrospraying was originally designed to more evenly and efficiently applying paint along with other coatings (Hudson et al., 2019). The principle of electrostatic spraying is to apply a positive charge on the item to be coated with a negative charge being applied to the spray. As the spray is delivered, the negatively charged sanitiser solution is attracted to the positive charges on the item, thereby leading to an even coverage. The approach has been used in the application of pesticides to growing crops where a more even coverage and less solution usage are positive attributes (Laryea, 2005).

Electro-spraying of electrolysed water supported the same log count reductions of *L. monocytogenes* compared with conventional spraying (Hsu et al., 2004). By applying electrostatic spraying, it is also possible to use sanitisers such as essential oils where the large-scale usage in wash tanks is unfeasible (Lee et al., 2021). In a further example, spinach and lettuce were electro-sprayed with grapeseed and malic acid and lactic acid. There was an initial 2-log reduction of *E. coli* O157:H7 followed by a further 2-log reduction during post-treatment storage (Ganesh et al., 2012). The authors noted the minimal volume required along with a low residual amount of the sanitising agent on the leafy greens, thereby exhibiting an antimicrobial effect with no negative effect on quality. In a comparative study, a malic acid–lactic acid combination was applied to cherry tomatoes via both conventional spraying and electro-spraying. No significant difference was found in terms of reducing *Salmonella*, although electro-sprayed cherry tomatoes had significantly better retention of quality during post-treatment storage (Massey et al., 2018). Electro-spraying will also make the application of other antimicrobials, such as bacteriophages, more feasible due to the low volumes that can be applied and the charges present on the proteins facilitating dispersion onto the fresh produce surface (Paczesny and Bielec, 2020).

Plasma-activated water (PAW) is another novel sanitiser receiving attention as an alternative to hypochlorite (Zeghioud et al., 2020). PAW is produced by introducing ionised gas, generated within a gas plasma unit, into water to produce a sanitiser containing reactive oxygen species (ROS) and/or reactive nitrogen species (Zeghioud et al., 2020). There are various forms of PAW reactors, with dielectric barrier discharge (DBD) being the most common. In this system, two closely spaced electrodes (0.3–1.0 cm) are polarised by a high alternating current to generate ionised plasma species in the headspace that subsequently are transferred into the water phase (Misra et al., 2011). The composition of the PAW is dependent on the gases within the headspace and aqueous solution. For example, a plasma generated from air produces PAW with hydrogen peroxide, peroxyacids, nitrous oxide and ozone (Misra et al.,

2011; Royintarat et al., 2020). The antimicrobial action of PAW can be enhanced by supplementing the water with hypochlorite or organic acids such as lactate or acetic acids (Qian et al., 2019; Liao et al., 2020). The general mode of action of PAW is thought to be direct oxidation by reactive radical species along with acidification of the cytoplasm by carboxylic acids. It has also been proposed that metal ions (Zn and Cu), originating from dissolution of the electrodes, could also contribute to the antimicrobial action by catalysing radical formation (Chen et al., 2018).

A PAW spray, referred to as a plasma jet, was reported to support a 5-log CFU reduction of *Salmonella* on tomatoes with no negative effect on fruit quality (Hou et al., 2021). Fresh-cut apples dipped in PAW reduced the total aerobic count by 1-log CFU, with residual antimicrobial activity throughout the 12-day storage period (Liu et al., 2020). In addition to antimicrobial activity, it has also been demonstrated that PAW can degrade pesticides and mycotoxins, although extended treatment times (>60 min) are required (Wang and Salvi, 2021). Attempts to scale up the production of PAW for use in commercial fresh produce processing have been met with variable results, partly due to the complexity of high-voltage reactors coupled with the ill-defined chemistry of solutions produced (Wang and Salvi, 2021). As PAW is unstable, it needs to be generated on-site, and similar to hypochlorite, the antimicrobial efficacy is decreased in the presence of organic loading (Zhao et al., 2020).

In a method related to PAW, hydroxyl-radical activated water is generated by passing a hydrogen peroxide solution containing ferric catalysts through a UV-C reactor. The activated water produced contains ROS and carboxylic acids that retain antimicrobial activity for 30 min post-generation (Wang et al., 2022). Studies performed with activated water have demonstrated that cross-contamination of lettuce batches by *E. coli* O157:H7, *Salmonella* and *L. monocytogenes* can be prevented, although log reductions on the leafy greens were not improved beyond standard hypochlorite washing (Wang et al., 2022). In a similar approach, it has been demonstrated that washing lettuce with peracetic acid along with the simultaneous irradiation with UV-C supported a 3-log CFU reduction of *Salmonella* (Lippman et al., 2020). It is believed that the generation of free radicals from photolytic degradation via UV-C resulted in synergistic antimicrobial action.

4 Advances in non-aqueous decontamination methods

Non-aqueous-based decontamination methods for fresh produce have been promoted due to the limited efficacy of the washing process along with the need to reduce water usage during processing (Pinela and Ferreira, 2017). Irradiation is the most extensively studied but is rarely applied commercially due to regulations along with negative consumer perception.

However, electron beams are considered to be a more positive form of irradiation and have been applied to inactivate pathogens associated with fresh produce.

Gas-phase treatments have a long historic use for decontaminating fresh produce, with methyl bromide, ethylene oxide, chlorine dioxide and ozone being the most widely applied (Bhilwadikar et al., 2019). The advances in gas-phase decontamination methods have focused on how treatments can be delivered without the associated hazards to workers, along with homogenous application. Novel technologies based on gas plasma or other free radical-based approaches will continue to be developed, with the anticipation of being a feasible alternative to the post-harvest washing process.

4.1 Electron beams

The gamma irradiation emitted by cobalt-60 and caesium-137 is known to be advantageous in treating fruit and vegetables to inactivate insects, retard ripening and reduce pathogen levels. In many respects, irradiation is a silver bullet in terms of fresh produce processing with positive aspects in both quality and safety, rather than one being at the expense of the other. However, the need to use central facilities adds to transportation costs and delays distribution to customers (Arvanitoyannis et al., 2009). As consumers have a negative perception of irradiation, any treated products must be identified via labelling.

The electron beam method is an alternative to gamma ray emitters and has several advantages as an irradiation treatment, i.e. the process can be performed on-site and is generally more acceptable by consumers (Magri and Teira-Esmatges, 2015). The beta rays emitted by electron beam devices are less energetic and hence have lower penetration along with delivering a lower dose per unit time. Consequently, there are no detectable by-products formed, unlike gamma irradiation, although this does not negate the need for irradiation labelling. Another consequence of the lower penetration of beta particles is that the produce needs to be thinly (2–3 cm) spread, although treatment times in the order of seconds are required to reduce pathogen levels. For example, electron beams delivered at 2.95 kGy could reduce the levels of feline calicivirus (surrogate for norovirus) by 1 log (Zhou et al., 2011). In comparison, vegetative pathogens such as *Salmonella* are more sensitive to viruses with a dose of 0.7 kGy required to support a log reduction (Pillai and Shayanfar, 2018). The need for preventative controls as part of the modern food safety regulations will likely restimulate interest in e-beams as a non-thermal approach to decontaminate fresh produce (Lung et al., 2015).

4.2 UV light

Germicidal UV light (UV-C) provides the advantage of being a chemical-free approach for surface disinfection that is compatible with continuous production (Singh et al., 2021). Despite the long history of UV surface disinfection, it has rarely been applied as an alternative to post-harvest washing as the fresh produce surface needs to be directly exposed to the UV photons. Consequently, the product has to be presented within a single layer and turned to expose undersides; however, UV is ultimately limited by the location of pathogens within cut edges, natural openings (e.g. stomata) and heterogeneous surface topography. Hence, UV can decrease pathogen levels, but extensive tailing occurs in inactivation curves (Koutchma, 2009).

Innovations in UV treatment have included assisted treatments along with the introduction of alternative UV sources. UV-assisted washing is a process in which lamps, or another UV source, are placed over the wash tank to inactivate pathogens released from fresh produce. Studies have illustrated that UV-assisted washing can increase the inactivation of murine norovirus on blueberries by 100-fold (Liu et al., 2015). A UV-C-assisted washing process was also reported to prevent cross-contamination of strawberries by *Salmonella* and *L. monocytogenes* (Ortiz-Sola et al., 2021). However, organic loading in water reduced the UV transmission, leading to a decrease in efficacy (Guo et al., 2017).

An interesting innovation in UV-based processing is microwave-assisted photolytic disinfection (MDEL). The method is based on using a quartz tube containing argon, mercury or cadmium gas that emits UV in the range of 185–254 nm, thereby generating ozone and UV-C photons. Consequently, there is a combination of microwave irradiation, ozone gas and UV-C to support the disinfection process. To date, MDEL has been primarily applied for wastewater treatment to reduce organic loading and pollutants, in addition to disinfection (Cheng et al., 2020). Although there are no published reports on the efficacy of UV microwaves to decontaminate fresh produce, the authors have demonstrated (unpublished results) that a 20-s treatment of frozen berries supported a 2.5-log CFU reduction of *L. monocytogenes* without defrosting the sample. To perform MDEL on an industrial scale could represent a challenge, given the need to shield from microwaves, in addition to UV light and ozone emissions.

Different light sources have been applied for surface disinfection, such as pulsed light and high-intensity lamps, although the limitation of shading is the same as for classic UV-C lamps (Singh et al., 2021). There has been interest in light-emitting diodes (LED)-based approaches given the advantages over mercury–UV (mercury-vapour) lamps (Green et al., 2020). Specifically, LEDs do not contain mercury, thereby reducing the risk of chemical contaminants and, from a performance aspect, can operate under a broad range of temperatures and be repeatedly switched on and off without negatively affecting working life (Chen et al., 2017). The advances in LED technology have improved the power output of lights, along with the range of wavelengths available, and costs have reduced compared with when first introduced.

LED light at longer wavelengths (>400 nm) demonstrates positive effects with regard to extending the shelf life of fresh produce and increasing the antioxidant content (Nassarawa et al., 2021). LED illumination of fresh produce at 461, 521 and 642 nm during storage resulted in an antimicrobial effect against *Salmonella* and *E. coli* via the generation of ROS. The effect of longer LED wavelengths was enhanced when combined with chilling, suggesting that the ability of pathogens to respond to the ROS burst is compromised at low temperatures (Chen et al., 2017).

UV-LEDs have been used for surface disinfection, and it was noted that different pathogens, e.g. *Listeria* versus *Salmonella*, have different susceptibilities depending on the wavelength applied (Green et al., 2018). Specifically, *Salmonella* was more sensitive to a wavelength of 268 nm with the maximal reduction of *Listeria* obtained at 259 nm (Green et al., 2018). The underlying reasons for the wavelength dependency on antimicrobial activity remain unclear, although this probably relates to the cellular composition that absorbs at specific wavelengths.

4.3 Gas-phase decontamination methods

Fumigation methods have been historically applied to inactivate pests and pathogens on fresh produce. The traditional fumigating agents were ethylene bromide and methyl bromide, which although effective, have raised food safety issues related to residues (Fields and White, 2002; Bononi et al., 2014), resulting in the use of alternatives such as nitric oxide, chlorine dioxide or ozone (Oh and Liu, 2020; Chen et al., 2020; Smith et al., 2015; Trinetta et al., 2011). The latest innovations within the area are related to delivery methods. Traditionally, fumigating gases were introduced into a sealed room or chamber containing the produce to be decontaminated (Table 1). The main challenge with such an approach is to achieve homogeneous distribution of the antimicrobial gas, with units exposed to the antimicrobial gas being overexposed while those deep within bins being underexposed (Perni et al., 2008; Tabakoglu and Karaca, 2015; Trinetta et al., 2010; Trinetta et al., 2011; Vurma et al., 2009; Skog and Chu, 2001). Such issues can be addressed by applying technologies such as the forced air approach, which streams antimicrobial gas through the bed of produce using the principle outlined by Darcy's law or the Klinkenberg effect. Specifically, the law establishes the diffusion coefficient of a fluid (e.g. antimicrobial gas) through a porous bed and is defined by the pressure differential across the

Table 1 Example of gas-phase, hydroxyl-/chloride-radical and gas plasma-based treatments for decontaminating fruit and vegetables

Microbe	Produce type	Process parameters	Log count reduction, log CFU	Source
Antimicrobial gas				
Salmonella	Carrots	Chlorine dioxide 4.1 mg/L for 30 min at 22°C	5.15	Sy et al. (2005)
L. monocytogenes	Tomatoes	Chlorine dioxide 0.5 mg/L for 12 min	>5.0	Bhagat et al. (2010)
Salmonella	Blueberries	Ozone 50 000 ppm for 64 min	3.00	Bialka and Demirci (2007)
E. coli O157:H7	Strawberries	Ozone 2–8 mg/L for 40 min	2.78	Han et al. (2002)
Hydroxyl-/chloride-radical process				
L. monocytogenes	Apples	H_2O_2: 6% Ozone: 11 mg UV-C: 54 mJ/cm^2	>4	Murray et al. (2018b)
E. coli O157:H7	Green onion	Chlorine: 100 ppm UV-C: 500 mJ/cm^2	>5	Durak et al. (2012)
Salmonella *L. monocytogenes*	Lemon	H_2O_2: 1.88% UV-C: 65 mJ/cm^2	>5	Hasani et al. (2019)
Salmonella	Lettuce	H_2O_2 1.5% UV-C: 37.8 mJ/cm^2	Surface: 4.12 Sub-surface: 2.84	Hadjok et al. (2008)
Salmonella	Lettuce	PAA: 100 ppm or chlorine: 100 ppm UV-C 450 mJ/cm^2	>4	Ge et al. (2013)
L. monocytogenes	Mushrooms	H_2O_2: 2% UV-C: 54 mJ/cm^2	1.2	Murray et al. (2015)
E. coli O157:H7	Mushrooms	H_2O_2: 3% UV-C: 450 mJ/cm^2	0.9	Guan et al. (2013)

(*Continued*)

Table 1 (*Continued*)

Microbe	Produce type	Process parameters	Log count reduction, log CFU	Source
Salmonella *L. monocytogenes* *E. coli* O157:H7	Spinach	ClO_2: 10 ppmv UV-C: 20 min @6W	>4	Park et al. (2018)
L. monocytogenes	Cantaloupe	H_2O_2: 1.5% Ozone: 11 mg UV-C: 60 mJ/cm^2	>5	Hasani et al (2019)
	Pepper		2.2	
	Courgette		1.4	
	Cucumber		3.4	
	Mango		4.8	
	Tomato		1.4	
	Peach		5.5	
	Avocado		5.7	
Gas plasma				
E. coli O104:H4	Apple	MW: air	4.6	Baier et al. (2015)
L. monocytogenes	Mango	APP: He + 0.5% O_2	2.5	Perni et al. (2008)
Salmonella	Lettuce	DBD: air	2.4	Ziuzina et al. (2015)
Salmonella	Lettuce	Microwave: N_2	1.5	Lee et al. (2015)
E. coli	Lettuce	DBD: air	3.3	Ziuzina et al. (2015)
Salmonella	Tomato	DBD: air	>5	Ziuzina et al. (2014)
L. monocytogenes	Tomato	DBD: air	>5	Ziuzina et al. (2014)
E. coli	Tomato	DBD: air	3.1	Ziuzina et al. (2014)

APP, atmospheric pressure plasma; PAA, peracetic acid.

bed. In practical terms, the theory suggests that ozone can be homogeneously distributed throughout the produce bed by modulating the air flow to establish a defined pressure differential that should represent a constant regardless of the bed geometry or size. Recent technology, using the forced air process, has been developed and commercialised, which utilises ozone gas and is known as the forced air–ozone reactor (Camargo et al., 2019; Murray et al., 2018b). A typical unit consists of a chamber into which the produce, in bins/containers, to be decontaminated is placed. The stacked bins are then wrapped in a curtain and the ozone is introduced at the top of the stack, with the air being drawn through the produce via a pump located at the exhaust port (Fig. 2). A laboratory-based unit was constructed that could treat 20 apples per batch, with the ozone being drawn through the fruit via an exhaust fan operating at an air velocity of 9.3 m/s. The ozone concentration measured after exiting the apple bed reached a steady state of 23 ppm. A 20-min ozone treatment supported a 2.12–3.07-log CFU reduction of *L. monocytogenes*, with no significant effect of apple position within the bed. In comparison, a commercial reactor introducing ozone (39 ppm) at a flow velocity of 0.022 m/s achieved a 2.5-log CFU reduction of *L. monocytogenes* within a 30-min treatment time and batch size of 570 kg (Camargo et al., 2019).

A further innovation is combining ozone gas treatment with vacuum cooling (Vurma et al., 2009). The principle of the approach is based on a common technique of removing field heat by applying a vacuum that results in sublimation of water, leading to the cooling of the fresh produce without the need to add ice or water. A further effect of applying a vacuum is that when

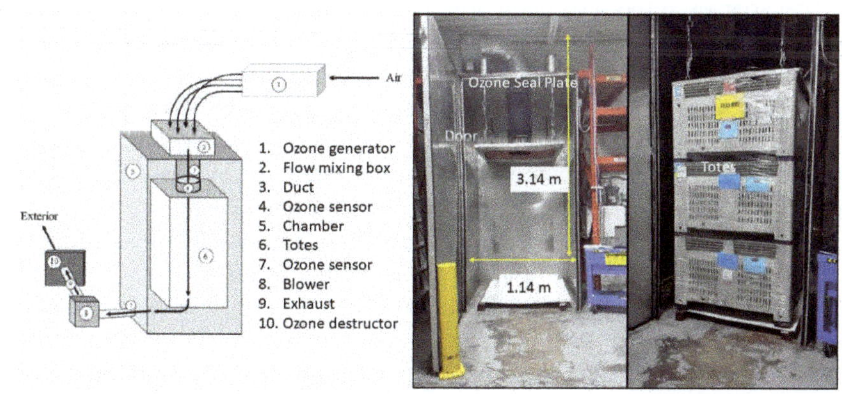

Figure 2 Forced air-ozone reactor to decontaminate fresh produce. The fruit or vegetables to be treated are packed into totes that are stacked and then wrapped with plastic film. The totes are positioned within the reactor and the ozone delivery plate lowered to make a seal. The chamber is then sealed with a stainless-steel door and the ozone:air mixture pulled through the blower with the air velocity controlled by the fan speed. The incoming and exiting ozone concentration is monitored via sensors positioned at the top and bottom of the reactor.

released, the air within stomata, cut edges along with other natural openings, is replaced by air/liquid on the surface (Chincinska, 2021). The ozone-vacuum treatment inactivated *E. coli* O157:H7 on and within baby spinach using 32–500 ppm ozone. It was found that the treatment supported a >1.5-log CFU reduction of *E. coli* O157:H7, although oxidative damage to leaves was observed even at the lowest concentration of ozone applied (Shynkaryk et al., 2016).

4.4 Gas plasma

In addition to producing PAW, the initial application of gas plasma was to deliver ionised gas directly to the surface being treated (Niemira, 2012). Early gas plasma reactors operated under vacuum and high temperatures, but units are now available that work under atmospheric pressure without generating excessive heat (Niemira, 2012). In the DBD format, the gas mixture can be air or a combination of gases such as oxygen, argon and/or nitrogen and is passed between high-voltage electrodes operating at >20 kV/cm^2. The gases become ionised with a mixture of radicals being produced, which include hydroxyl-radicals, ozone, hydrogen peroxide and nitrous oxide, amongst others, in addition to UV photons (Pankaj and Keener, 2017). The plasma can be transferred to a container holding the sample or generated within packs (Rana et al., 2020). In addition, it is possible to deliver the ionised gas directly via a plasma jet. The mode of inactivation by gas plasma is through UV photons that damage nucleic acids along with the oxidation of cell structures via the radical species. It has also been noted that acidic species (for example, nitric acid) provide a residual antimicrobial effect that can reduce microbial levels during post-treatment storage (Rossow et al., 2018).

There have been numerous studies on decontaminating fruit and vegetables using gas plasma, with varying efficacy being reported (Table 1). In general, the antimicrobial efficacy is dependent on the gas composition, treatment time and produce type, for example, gases high in oxygen and nitrogen produce ROS and nitric oxide synthase with high efficacy. The inactivation kinetics follows a linear decline, and hence extended treatment times support higher log reductions. Produce with a high antioxidant content and rough surface, for example strawberries, require longer treatment times compared with produce such as lettuce. The challenges with gas plasma include the need for high-voltage units, with the generated plasma being focused directly onto the produce being treated. In addition, treatment times in the order of minutes are required, which may compromise commercial feasibility. Toxicology assessment of the by-products generated on the fresh produce surface is required, along with balancing decontamination efficacy and any negative effect on quality (Niveditha et al., 2021).

4.5 Gas-phase hydroxyl-radical process

Gas-phase hydroxyl-radicals are generated from the UV-C-mediated degradation of hydrogen peroxide vapour and ozone gas. In nature, the reaction occurs within clouds, whereby the UV-C from the sun degrades ozone to oxygen radicals and subsequently hydrogen peroxide to form hydroxyl-radicals (Warriner et al., 2021). It is thought the hydroxyl-radicals degrade pollutants within the atmosphere. The process is related to the AOP applied in wastewater treatment to degrade pollutants in water such as pesticides, drugs and those not removed by standard treatments.

The chemistry of hydroxyl-radical generation from the UV-C degradation of hydrogen peroxide and ozone is not fully understood, given the transient nature of the intermediates. Nevertheless, it has been proposed that reactions follow the Haber–Weiss process that is initiated by the Fenton reaction referred to previously. The Haber–Weiss process is separated into three parts consisting of initiation, propagation and finally termination (Fig. 3). The initiation stage forms the hydroxyl-radicals that subsequently react with hydrogen peroxide and oxygen radicals to form additional hydroxyl-radicals. The process continues and can become self-sustaining, with an increase in hydrogen peroxide being observed followed by degradation to hydroxyl-radicals. In addition to hydroxyl-radicals, it is also thought that a range of other radicals can be formed such as trioxidane that have a transient existence (Kehrer, 2000).

In the event of excess hydroxyl-radicals, there are self-neutralisation reactions whereby water is formed. In aqueous solution, the optimum hydrogen peroxide level is 1–2% v/v; levels below this concentration generate negligible hydroxyl-radicals, but higher levels result in termination reactions (Fan and Song, 2020; Bayliss and Waites, 1979). However, within the gas phase, there is no evidence of an optimum hydrogen peroxide concentration, although above 6% H_2O_2, the primary mode of microbe inactivation is via direct action of the hydrogen peroxide rather than hydroxyl-radical generation (Hasani et al., 2020).

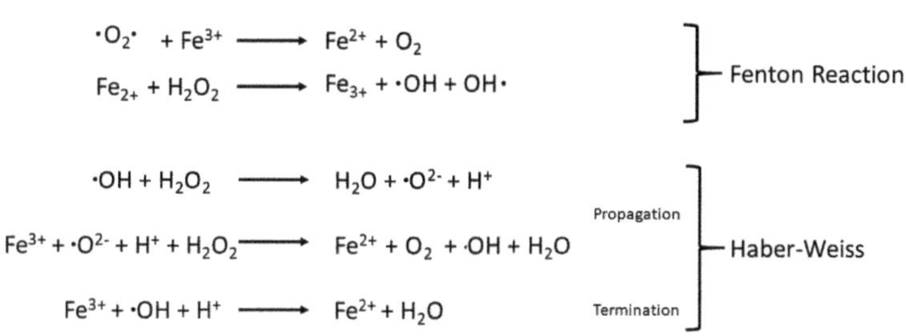

Figure 3 Generation of hydroxyl-radicals via the Fenton reaction and Haber–Weiss reaction. The Fenton reaction initiates the generation of hydroxyl-radicals in the presence of a ferric catalyst. The hydroxyl-radicals are propagated via the Haber–Weiss process with termination when an excess of radicals results in self-neutralisation.

The basic construction of a gas-phase hydroxyl-radical reactor consists of UV-C lamps, with hydrogen peroxide being misted or vaporised into the chamber. The ozone can be generated *in situ* via UV lamps emitting at 174 nm or via an external generator. The process can be performed within a batch or a continuous system and can be readily scaled up (Warriner et al., 2021).

The gas-phase hydroxyl-radical treatment has been applied for carton packaging sterilisation since the 1980s in a process developed by Elopak Ltd (Stannard et al., 1983). The application for decontaminating food surfaces was initially developed in 2006 using a batch process based on a UV-C lamp under which the sample was placed and simultaneously sprayed with hydrogen peroxide (Hadjok et al., 2008). It was found that a 30-s treatment using 1.5% v/v hydrogen peroxide introduced at 50°C supported a >4-log CFU reduction in *Salmonella* on the surface and 2-log CFU reduction of internalised populations (vacuum infiltrated) in shredded lettuce (Hadjok et al., 2008). Importantly, the hydroxyl-radical treatment did not result in negative changes in quality. In comparison, washing inoculated lettuce samples in 100 ppm sodium hypochlorite supported a <2-log CFU reduction with a negligible decrease in internalised populations. Studies have also illustrated that the hydroxyl-radical treatment can support a >4-log reduction of norovirus along with the MS2 coliphage surrogate (Li et al., 2011; Xie et al., 2008).

A comparative study has shown that the efficacy of the hydroxyl-radical treatment is dependent on the produce type, in addition to the operating parameters (Hasani et al., 2019). It was found that cantaloupes, peaches, avocados, apples, peppers, mangoes and cucumbers were effectively decontaminated by applying a UV–hydrogen peroxide–ozone treatment. In comparison, *Listeria* species on courgettes and tomatoes were reduced to a lower extent (1 -log CFU reduction) (Table 1). The underlying reasons for the efficacy of the hydroxyl-radical treatment being produce dependent remain unknown but is believed to be associated with the hydrophobic nature of the surface along with the antioxidant content.

In addition to pathogen reduction, the hydroxyl-radical process has also been demonstrated to extend the shelf life of fruit and vegetables (Warriner and Hasani, 2020b). For example, strawberries treated with the hydroxyl-radical process retained quality over a 15-day storage period (Fig. 4); in comparison, non-treated controls exhibited softening and internal browning. A similar effect was seen for other soft fruit (blackberries, blueberries and raspberries) treated with gas-phase hydroxyl-radicals and then stored at 4°C for 16 days.

Lemons inoculated with *Penicillium digitatum* using a probe applicator that drove spores of the mould 2 mm into the fruit peel were treated with the hydroxyl-radical process and then held at room temperature for 10 days. Controls exhibited extensive, confluent growth of mould in contrast to those treated with the hydroxyl-radical process that exhibited no or minor growth. In

Figure 4 Strawberries treated with the hydroxyl-radical process and then stored aerobically at 4°C for 15 days. Whole strawberries were passed through a hydroxyl-radical process operating with a 1.5% v/v hydrogen peroxide flow rate of 40 mL/min and 12 ppm ozone with a UV-C treatment time of 30 s (114 mJ/cm^2). At the end of the 15-day shelf life, the strawberries were halved to evaluate the degree of internal browning.

a further example, avocados (n = 18) treated with the hydroxyl-radical process retained quality over a 10-day storage period. In comparison, only 3 of a batch of 18 fruit of non-treated controls remained acceptable at the end of the storage period. It was interesting to note that the main symptom of spoilage of avocados was softening and discolouration of the inner flesh. Given that the hydroxyl-radical process is a surface treatment, it is likely that the physiology of the avocado had been changed. A similar effect has been observed with soft fruit treated with gas plasma, whereby the generation of free radicals stimulates the synthesis of antioxidant compounds that provide protection during post-treatment storage (Pan et al., 2019).

5 Considerations for developing a standard validation method for fresh produce decontamination methods

5.1 Developing a standard method for fresh produce-processing verification

There are a number of publications on decontamination methods for fruit and vegetables, but one of the key challenges is how to compare the results of different studies. This is primarily due to the different strains, methodology and reactor designs. To address this issue, there have been attempts to develop a standard methodology to validate and verify decontamination methods. Standard methods have previously been proposed in performing validation trials, although these have been generic in nature (Natl Advisory Comm Microbiol, 2010; Ceylan et al., 2021).

A working group has proposed different options for validating wash processes to demonstrate that the working parameters can prevent cross-contamination (Gombas et al., 2017). Three options are provided in the Gombas et al. (2017) outline, with the first using a surrogate to mimic the fate of pathogens. The second option is to measure the concentration of sanitiser under the worst-case scenario, for example, organic loading (Gombas et al., 2017). The third option, like option 2, sets a minimum sanitiser concentration to be maintained throughout the tank under a range of processing conditions. The guidance document aimed to help the produce industry develop protocols to validate commercial post-harvest wash processes to prevent cross-contamination between batches. However, the guidance did more to highlight the complexity of defining the wash process and which parameters should be considered. The challenges identified were finding surrogates that could be applied to represent the pertinent pathogen. However, it was unclear what phenotype the surrogate should represent of the pathogens of interest, for example, it was suggested that the surrogate should have higher resistance to chlorine compared with the pathogen. Yet, this alone would be inadequate given the transfer between produce would also need to be considered along with tolerance to the sanitiser and expression of stress responses under conditions within the wash tank. There were also issues with the surrogate approach in terms of what level of inactivation (log reduction) should be designated and within what time period.

As an alternative to surrogates, Gombas et al. (2017) suggested maintaining the sanitiser above a critical limit could be used in post-harvest wash validation. However, the working group came across the challenge of identifying such critical limits given the broad range (3 ppm to >20 ppm) reported in the literature. However, a greater challenge was to identify a worst-case scenario in terms of wash water characteristics and how to account for the dynamics of chlorine sequestering along with the generation of disinfection by-products. The guidance document concluded that the validation of post-harvest wash processes is challenging without giving assurance that any of the options outlined would be sufficient to provide evidence of a validated process. Ultimately, the guidance document illustrated the lack of control of the post-harvest wash process and need for alternative decontamination approaches.

Alternative decontamination methods to post-harvest washing are generally easier to validate, given these represent less complex, more defined, systems. For example, if a gas-phase hydroxyl-radical process is considered, then the operating parameters that define the antimicrobial action can be defined as UV-C dose, hydrogen peroxide and ozone concentrations. Each parameter can be monitored, and hence critical limits applied, thereby representing a critical control point/preventative control in fresh produce processing. Being able to define the monitoring parameters appears to be a prerequisite for developing a standard validation method.

Additional parameters for developing a standard method include selecting the relevant pathogens that are associated with the produce. From this point, it is possible to identify surrogates that should have the same resistance to the decontamination method as the identified pathogen. It must also be appreciated that surrogates would be readily available and cultured under standard laboratory conditions. The working parameters that achieve the Food Safety Objective can then be identified along with establishing the worst-case scenario (Ceylan et al., 2021). Although generic outlines are useful, standardised protocols that are commodity specific need to be established. In the case of fruit and vegetables, the lethality of the treatment against relevant pathogens and the effect on quality metrics need to be considered. This is often overlooked in validation trials, which results in a disconnect between operating parameters optimised in the laboratory compared with those applied on an industrial scale.

A key challenge in developing a standardised validation protocol is agreement between different groups of academics, certification bodies and, most importantly, government regulators. The elements of a standard protocol would be to select pathogens derived from clinical cases or outbreaks linked to fresh produce. The inoculation method should also be considered in relation to submersion, spray or spot inoculation. Vacuum infiltration should also be considered when focusing on worst-case scenarios. An additional aspect is to define the log count reduction, which is traditionally set at 5-log CFU; however, a more informative metric is the log count reduction obtained using a reference treatment such as post-harvest washing in sanitiser solutions. If this is coupled with shelf-life assessment through sensory or measurements of colour and texture, then operational parameters can be identified that can be transferred to a commercial process. The development of an international standard should be a global effort and coordinated by a body such as Codex Alimentarius.

5.2 Pathogen resistance and tolerance to decontamination interventions

In pathogen challenge testing of decontamination technologies, it is common practice to cultivate bacteria under optimal conditions and then prepare cell suspensions that are subsequently inoculated onto the fresh produce sample. Clearly, this is far removed from the natural environment where the cells could be stressed, sub-lethally injured and have intrinsically higher resistance. To recreate such conditions, the culture is harvested in the stationary phase rather than the exponential growth phase, and glucose is added to the medium to induce acid resistance (Bayles, 2004). The cell suspensions can also be held for a period at 4°C to enhance resistance. For example, *L. monocytogenes* derived from 7-day-aged cultures exhibited significant resistance to ozone gas compared with cells held for a shorter period (Sheng et al., 2018). Cell survival can be attributed to the induction of stress proteins that stabilise cellular structures against a broad range of stresses (Bansal et al., 2018). In addition, cells can also enter a dormant state known as persisters or viable but non-culturable (VBNC) that exhibit high stress tolerance to actively metabolizing cells (Wood et al., 2013). While in the dormant state, the cells exhibit a low metabolic state with the aggregation of ribosomes and other proteins to enhance resistance. The differentiation of persisters and VBNC is open to debate, with one school of thought that both represent the same state, while others suggest these are distinct (Ayrapetyan et al., 2015). Historically, persisters have been defined as a sub-population of cells that survive exposure to antibiotic treatment. In turn, VBNC are defined as cells that cannot be cultured in a standard laboratory medium but can become activated under certain conditions, for example, within the gastrointestinal tract (Highmore et al., 2018; Arvaniti et al., 2021). There is also anecdotal evidence that *E. coli* O157:H7 can enter a dormant state on lettuce but then revive over storage as the plant tissue breaks down; however, if such recovery is due to revival from the dormant state remains to be elucidated.

Critically, dormant cells, regardless of persisters or VBNC, exhibit enhanced tolerance to stress and cannot be readily cultivated (false negative in diagnostics tests) but can cause illness or lead to latent infections within susceptible hosts (Wood et al., 2013).

The relevance of dormant cells in relation to fresh produce safety remains to be fully elucidated. However, there is evidence that enteric pathogens such as toxigenic *E. coli* typically die off in the soil environment, but a small population persists over months. The induction of VBNC in *L. monocytogenes* subjected to chlorinated washes has also been demonstrated and shown to lose culturability but retain infectivity (Arvaniti et al., 2021; Highmore et al., 2018).

A further challenge is to differentiate dormant cells from those that would eventually become non-viable. It was anticipated that viability staining and molecular methods could be used; however, this is not the case (Kim et al., 2018; Song and Wood, 2021). Consequently, there is active research into additives and culture methods that can enhance the recovery of dormant cells such as the inclusion of carbohydrates, esters, alcohols and signal molecules (Chen et al., 2021). By being able to detect and enumerate dormant cells will enable reliable pathogen screening diagnostics, facilitate risk analysis and, ultimately, pathogen pre harvest and post-harvest controls.

6 Conclusion and future trends in research

With the increase in high-intensity animal production and pressure on water sources, in combination with climate change, it is believed that outbreaks linked to fresh produce will continue to rise. The pressure on food resources will also necessitate reducing food waste, requiring the shelf life of produce to be extended. Consequently, the post-harvest decontamination treatment of fruit and vegetables will become increasingly significant in relation to inactivating pathogens and spoilage microbes. Treatments based on post-harvest washing have limited efficacy, and alternative methods are required. To this end, technologies based on electrostatic spraying of sanitisers could be introduced along with novel technologies based on gas plasma or hydroxyl-radical treatments. Such approaches will provide more effective treatments with the potential to alter the physiology of plants to extend shelf life. Immediate research is necessary to develop standardised validation protocols to enable different studies to be compared to provide both industry and regulator assessments of treatment suitability. Ultimately, this will be required to demonstrate preventative controls within food safety management systems in the fresh produce sector.

7 Where to look for further information

- Center for Produce Safety https://www.centerforproducesafety.org/.
- Fresh Produce Safety Center Australia and New Zealand https://fpsc-anz.com/. - the FPSCAZ is a consortium of academics, government and industry who have developed guidelines on fresh produce safety. The association embrances new technologies that can be applied in the fresh fruit & vegetable sector with respect to enhancing safety and quality.
- Clean Works Corporation https://cleanworkscorp.com/. Clean Works have commercialized hydroxyl-radical process for decontaminating a broad range of fruit and vegetable types.

7.1 Further reading

- Innovative Food Processing Technologies: A comprehensive review. Ed Knoerzer and Muthukumarappan. Academic Press, New York.
- Global Safety of Fresh Produce: A handbook of best practice, innovation commercial solutions and case studies. Ed J. Hoorfar, Academic Press, New York.
- Guidance for Industry: Guide to minimize microbial food safety hazards of fresh cut fruit and vegetables. https://www.fda.gov/regulatory-information/search-fda-guidance-documents/guidance-industry-guide-minimize-microbial-food-safety-hazards-fresh-cut-fruits-and-vegetables.

7.2 Key journals and conferences

- Food Control.
- International Journal of Food Microbiology.
- Journal of Food Protection.
- Food Microbiology.
- Post-harvest Biology and Technology.
- Center for Fresh Produce Safety – Annual symposium.
- International Journal for Food Protection Annual symposium.

8 References

Abnavi, M. D., Kothapalli, C. R. and Srinivasan, P. 2021. Total amino acids concentration as a reliable predictor of free chlorine levels in dynamic fresh produce washing process. *Food Chemistry* 335, 127651.

Ali, A., Yeoh, W. K., Forney, C. and Siddiqui, M. W. 2018. Advances in postharvest technologies to extend the storage life of minimally processed fruits and vegetables. *Critical Reviews in Food Science and Nutrition* 58(15), 2632–2649.

Arvaniti, M., Tsakanikas, P., Papadopoulou, V., Giannakopoulou, A. and Skandamis, P. 2021. *Listeria monocytogenes* sublethal injury and viable-but-nonculturable state induced by acidic conditions and disinfectants. *Microbiology Spectrum* 9(3), e0137721.

Arvanitoyannis, I. S., Stratakos, A. C. and Tsarouhas, P. 2009. Irradiation applications in vegetables and fruits: A review. *Critical Reviews in Food Science and Nutrition* 49(5), 427–462.

Ayrapetyan, M., Williams, T. C. and Oliver, J. D. 2015. Bridging the gap between viable but non-culturable and antibiotic persistent bacteria. *Trends in Microbiology* 23(1), 7–13.

Baier, M., Janssen, T., Wieler, L. H., Ehlbeck, J., Knorr, D. and Schluter, O. 2015. Inactivation of Shiga toxin-producing *Escherichia coli* O104:H4 using cold atmospheric pressure plasma. *Journal of Bioscience and Bioengineering* 120(3), 275–279.

Banach, J. L., Sampers, I., Van Haute, S. and Van Der Fels-Klerx, H. J. 2015. Effect of disinfectants on preventing the cross-contamination of pathogens in fresh produce washing water. *International Journal of Environmental Research and Public Health* 12(8), 8658–8677.

Bansal, M., Nannapaneni, R., Sharma, C. S. and Kiess, A. 2018. *Listeria monocytogenes* response to sublethal chlorine induced oxidative stress on homologous and heterologous stress adaptation. *Frontiers in Microbiology* 9, 2050.

Bayles, D. O. 2004. Changes in heat resistance resulting from pH and nutritional shifts of acid-adapted and non-acid-adapted *Listeria monocytogenes*. *Journal of Food Protection* 67(2), 316–321.

Bayliss, C. E. and Waites, W. M. 1979. The combined effect of hydrogen peroxide and ultraviolet irradiation on bacterial spores. *Journal of Applied Bacteriology* 47(2), 263–269.

Bertoldi, B., Bardsley, C. A., Pabst, C. R., Baker, C. A., Gutierrez, A., De, J., Luo, Y. and Schneider, K. R. 2022. Influence of free chlorine and contact time on the reduction of *Salmonella* cross-contamination of tomatoes in a model flume system. *Journal of Food Protection* 85(1), 22–26.

Beuchat, L. R., Nail, B. V., Adler, B. B. and Clavero, M. R. S. 1998. Efficacy of spray application of chlorinated water in killing pathogenic bacteria on raw apples, tomatoes, and lettuce. *Journal of Food Protection* 61(10), 1305–1311.

Bhagat, A., Mahmoud, B. S. M. and Linton, R. H. 2010. Inactivation of *Salmonella enterica* and *Listeria monocytogenes* inoculated on hydroponic tomatoes using chlorine dioxide gas. *Foodborne Pathogens and Disease* 7(6), 677–685.

Bhilwadikar, T., Pounraj, S., Manivannan, S., Rastogi, N. K. and Negi, P. S. 2019. Decontamination of microorganisms and pesticides from fresh fruits and vegetables: A comprehensive review from common household processes to modern techniques. *Comprehensive Reviews in Food Science and Food Safety* 18(4), 1003–1038.

Bialka, K. L. and Demirci, A. 2007. Decontamination of *Escherichia coli* O157: H7 and *Salmonella enterica* on blueberries using ozone and pulsed UV-light. *Journal of Food Science* 72(9), M391–M396.

Bononi, M., Quaglia, G. and Tateo, F. 2014. Identification of ethylene oxide in herbs, spices and other dried vegetables imported into Italy. *Food Additives and Contaminants. Part A, Chemistry, Analysis, Control, Exposure and Risk Assessment* 31(2), 271–275.

Camargo, J. A., Murray, K., Warriner, K. and Lubitz, W. 2019. Characterization of efficacy and flow in a commercial scale forced air ozone reactor for decontamination of apples. *LWT-Food Science and Technology* 113(8), 108325.

Ceylan, E., Amezquita, A., Anderson, N., Betts, R., Blayo, L., Garces-Vega, F., Gkogka, E., Harris, L. J., McClure, P., Winkler, A. and Den Besten, H. M. W. 2021. Guidance on validation of lethal control measures for foodborne pathogens in foods. *Comprehensive Reviews in Food Science and Food Safety* 20(3), 2825–2881.

Chen, H., Green, A., Martz, K., Wu, X. Y., Alzahrani, A. and Warriner, K. 2021. The progress of type II persisters of *Escherichia coli* O157:H7 to a non-culturable state during prolonged exposure to antibiotic stress with revival being aided through acid-shock treatment and provision of methyl pyruvate. *Canadian Journal of Microbiology* 67(7), 518–528.

Chen, J., Loeb, S. and Kim, J. H. 2017. LED revolution: Fundamentals and prospects for UV disinfection applications. *Environmental Science: Water Research and Technology* 3(2), 188–202.

Chen, M., Chen, X., Ray, S. and Yam, K. 2020. Stabilization and controlled release of gaseous/volatile active compounds to improve safety and quality of fresh produce. *Trends in Food Science and Technology* 95, 33–44.

Chen, T. P., Liang, J. F. and Su, T. L. 2018. Plasma-activated water: Antibacterial activity and artifacts? *Environmental Science and Pollution Research International* 25(27), 26699–26706.

Chen, X. and Hung, Y. C. 2017. Effects of organic load, sanitizer pH and initial chlorine concentration of chlorine-based sanitizers on chlorine demand of fresh produce wash waters. *Food Control* 77, 96–101.

Cheng, G., Li, Z., Sun, L. M., Li, Y. Z. and Fu, J. 2020. Application of microwave/electrodeless discharge ultraviolet/ozone sterilization technology in water reclamation. *Process Safety and Environmental Protection* 138, 148–156.

Chincinska, I. A. 2021. Leaf infiltration in plant science: Old method, new possibilities. *Plant Methods* 17(1), 83.

Davidson, G. R., Kaminski-Davidson, C. N. and Ryser, E. T. 2017. Persistence of *Escherichia coli* O157:H7 during pilot-scale processing of iceberg lettuce using flume water containing peroxyacetic acid-based sanitizers and various organic loads. *International Journal of Food Microbiology* 248, 22-31.

Dos Santos, N. D., Teixeira, L. A., Zhou, Q. Z., Burke, G. and Campos, L. C. 2022 Fenton pre-oxidation of natural organic matter in drinking water treatment through the application of iron nails. *Environmental Technology* 14, 1-14.

Durak, M. Z., Churey, J. J., Gates, M., Sacks, G. L. and Worobo, R. W. 2012. Decontamination of green onions and baby spinach by vaporized ethyl pyruvate. *Journal of Food Protection* 75(6), 1012–1022.

Fan, X. T. and Song, Y. Y. 2020. Advanced oxidation process as a postharvest decontamination technology to improve microbial safety of fresh produce. *Journal of Agricultural and Food Chemistry* 68(46), 12916–12926.

Feliziani, E., Lichter, A., Smilanick, J. L. and Ippolito, A. 2016. Disinfecting agents for controlling fruit and vegetable diseases after harvest. *Postharvest Biology and Technology* 122, 53–69.

Fields, P. G. and White, N. D. G. 2002. Alternatives to methyl bromide treatments for stored-product and quarantine insects. *Annual Review of Entomology* 47, 331–359.

Ganesh, V., Hettiarachchy, N. S., Griffis, C. L., Martin, E. M. and Ricke, S. C. 2012. Electrostatic spraying of food-grade organic and inorganic acids and plant extracts to decontaminate *Escherichia coli* O157:H7 on spinach and iceberg lettuce. *Journal of Food Science* 77(7), M391–M396.

Ge, C., Bohrerova, Z. and Lee, J. 2013. Inactivation of internalized *Salmonella* Typhimurium in lettuce and green onion using ultraviolet C irradiation and chemical sanitizers. *Journal of Applied Microbiology* 114(5), 1415–1424.

Gerdes, M. E., Cruz-Cano, R., Solaiman, S., Ammons, S., Allard, S. M., Sapkota, A. R., Micallef, S. A. and Goldstein, R. E. R. 2022. Impact of irrigation water type and sampling frequency on microbial water quality profiles required for compliance with US Food Safety Modernization Act Produce Safety Rule standards. *Environmental Research* 205, 112480.

Giannakis, S. 2019. A review of the concepts, recent advances and niche applications of the (photo) Fenton process, beyond water/wastewater treatment: Surface functionalization, biomass treatment, combatting cancer and other medical uses. *Applied Catalysis B: Environmental* 248, 309–319.

Gokarn, S. and Kuthambalayan, T. S. 2019. Creating sustainable fresh produce supply chains by managing uncertainties. *Journal of Cleaner Production* 207, 908–919.

Gombas, D., Luo, Y., Brennan, J., Shergill, G., Petran, R., Walsh, R., Hau, H., Khurana, K., Zomorodi, B., Rosen, J., Varley, R. and Deng, K. 2017. Guidelines to validate control of cross-contamination during washing of fresh-cut leafy vegetables. *Journal of Food Protection* 80(2), 312-330.

Goss, M. J., Tubeileh, A. and Goorahoo, D. 2013. A review of the use of organic amendments and the risk to human health. *Advances in Agronomy* 120(120), 275–379.

Green, A., Popovic, V., Pierscianowski, J., Biancaniello, M., Warriner, K. and Koutchma, T. 2018. Inactivation of *Escherichia coli*, *Listeria* and *Salmonella* by single and multiple wavelength ultraviolet-light emitting diodes. *Innovative Food Science and Emerging Technologies* 47, 353-361.

Green, A., Popovic, V., Warriner, K. and Koutchma, T. 2020. The efficacy of UVC LEDs and low pressure mercury lamps for the reduction of *Escherichia coli* O157:H7 and *Listeria monocytogenes* on produce. *Innovative Food Science and Emerging Technologies* 64, 10.

Guan, W., Fan, X. and Yan, R. 2013. Effect of combination of ultraviolet light and hydrogen peroxide on inactivation of *Escherichia coli* O157:H7, native microbial loads, and quality of button mushrooms. *Food Control* 34(2), 554–559.

Guo, S. H., Huang, R. Z. and Chen, H. Q. 2017. Application of water-assisted ultraviolet light in combination of chlorine and hydrogen peroxide to inactivate *Salmonella* on fresh produce. *International Journal of Food Microbiology* 257, 101–109.

Gutierrez-Rodriguez, E. and Adhikari, A. 2018. Preharvest farming practices impacting fresh produce safety. *Microbiology Spectrum* 6(2), 20.

Hadjok, C., Mittal, G. S. and Warriner, K. 2008. Inactivation of human pathogens and spoilage bacteria on the surface and internalized within fresh produce by using a combination of ultraviolet light and hydrogen peroxide. *Journal of Applied Microbiology* 104(4), 1014–1024.

Han, Y., Floros, J. D., Linton, R. H., Nielsen, S. S. and Nelson, P. E. 2002. Response surface modeling for the inactivation of *Escherichia coli* O157:H7 on green peppers (*Capsicum annuum*) by ozone gas treatment. *Journal of Food Science* 67(3), 1188–1193.

Hasani, M., Chudyk, J., Murray, K., Lim, L. T., Lubitz, D. and Warriner, K. 2019. Inactivation of *Salmonella*, *Listeria monocytogenes*, *Aspergillus* and *Penicillium* on lemons using advanced oxidation process optimized through response surface methodology. *Innovative Food Science and Emerging Technologies* 54, 182–191.

Hasani, M., Wu, F., Hu, K., Farber, J. and Warriner, K. 2020. Inactivation of *Salmonella* and *Listeria monocytogenes* on dried fruit, pistachio nuts, cornflakes and chocolate crumb using a peracetic acid-ethanol based sanitizer or advanced oxidation process. *International Journal of Food Microbiology* 333, 108789.

Highmore, C. J., Warner, J. C., Rothwell, S. D., Wilks, S. A. and Keevil, C. W. 2018. Viable-but-nonculturable *Listeria monocytogenes* and *Salmonella enterica* serovar Thompson induced by chlorine stress remain infectious. *mBio* 9(2), 12.

Hoff, C., Higa, J., Patel, K., Gee, E., Wellman, A., Vidanes, J., Holland, A., Kozyreva, V., Zhu, J., Mattioli, M., Roundtree, A., Mcfadden, K., Whitlock, L., Wise, M., Gieraltowski, L. and Schwensohn, C. 2021. An outbreak of *Escherichia coli* O157:H7 infections linked to romaine lettuce exposure United States, 2019. *MMWR-Morbidity and Mortality Weekly Report* 70(18), 689–690.

Hou, C. Y., Lai, Y. C., Hsiao, C. P., Chen, S. Y., Liu, C. T., Wu, J. S. and Lin, C. M. 2021. Antibacterial activity and the physicochemical characteristics of plasma activated water on tomato surfaces. *LWT* 149, 111879.

Hsu, S. Y., Kim, C., Hung, Y. C. and Prussia, S. E. 2004. Effect of spraying on chemical properties and bactericidal efficacy of electrolysed oxidizing water. *International Journal of Food Science and Technology* 39(2), 157–165.

Hudson, J. C., Tolen, T. N., Kirsch, K. R., Acuff, G., Taylor, T. M., Lucia, L. M. and Castillo, A. 2019. Comparison of antimicrobial treatments applied via conventional or handheld electrostatic spray to reduce shiga toxin-producing *Escherichia coli* on chilled beef outside rounds. *Journal of Food Protection* 82(5), 862–868.

Jongman, M. and Korsten, L. 2017. Assessment of irrigation water quality and microbiological safety of leafy greens in different production systems. *Journal of Food Safety* 37(3), 12.

Kehrer, J. P. 2000. The Haber-Weiss reaction and mechanisms of toxicity. *Toxicology* 149(1), 43–50.

Kim, J. S., Chowdhury, N., Yamasaki, R. and Wood, T. K. 2018. Viable but non-culturable and persistence describe the same bacterial stress state. *Environmental Microbiology* 20(6), 2038–2048.

Kongkaew, S., Joonyong, K., Kanatharana, P., Thavarungkul, P. and Limbut, W. 2021. Fabrication and characterization of Prussian Blue screen-printed working electrode and their application for free chlorine monitoring in swimming pool water. *Electrochimica Acta* 388, 138558.

Koutchma, T. 2009. Advances in ultraviolet light technology for non-thermal processing of liquid foods. *Food and Bioprocess Technology* 2(2), 138–155.

Laryea, G. N. 2005. Effect of fan speed and electrostatic charge on deposition of orchard canopy sprays. *Atomization and Sprays* 15(2), 133–144.

Lee, C. H., Woo, H. J., Kang, J. H. and Song, K. B. 2021. Electrostatic spraying of passion fruit (*Passiflora edulis* L.) peel extract for inactivation of *Escherichia coli* O157:H7 and *Listeria monocytogenes* on fresh-cut lollo rossa and beetroot leaves. *Food and Bioprocess Technology* 14(5), 898–908.

Lee, H., Kim, J. E., Chung, M. S. and Min, S. C. 2015. Cold plasma treatment for the microbiological safety of cabbage, lettuce, and dried figs. *Food Microbiology* 51, 74–80.

Li, D., Baert, L., De Jonghe, M., Van Coillie, E., Ryckeboer, J., Devlieghere, F. and Uyttendaele, M. 2011. Inactivation of Murine Norovirus 1, Coliphage phi X174, and *Bacillus fragilis* phage B40-8 on surfaces and fresh-cut iceberg lettuce by hydrogen peroxide and UV light. *Applied and Environmental Microbiology* 77(4), 1399–1404.

Li, H. P., Tajkarimi, M. and Osburn, B. I. 2008. Impact of vacuum cooling on *Escherichia coli* O157:H7 infiltration into lettuce tissue. *Applied and Environmental Microbiology* 74(10), 3138–3142.

Liao, X. Y., Xiang, Q. S., Cullen, P. J., Su, Y. A., Chen, S. G., Ye, X. Q., Liu, D. H. and Ding, T. 2020. Plasma-activated water (PAW) and slightly acidic electrolyzed water (SAEW) as beef thawing media for enhancing microbiological safety. *LWT* 117, 6.

Lippman, B., Yao, S. Y., Huang, R. Z. and Chen, H. Q. 2020. Evaluation of the combined treatment of ultraviolet light and peracetic acid as an alternative to chlorine washing for lettuce decontamination. *International Journal of Food Microbiology* 323, 108590.

Liu, C. H., Chen, C., Jiang, A. L., Sun, X. Y., Guan, Q. X. and Hu, W. Z. 2020. Effects of plasma-activated water on microbial growth and storage quality of fresh-cut apple. *Innovative Food Science and Emerging Technologies* 59, 102256.

Liu, C. H., Li, X. H. and Chen, H. Q. 2015. Application of water-assisted ultraviolet light processing on the inactivation of murine Norovirus on blueberries. *International Journal of Food Microbiology* 214, 18–23.

Lopez-Galvez, F., Tudela, J. A., Allende, A. and Gil, M. I. 2019. Microbial and chemical characterization of commercial washing lines of fresh produce highlights the need for process water control. *Innovative Food Science and Emerging Technologies* 51, 211–219.

Lung, H. M., Cheng, Y. C., Chang, Y. H., Huang, H. W., Yang, B. B. and Wang, C. Y. 2015. Microbial decontamination of food by electron beam irradiation. *Trends in Food Science and Technology* 44(1), 66–78.

Luo, Y., Nou, X., Yang, Y., Alegre, I., Turner, E., Feng, H., Abadias, M. and Conway, W. 2011. Determination of free chlorine concentrations needed to prevent *Escherichia coli* O157:H7 cross-contamination during fresh-cut produce wash. *Journal of Food Protection* 74(3), 352–358.

Luo, Y. G., Zhou, B., Van Haute, S., Nou, X. W., Zhang, B. C., Teng, Z., Turner, E. R., Wang, Q. and Millner, P. D. 2018. Association between bacterial survival and free chlorine concentration during commercial fresh-cut produce wash operation. *Food Microbiology* 70, 120–128.

Lyons, S. M., Harrison, M. A. and Law, S. E. 2011. Electrostatic application of antimicrobial sprays to sanitize food handling and processing surfaces for enhanced food safety. In: *13th International Conference on Electrostatics*, April 10–14, 2011. Bangor, Wales, Bristol: IOP Publishing Ltd.

Ma, L., Zhang, M., Bhandari, B. and Gao, Z. X. 2017. Recent developments in novel shelf life extension technologies of fresh-cut fruits and vegetables. *Trends in Food Science and Technology* 64, 23–38.

Ma, Q. M., Davidson, P. M. and Zhong, Q. X. 2020. Properties and potential food applications of lauric arginate as a cationic antimicrobial. *International Journal of Food Microbiology* 315, 108417.

Magri, A. and Teira-Esmatges, M. R. 2015. Assessment of a composting process for the treatment of beef cattle manure. *Journal of Environmental Science and Health: Part B, Pesticides, Food Contaminants, and Agricultural Wastes* 50(6), 430–438.

Massey, L. M., Hettiarachchy, N. S., Horax, R., Rayaprolu, S. J., Kumar-Phillips, G., Martin, E. M. and Ricke, S. C. 2018. Efficacy of organic acid electrostatic spray for decontaminating *Salmonella* on cantaloupe cubes and cherry tomatoes. *Journal of Food Processing and Preservation* 42(10), 13748.

Meireles, A., Giaouris, E. and Simoes, M. 2016. Alternative disinfection methods to chlorine for use in the fresh-cut industry. *Food Research International* 82, 71–85.

Miller, F. A., Silva, C. L. M. and Brandao, T. R. S. 2013. A review on ozone-based treatments for fruit and vegetables preservation. *Food Engineering Reviews* 5(2), 77–106.

Misra, N. N., Tiwari, B. K., Raghavarao, K. S. M. S. and Cullen, P. J. 2011. Nonthermal plasma inactivation of food-borne pathogens. *Food Engineering Reviews* 3(3–4), 159–170.

Mokhtari, A., Pang, H., Farakos, S. S., Davidson, G. R., Williams, E. N. and Van Doren, J. M. 2021. Evaluation of potential impacts of free chlorine during washing of fresh-cut leafy greens on *Escherichia coli* O157:H7 cross-contamination and risk of illness. *Risk Analysis*, September 15, 34528270.

Mundi, G., Zytner, R. G., Warriner, K., Bonakdari, H. and Gharabaghi, B. 2021. Machine learning models for predicting water quality of treated fruit and vegetable Wastewater. *Water* 13(18), 2485.

Mundi, G. S., Zytner, R. G., Warriner, K. and Gharabaghi, B. 2017. Predicting fruit and vegetable processing wash-water quality. *Water Science and Technology: A Journal of the International Association on Water Pollution Research* 2017(1), 256–269.

Murray, K., Aldossari, H., Wu, F. and Warriner, K. 2018a. Dynamic changes in free-chlorine levels within a commercial post-harvest wash and prevention of cross-contamination between shredded lettuce batches. *Food Control* 85, 127–134.

Murray, K., Moyer, P., Wu, F., Goyette, J. B. and Warriner, K. 2018b. Inactivation of *Listeria monocytogenes* on and within apples destined for caramel apple production by using sequential forced air ozone gas followed by a continuous advanced oxidative process treatment. *Journal of Food Protection* 81(3), 357–364.

Murray, K., Wu, F., Aktar, R., Namvar, A. and Warriner, K. 2015. Comparative study on the efficacy of bacteriophages, sanitizers, and UV light treatments to control

Listeria monocytogenes on sliced mushrooms (*Agaricus bisporus*). *Journal of Food Protection* 78(6), 1147–1153.

Murray, K., Wu, F., Shi, J., Jun Xue, S. and Warriner, K. 2017. Challenges in the microbiological food safety of fresh produce: Limitations of post-harvest washing and the need for alternative interventions. *Food Quality and Safety* 1(4), 289–301.

Nassarawa, S. S., Abdelshafy, A. M., Xu, Y. Q., Li, L. and Luo, Z. S. 2021. Effect of light-emitting diodes (LEDs) on the quality of fruits and vegetables during postharvest period: A review. *Food and Bioprocess Technology* 14(3), 388–414.

National Advisory Committee on Microbiological Criteria for Foods. 2010. Parameters for determining inoculated pack/challenge study protocols. *Journal of Food Protection* 73(1), 140–202.

Niemira, B. A. 2012. Cold plasma decontamination of foods. In: Doyle, M. P. and Klaenhammer, T. R. (Eds.), *Annual Review of Food Science and Technology* (Vol. 3). Palo Alto: Annual Reviews.

Niveditha, A., Pandiselvam, R., Prasath, V. A., Singh, S. K., Gul, K. and Kothakota, A. 2021. Application of cold plasma and ozone technology for decontamination of *Escherichia coli* in foods - A review. *Food Control* 130, 108338.

Nubling, S., Hagele, F., Wohlt, D., Graf, B., Schweiggert, R. M., Carle, R., Schmidt, H. and Weiss, A. 2017. Effects of Quillaja Saponaria extract and N-alpha-lauroyl-L-arginine ethyl ester on reducing selected foodborne pathogens in vitro and maintaining quality of fresh-cut endive (*Cichorium endivia* L.) at pilot plant scale. *Food Control* 73, 393–400.

Oh, S. and Liu, Y. B. 2020. Effectiveness of nitrogen dioxide fumigation for microbial control on stored almonds. *Journal of Food Protection* 83(4), 599–604.

Olaimat, A. N. and Holley, R. A. 2012. Factors influencing the microbial safety of fresh produce: A review. *Food Microbiology* 32(1), 1–19.

Ortiz-Sola, J., Abadias, I., Colas-Meda, P., Anguera, M. and Vinas, I. 2021. Inactivation of *Salmonella enterica*, *Listeria monocytogenes* and murine Norovirus (MNV-1) on fresh strawberries by conventional and water-assisted ultraviolet light (UV-C). *Postharvest Biology and Technology* 174, 8.

Paczesny, J. and Bielec, K. 2020. Application of bacteriophages in nanotechnology. *Nanomaterials* 10(10).

Pan, Y. Y., Cheng, J. H. and Sun, D. W. 2019. Cold plasma-mediated treatments for shelf Life extension of fresh produce: A review of recent research developments. *Comprehensive Reviews in Food Science and Food Safety* 18(5), 1312–1326.

Pankaj, S. K. and Keener, K. M. 2017. Cold plasma: Background, applications and current trends. *Current Opinion in Food Science* 16, 49–52.

Park, S. H., Kang, J. W. and Kang, D. H. 2018. Inactivation of foodborne pathogens on fresh produce by combined treatment with UV-C radiation and chlorine dioxide gas, and mechanisms of synergistic inactivation. *Food Control* 92, 331–340.

Parker, J. S., Wilson, R. S., Lejeune, J. T. and Doohan, D. 2012. Including growers in the "food safety" conversation: Enhancing the design and implementation of food safety programming based on farm and marketing needs of fresh fruit and vegetable producers. *Agriculture and Human Values* 29(3), 303–319.

Perni, S., Shama, G. and Kong, M. G. 2008. Cold atmospheric plasma disinfection of cut fruit surfaces contaminated with migrating microorganisms. *Journal of Food Protection* 71(8), 1619–1625.

Petri, E., Virto, R., Mottura, M. and Parra, J. 2021. Comparison of peracetic acid and chlorine effectiveness during sresh-cut vegetable processing at industrial scale. *Journal of Food Protection* 84(9), 1592–1602.

Pillai, S. D. and Shayanfar, S. 2018. Electron beam processing of fresh produce - A critical review. *Radiation Physics and Chemistry* 143, 85–88.

Pinela, J. and Ferreira, I. C. 2017. Nonthermal physical technologies to decontaminate and extend the shelf-life of fruits and vegetables: trends aiming at quality and safety. *Critical Reviews in Food Science and Nutrition* 57(10), 2095–2111.

Qi, H. and Hung, Y. C. 2019. Effectiveness of activated persulfate in removal of foodborne pathogens from romaine lettuce. *Food Control* 106, 106708.

Qi, H., Wang, L., Huang, Q. G. and Hung, Y. C. 2020. Effect of organic load on the efficacy of activated persulfate in inactivating *Escherichia coli* O157:H7 and the production of halogenated by-products. *Food Control* 114, 107218.

Qian, J., Zhuang, H., Nasiru, M. M., Muhammad, U., Zhang, J. and Yan, W. 2019. Action of plasma-activated lactic acid on the inactivation of inoculated *Salmonella* enteritidis and quality of beef. *Innovative Food Science and Emerging Technologies* 57, 102196.

Rana, S., Mehta, D., Bansal, V., Shivhare, U. S. and Yadav, S. K. 2020. Atmospheric cold plasma (ACP) treatment improved in-package shelf-life of strawberry fruit. *Journal of Food Science and Technology* 57(1), 102–112.

Rock, C. M., Brassill, N., Dery, J. L., Carr, D., Mclain, J. E., Bright, K. R. and Gerba, C. P. 2019. Review of water quality criteria for water reuse and risk-based implications for irrigated produce under the FDA food safety modernization act, produce safety rule. *Environmental Research* 172, 616–629.

Rossow, M., Ludewig, M. and Braun, P. G. 2018. Effect of cold atmospheric pressure plasma treatment on inactivation of *Campylobacter jejuni* on chicken skin and breast fillet. *LWT* 91, 265–270.

Rouge, V., Von Gunten, U. and Allard, S. 2020. Efficiency of pre-oxidation of natural organic matter for the mitigation of disinfection byproducts: Electron donating capacity and UV absorbance as surrogate parameters. *Water Research* 187(10), 116418.

Royintarat, T., Choi, E. H., Boonyawan, D., Seesuriyachan, P. and Wattanutchariya, W. 2020. Chemical-free and synergistic interaction of ultrasound combined with plasma-activated water (PAW) to enhance microbial inactivation in chicken meat and skin. *Scientific Reports* 10(1), 1559.

Sharma, M. and Reynnells, R. 2016. Importance of soil amendments: Survival of bacterial pathogens in manure and compost used as organic fertilizers. *Microbiology Spectrum* 4(4), 27726763.

Sheng, L. N., Hanrahan, I., Sun, X. F., Taylor, M. H., Mendoza, M. and Zhu, M. J. 2018. Survival of *Listeria innocua* on Fuji apples under commercial cold storage with or without low dose continuous ozone gaseous. *Food Microbiology* 76, 21–28.

Shynkaryk, M. V., Pyatkovskyy, T. I., Yousef, A. E. and Sastry, S. K. 2016. Gaseous ozone treatment of baby spinach within the existing production chain for inactivation of *Escherichia coli* O157:H7. *Journal of Food Engineering* 191, 10–18.

Singh, H., Bhardwaj, S. K., Khatri, M., Kim, K. H. and Bhardwaj, N. 2021. UVC radiation for food safety: An emerging technology for the microbial disinfection of food products. *Chemical Engineering Journal* 417, 128084.

Skog, L. J. and Chu, C. L. 2001. Effect of ozone on qualities of fruits and vegetables in cold storage. *Canadian Journal of Plant Science* 81(4), 773–778.

Smith, D. J., Ernst, W. and Herges, G. R. 2015. Chloroxyanion residues in cantaloupe and tomatoes after chlorine dioxide gas sanitation. *Journal of Agricultural and Food Chemistry* 63(43), 9640–9649.

Solomon, E. B. and Matthews, K. R. 2006. Interaction of live and dead *Escherichia coli* O157:H7 and fluorescent microspheres with lettuce tissue suggests bacterial processes do not mediate adherence. *Letters in Applied Microbiology* 42(2), 88–93.

Song, S. and Wood, T. K. 2021. 'Viable but non-culturable cells' are dead. *Environmental Microbiology* 23(5), 2335–2338.

Soon, J. M. and Baines, R. N. 2012. Food safety training and evaluation of handwashing intention among fresh produce farm workers. *Food Control* 23(2), 437–448.

Stannard, C. J., Abbiss, J. S. and Wood, J. M. 1983. Combined treatment with hydrogen peroxide and ultra-violet irradiation to reduce microbial contamination levels in pre-formed food packaging cartons. *Journal of Food Protection* 46(12), 1060–1064.

Sy, K. V., Murray, M. B., Harrison, M. D. and Beuchat, L. R. 2005. Evaluation of gaseous chlorine dioxide as a sanitizer for killing *Salmonella*, *Escherichia coli* O157:H7, *Listeria monocytogenes*, and yeasts and molds on fresh and fresh-cut produce. *Journal of Food Protection* 68(6), 1176–1187.

Tabakoglu, N. and Karaca, H. 2015. Enhanced degradation of azoxystrobin in grapes and model systems by ozone fumigation during Storage. *Ozone: Science and Engineering* 37(6), 479–488.

Tak, S. and Vellanki, B. P. 2020. Comparison of O-3-BAC, UV/H2O2-BAC, and O-3/H_2O_2-BAC treatments for limiting the formation of disinfection byproducts during drinking water treatment in India. *Journal of Environmental Chemical Engineering* 8(5), 104434.

Teng, Z., Luo, Y. G., Alborzi, S., Zhou, B., Chen, L., Zhang, J. L., Zhang, B. C., Millner, P. and Wang, Q. 2018. Investigation on chlorine-based sanitization under stabilized conditions in the presence of organic load. *International Journal of Food Microbiology* 266, 150–157.

Tirado, M. C., Clarke, R., Jaykus, L. A., Mcquatters-Gollop, A. and Franke, J. M. 2010. Climate change and food safety: A review. *Food Research International* 43(7), 1745–1765.

Tomas-Callejas, A., Lopez-Velasco, G., Valadez, A. M., Sbodio, A., Artes-Hernandez, F., Danyluk, M. D. and Suslow, T. V. 2012. Evaluation of current operating standards for chlorine dioxide in disinfection of dump tank and flume for fresh tomatoes. *Journal of Food Protection* 75(2), 304–313.

Trajer, J., Winiczenko, R. and Drozdz, B. 2021. Analysis of water consumption in fruit and vegetable processing plants with the use of artificial intelligence. *Applied Sciences* 11(21), 10167.

Trinetta, V., Morgan, M. T. and Linton, R. H. 2010. Use of high-concentration-short-time chlorine dioxide gas treatments for the inactivation of *Salmonella enterica* spp. inoculated onto roma tomatoes. *Food Microbiology* 27(8), 1009–1015.

Trinetta, V., Vaidya, N., Linton, R. and Morgan, M. 2011. Evaluation of chlorine dioxide gas residues on selected food produce. *Journal of Food Science* 76(1), T11–T15.

Tudela, J. A., Lopez-Galvez, F., Allende, A. and Gil, M. I. 2019. Chlorination management in commercial fresh produce processing lines. *Food Control* 106, 106760.

Turner, K., Moua, C. N., Hajmeer, M., Barnes, A. and Needham, M. 2019. Overview of leafy greens-related food safety incidents with a California link: 1996 to 2016. *Journal of Food Protection* 82(3), 405–414.

Van Haute, S., Luo, Y., Sampers, I., Mei, L., Teng, Z., Zhou, B., Bornhorst, E. R., Wang, Q. and Millner, P. 2018. Can UV absorbance rapidly estimate the chlorine demand in wash water during fresh-cut produce washing processes? *Postharvest Biology and Technology* 142, 19-27.

Van Haute, S., Zhou, B., Luo, Y. G., Sampers, I., Vanhaverbeke, M. and Millner, P. 2019. The use of redox potential to estimate free chlorine in fresh produce washing operations: Possibilities and limitations. *Postharvest Biology and Technology* 156, 110957.

Varanasi, L., Coscarelli, E., Khaksari, M., Mazzoleni, L. R. and Minakata, D. 2018. Transformations of dissolved organic matter induced by UV photolysis, hydroxyl radicals, chlorine radicals, and sulfate radicals in aqueous-phase UV-Based advanced oxidation processes. *Water Research* 135, 22–30.

Vurma, M., Pandit, R. B., Sastry, S. K. and Yousef, A. E. 2009. Inactivation of *Escherichia coli* O157:H7 and natural microbiota on spinach leaves using gaseous ozone during vacuum cooling and simulated transportation. *Journal of Food Protection* 72(7), 1538–1546.

Wang, H., Hasani, M., Wu, F., Prossor, R., Basto-Machado, G. B. and Warriner, K. 2022. Hydroxyl-radical activated water for inactivation of *Escherichia coli* O157:H7, *Salmonella* and *Listeria monocytogenes* on germinating mung beans. *International Journal of Food Microbiology* 367, 109587.

Wang, H. R., Hasani, M., Wu, F. and Warriner, K. 2022. Pre-oxidation of spent lettuce wash water by continuous advanced oxidation process to reduce chlorine demand and cross-contamination of pathogens during post-harvest washing. *Food Microbiology* 103, 103937.

Wang, J. Y., Lei, Y., Yu, Y. G., Yin, L. B. and Zhang, Y. Y. 2021. Use of acetic acid to partially replace lactic acid for decontamination against *Escherichia coli* O157:H7 in fresh produce and mechanism of action. *Foods* 10(10), 2406.

Wang, J. Y., Tao, D. B., Wang, S., Li, C., Li, Y. R., Zheng, F. G. and Wu, Z. X. 2019. Disinfection of lettuce using organic acids: An ecological analysis using 16S rRNA sequencing. *RSC Advances* 9(30), 17514–17520.

Wang, Q. Y. and Salvi, D. 2021. Recent progress in the application of plasma-activated water (PAW) for food decontamination. *Current Opinion in Food Science* 42, 51–60.

Warriner, K. and Hasani, M. 2020a. Post-harvest risk management of biological hazards encountered in horticultureal produce. In: Olatunji, C., East, A. R., Delong, J., Prange, R. and Brandenburg, J. S. (Eds.), *Advances in Postharvest Management of Horticulture Produce* 1, 251–266. Burleigh Dodds Science Publishing.

Warriner, K. and Hasani, M. 2020b. UV technology lights the way to plant vaccines: Boosting plant defenses to extend shelf life of fruit and vegetables. *UV Solutions,* Vol. 5. https://uvsolutionsmag.com/articles/2020/uv-technology-lights-the-way-to-plant-vaccines-boosting-plant-defenses-to-extend-shelf-life-of-fruits-and-vegetables/. Accessed 5 January 2022.

Warriner, K., Wang, H. and Hasani, M. 2021. Gas-phase advanced oxidation process for surface disinfection of foods and food contact surfaces. In: Knoerzer, K. and Muthukumarappan, K. (Eds.), *Innovative Food Processing Technologies: A Comprehensive Review*. New York: Academic Press, Vol. 1, pp. 317-334.

Wood, T. K., Knabel, S. J. and Kwan, B. W. 2013. Bacterial persister cell formation and dormancy. *Applied and Environmental Microbiology* 79(23), 7116-7121.

Xiang, Y. Y., Gonsior, M., Schmitt-Kopplin, P. and Shang, C. 2020. Influence of the UV/H_2O_2 advanced oxidation process on dissolved organic matter and the connection between elemental composition and disinfection byproduct formation. *Environmental Science and Technology* 54(23), 14964–14973.

Xie, Y., Hajdok, C., Mittal, G. S. and Warriner, K. 2008. Inactivation of MS2 F (+) coliphage on lettuce by a combination of UV light and hydrogen peroxide. *Journal of Food Protection* 71(5), 903–907.

Zeghioud, H., Nguyen-Tri, P., Khezami, L., Amrane, A. and Assadi, A. A. 2020. Review on discharge plasma for water treatment: Mechanism, reactor geometries, active species and combined processes. *Journal of Water Process Engineering* 38, 101664.

Zhao, Y. M., Patange, A., Sun, D. W. and Tiwari, B. 2020. Plasma-activated water: Physicochemical properties, microbial inactivation mechanisms, factors influencing antimicrobial effectiveness, and applications in the food industry. *Comprehensive Reviews in Food Science and Food Safety* 19(6), 3951–3979.

Zhou, B., Luo, Y. G., Nou, X. W. and Millner, P. 2014. Development of an algorithm for feed-forward chlorine dosing of lettuce wash operations and correlation of chlorine profile with *Escherichia coli* O157:H7 inactivation. *Journal of Food Protection* 77(4), 558–566.

Zhou, F. H., Harmon, K. M., Yoon, K. J., Olson, D. G. and Dickson, J. S. 2011. Inactivation of feline calicivirus as a surrogate for Norovirus on lettuce by electron beam irradiation. *Journal of Food Protection* 74(9), 1500–1503.

Ziuzina, D., Han, L., Cullen, P. J. and Bourke, P. 2015. Cold plasma inactivation of internalised bacteria and biofilms for *Salmonella enterica* serovar Typhimurium, *Listeria monocytogenes* and *Escherichia coli*. *International Journal of Food Microbiology* 210, 53–61.

Ziuzina, D., Petil, S., Cullen, P. J., Keener, K. M. and Bourke, P. 2014. Atmospheric cold plasma inactivation of *Escherichia coli*, *Salmonella enterica* serovar Typhimurium and *Listeria monocytogenes* inoculated on fresh produce. *Food Microbiology* 42, 109–116.

Chapter 10

Developments in packaging techniques and their assessment for assuring the safety of fresh produce

Jinhe Bai, Gabriela Maria Olmedo and Xiuxiu Sun, USDA-ARS, USA

1 Introduction

The packaging of fruits and vegetables plays a crucial role in their journey from farm to table. Fresh produce is highly perishable and undergoes various metabolic processes such as ripening and aging. Also, during this journey, fresh produce can become contaminated with microorganisms. Since fresh produce is often consumed raw, there is a risk of foodborne illnesses if the produce is contaminated with pathogens such as *Salmonella*, *Escherichia coli* O157:H7 and O104:H4, and *Listeria monocytogenes*. Packaging is used to slow down the aging process and maintain the quality of the produce. Moreover, packaging provides protection from contamination to fresh produce and the incorporation of antimicrobial compounds into packaging further enhances food safety.

Traceability is a critical component of food safety and packaging plays an important role in this regard. It helps to ensure that all stakeholders along the supply chain comply with food safety requirements.

http://dx.doi.org/10.19103/AS.2023.0121.17
Published by Burleigh Dodds Science Publishing Limited, 2023.

2 Packaging for physical protection

Packaging serves to safeguard produce from physical harm throughout storage, transportation, and marketing. It is crucial for preventing mechanical damage caused by shock, vibration, compression, and bruising. To achieve this, the packaging must be strong enough to withstand handling without sustaining damage. Technological advancements have resulted in the evolution of packaging materials and designs over time. For instance, while woven baskets were once used to transport produce, wooden or plastic bins with up to 1000 kg capacities are now common for storing fruit like apples.

Research conducted almost a century ago reported that wooden bins can be contaminated with fragments of decay fungi, resulting in fruit surface contamination (Heald and Ruehle, 1931). These fungi include *Alternaria* spp. (69–73% of bins, same as below), *Cladosporium* spp. (25–46%), *Mucor piriformis* (19–21%), and *Penicillium* spp. (77–83%) in the 1988 and 1989 seasons in Oregon (Spotts and Cervantes, 1994). Studies have been conducted in various regions and years to disinfect the bins, such as the use of 150 ppm of NaOCl to reduce 90% of *Penicillium expansum* (Baker and Heald, 1932) or their exposure to steam for 2 min (Wellman and Heald, 1938). In the 1970s and 1980s, many fungicides and chemicals, including dichloran, benomyl, thiabendazole, captan, glyodin, zineb, darisan, hitor, and combinations of these agents, were used to control fungi contamination to some extent (Spotts and Cervantes, 1994).

Plastic bins are often preferred over wooden ones because they are lighter, do not absorb water or chemicals, allow for better air circulation, are easier to clean and sanitize, and are more effective in killing *P. expansum* spores when disinfected with NaOCl (Spotts and Cervantes, 1994). Another disadvantage of wood bins is that they break easily, so the sharp broken pieces can cause cuts, punctures, or bruises on the fruit (Timm et al., 1998). However, it is important to note that the sanitation problems associated with wooden surfaces are primarily due to difficulties in cleaning, not because of their rough or porous nature (López-Gálvez et al., 2021). In fact, the porous nature of wood often creates unfavorable conditions for microorganisms, and wood has the unique ability to produce antimicrobial components that can inhibit or limit the growth of pathogenic microorganisms (Aviat et al., 2016). Figure 1 depicts a blueberry pallet and enhanced packaging.

Throughout the various stages of the fruit and vegetable supply chain, including harvesting, handling, packaging, transportation, and retail, packing boxes or crates made of plastic, corrugated fiberboard, or wood are commonly utilized. The transfer of microorganisms between packaging surfaces and produce, as well as vice versa, is impacted by multiple factors such as the microbe type, inoculation method, contamination level, surface and produce type, temperature, moisture, contact duration, ventilation, and pressure (Villegas

Published by Burleigh Dodds Science Publishing Limited, 2023.

Figure 1 Blueberry pallet for storage and transportation. Berries are packaged in clear plastic clamshells, with proper ventilation to maintain freshness without dehydration or anaerobic conditions. These clamshells are then grouped in sets of 6–12 and placed in a corrugated fiberboard flat, also designed with ventilation. The flats are then stacked on top of a wood pallet. To ensure protection during transportation, corrugated fiberboard covers are added to cover several layers of the pallet, along with corrugated fiberboard corner reinforcements and bindings.

et al., 2020; Newman et al., 2017; López-Gálvez et al., 2021). Several studies have suggested that bacterial biofilms tend to form more readily on hydrophilic surfaces, such as glass and stainless steel, than on hydrophobic surfaces, such as plastic or Buna N rubber (Abeysundara et al., 2018). For example, Patrignani et al. (2016) showed a higher transfer of bacteria from reusable plastic crates to peaches than from single-use cardboard, hypothesizing that this difference was caused by the higher entrapment capability of cardboard. Similarly, Aviat et al. (2020) found a higher transfer of *E. coli* to apples from polypropylene (PP) surfaces compared to wood and cardboard surfaces and attributed this to the ability of microorganisms to form biofilms on plastic surfaces. Siroli et al. (2017) also confirmed that the risk of spoilage and pathogenic microbial cross-contamination is higher via plastic surfaces than via cardboard surfaces. While reusable plastic containers are increasingly being used to replace single-use cardboard and wooden containers for environmental sustainability, several studies have raised concerns about their hygienic status and potential role as a source of microbiological contamination (López-Gálvez et al., 2021; Abejón et al., 2020; Zhu et al., 2020).

To prevent physical damage to fruits, cushion, or filler materials, such as styrofoam fruit nets, modeled fruit trays made of pulp fiber, expanded polyethylene foam (EPE), styrofoam, plastic (polypropylene (PP), polystyrene

Published by Burleigh Dodds Science Publishing Limited, 2023.

Figure 2 Examples of cushion and filler. (Left) Apples in expanded polyethylene (EPE) foam nets. (Right) Pears on a pulp fiber tray.

(PS), oriented polystyrene (OPS), polyethylene terephthalate (PET)), and paper wrap are frequently used in addition to outer packaging (Fig. 2) (Berardinelli et al., 2001; Patil et al., 2011).

Over the course of several decades, apples and pears have been wrapped in individual paper packaging that contains a combination of petroleum oils, fungicides, antioxidants, and minerals. The purpose of these wraps is to address a variety of issues related to postharvest decay and disorders. Originally, paper wraps were used to protect the fruit during transport and to prevent the spread of decay. In the early 1900s, it was suggested that oil-impregnated paper could help reduce superficial scald, a physiological disorder affecting the fruit (Brooks, 1920). Later, Cooley and Crenshaw (1931) recommended adding copper sulfate to the paper wrap to prevent rot from spreading. In 1957, Smock (1957) suggested the use of ethoxyquin or diphenylamine to control superficial scald in apple and pear fruit. Since then, incorporating ethoxyquin into paper wraps has been found to be effective in controlling superficial scald in pears (Drake et al., 2004). Additionally, paper wraps containing potassium sorbate and potassium metabisulfite have also been shown to inhibit the growth of *Candida pelliculosa* in peaches and plums (Rudra et al., 2013).

3 Modified atmosphere packaging

Modified atmosphere packaging (MAP) is the most commonly used method for preserving the freshness of fruits and vegetables. It involves modifying the gas composition within the packaging to reduce water loss, slow down senescence, and prevent microbial contamination. The gas mixture in the packaging can be regulated through macro- or micro-perforation and film gas permeability.

After harvest, fruits and vegetables continue to respire by two types of respiration: aerobic and anaerobic. Both pathways produce carbon dioxide

(CO_2), which can impact produce quality, shelf life, and marketability (Ayhan, 2019). The aim of MAP is to slow down the respiration process, inhibit spoilage, and keep food attractive and edible for a longer period. MAP systems involve the management of the product surrounding the atmosphere and package gas permeability. When gases, conditions, and packaging materials are appropriately chosen, products exhibit a reduction in physiological and biochemical processes of ripening, disease incidence, senescence, and ethylene production. This helps to keep quality, freshness, and microbial safety (Ghidelli and Pérez-Gago, 2018). Generally, MAP provides an environment with low oxygen (O_2) and high CO_2 levels (Qu et al., 2022). The most suitable combination of gases and their concentrations should be adjusted for each specific product (Belay et al., 2019; Zhang et al., 2015; Ghidelli and Pérez-Gago, 2018). For instance, Table 1 shows how different O_2 and carbon combinations at several storage temperatures impact shelf life of whole and fresh-cut produce.

3.1 History and scope of modified atmosphere packing concept

The practice of modifying product surroundings to maintain or extend the quality of food can be traced back at least 2000 years when underground storage of food provoked an unintentional but effective modified atmosphere. There is also evidence of the practice of sealing food in clay containers together with fresh leaves by ancient civilizations in China and Greece. This practice helped to delay fruit ripening by quickly modifying the atmosphere within the container (Beaudry, 2007; Floros and Matsos, 2005). However, the first studies on the effect of the atmosphere on fruit ripening were not performed until 1820. Before 1930, it was demonstrated that lamb, pork, and fish maintained freshness for significantly longer periods in a saturated CO_2 atmosphere compared with storage in air at cold temperatures (Robertson, 2005). After decades of research, MAP packages were introduced in Mexican stores for fish and bacon during the 1970s for the first time (Blakistone, 1999). Since then, MAP technology has been thoroughly developed and improved, gaining acceptance among consumers.

Within the wide range of food preservative alternatives, MAP emerges as a proposal with many advantages and disadvantages to consumers and producers. For consumers, it offers products with an extended shelf life maintaining organoleptic and nutritional quality. At the same time, it reduces the need for chemical preservatives, leading to more natural products. For producers, shelf life extension achieved by MAP enables more routes and destinations to be exploited. At the same time, this technology exhibits several disadvantages, most of them related to transport and logistics.

Table 1 Recent modified atmosphere packaging applications in whole and fresh-cut fruits and vegetables

Fruit	Gas combination		Temperature (°C)	Shelf life (days)	Reference
	O_2	CO_2			
Green asparagus	10%	5%	2	16–20	An et al. (2006)
	18%	4%	4	30	Toscano et al. (2021)
Broccoli	3%	8%	3	35	Tano et al. (2007)
	30%	70%	10 ± 1	24	Wang et al. (2020)
Cauliflower	3%	5%	4 ± 1	30	Alden et al. (2019)
Carrot	14 ± 1%	6.7 ± 0.6%	4 ± 1	14	Cozzolino et al. (2021)
Sliced carrot	5%	5%	5 ± 2	13	Alasalvar et al. (2005)
Celery	18 kPa	3 kPa	4	>10	
Celery sticks	6 kPa	7 kPa	5	15	Gómez and Artés (2005)
Fresh-cut peppers	80/50 kPa	15 kPa	5	9–10	Conesa et al. (2007)
Cucumber	5%	5%			
Sea asparagus	4%	5%	2	28–32	Lu et al. (2010)
Summer truffles	7%	15%	4	21	Rivera et al. (2010)
White mushrooms	5%	4%	4 ± 1	15	Tao et al. (2006)
Shiitake mushroom	2%	10%	4	17	Ye et al. (2012)
Agrocybe chaxingu	5%	10%	3 ± 1	20	Jiang et al. (2010)
Kohlrabi	5 kPa	10–15 kPa	0	14	Escalona et al. (2007a)
Fresh-cut kohlrabi	7 kPa	9 kPa	0	14	Escalona et al. (2007b)
Spinach	10%	9%	7	7	Tudela et al. (2013)
Fresh-cut apple	5%	5%	5 ± 2	42	Kumar and Sethi (2021)
Fresh-cut cantaloupe	4 kPa	10 kPa	5	9	Bai et al. (2001)
Fresh-cut honeydew	5 kPa	5 kPa	5	11	Bai et al. (2003)

Fresh-cut melon	70%	0	5	10-14	Oms-Oliu et al. (2008a)
Fresh-cut jackfruit	3%	6%	6	50	Saxena et al. (2013)
Jujube	14.5 kPa	3.86 kPa	5	49	Reche et al. (2019)
Plum	10%	4%	0	56	Wang et al. (2021)
Peach	10%	0	4	90	Oliveira et al. (2015)
Cherry	16%	20%	5	21	Cozzolino et al. (2019)
	5%	10%	0 ± 1	63	Yang et al. (2019)
Apricot	5%	10%	4	42	Muftuoğlu et al. (2012)
Grape	2%	5%	0 ± 0.5	>45	Admane et al. (2018)
Persimmon	20 kPa	2 kPa	5	30	Liamnimitr et al. (2018)
Papaya	3-5%	6-9%	10	25	Gonzalez-Aguilar et al. (2003)
Fresh-cut papaya	5%	10%	5	25	Waghmare and Annapure (2013)
Pomegranate	5%	10%	6	100	Selcuk and Erkan (2014)
Fresh-cut pear	2.5%	7%	4	14	Oms-Oliu et al. (2008b)
Blueberry	10 kPa	40 kPa	4	15	Pinto et al. (2020)
	4.5-6%	13-15%	0	45	Rodriguez and Zoffoli (2016)
Raspberry	3%	5%	7	7	Siro et al. (2006)
Strawberry	5-10%	10-15%	4 ± 1	18	Zhao et al. (2019)
	3-7%	1.2-4.2%	5	22	Yun et al. (2017)

MAP involves larger package volumes requiring more space in containers for transportation and retail. Also, MAP involves the use of specialized and expensive equipment and trained operators and the need to determine the best gas formulation for each product (Floros and Matsos, 2005). For this reason, R&D departments throughout the world are constantly searching for ways to reduce the initial implementation costs of this type of technology (Panea and Ripoll, 2019).

It is worth noting that, despite sometimes being used synonymously, controlled atmosphere packaging (CAP) and MAP are different technologies. In MAP, the product surrounding atmosphere is altered initially and then allowed to change over time, in response to product and microbial metabolism. On the other hand, in CAP, often included in the category of intelligent packaging, the atmosphere is initially altered and then maintained during the life of the package. Atmosphere deviations are monitored and can be adjusted using scavengers, sensors, and other specific tools (Brody, 2005). The controlled release will be deeply analyzed in the 'active packaging' section. Here, MAP types, packaging materials, methods to obtain the desired atmosphere, cases of success in fruit and vegetable preservation, and synergy with other technologies will be addressed.

3.2 Types of modified atmosphere packing

The atmosphere within the package can be modified passively or actively (Ghidelli and Pérez-Gago, 2018).

In passive MAP (pMAP), atmosphere modification occurs as a consequence of product respiration and/or the metabolism of microorganisms associated with food and the permeation of gases through packaging materials. This method is commonly used for fresh respiring fruits and vegetables. On the other hand, active MAP (aMAP) consists of gas flushing or displacing air with a controlled and desired mixture of gases to modify the atmosphere within the package before sealing. Among numerous differences between these two methods, the main disadvantage of the passive atmosphere modification method is that the desired atmosphere is achieved slowly, which sometimes results in detrimental effects on product quality. Active modification of the atmosphere is exempt from this problem since atmosphere modification is practically instantaneous (Floros and Matsos, 2005).

The choice between active and passive methods depends on the type of product. For some products, pMAP is more appropriate than aMAP, and vice versa. In the case of fresh and minimally processed fruits and vegetables, regardless of which MAP method is intended to use, the equilibrated internal atmosphere depends on the equilibrium between the respiration rate of produce and permeability of the packaging materials; that is, it always results in pMAP (Mullan and McDowell, 2011).

3.3 Packaging materials: criteria and requisites

Selection and design of the most appropriate packaging materials is essential to maintain the quality and safety of food, with flexible and semi-rigid plastics and plastic laminates being the most common ones. Among the properties that make a plastic suitable for MAP, light weight, good clarity, heat sealing, easy forming, and strength are the most important. MAP films are usually multilayer structures formed from several layers of different plastics. This allows for designing a material that matches the requirements of a MAP system. The most common materials used for this purpose are polyethylene (PE), PP, polyamide (PA/nylons), PET, polyvinyl chloride (PVC), polyvinylidene chloride, ethylene vinyl alcohol unplasticized PVC, and expanded PS. The forms of flexible packages include bags, pouches, pillow packs, and top webs in sealed tray systems. On the other hand, rigid and semi-rigid packages are represented by base trays, dishes, cups, and tubs.

The selection of plastic packaging materials must be performed while considering several factors:

- Food contact approval by the agencies that monitor public health in the market to which the products are intended to be commercialized: Suppliers ought to be able to demonstrate that the package does not transfer components from the packaging to the food product in amounts that could harm the consumer.
- Material transmission rates for gases and water, defined as the quantity of gas (or other permeant) passing through a material of a certain area over time. In general, the lower the transmission rate value, the better the gas barrier properties of the material. This measure must be considered together with temperature and relative humidity (RH). No single commercially available plastic provides a total barrier to gases and vapors, and therefore, the materials are selected on the basis of product type, desired shelf life, gas composition, availability, and cost.
- Other: Optical properties (gloss, transparency, color); antifogging properties (condensation on the inner surface leads to poor product visibility and an aesthetically unpleasing appearance of the pack); mechanical properties (such as resistance to tearing and puncture, good machine handling characteristics, ability of layers to bond effectively during the packaging and subsequent storage and handling); and heat sealing properties (Mullan and McDowell, 2011).

3.4 Creation of a modified atmosphere: methods, gases selection and measurements

There are two main techniques to create a modified atmosphere in an active way. One of them is gas flushing. This method consists of a continuous gas stream that flushes air out of the package prior to sealing. The main advantage of this method is that it consists of a simple and rapid operation, and, in consequence, machines pack at a high rate. Gas flushing is not suitable for oxygen-sensitive food products since the method is not completely effective at flushing air out. The other aMAP method is vacuum - gas flushing, which consists of a two-step process: (1) the evacuation stage, where air is pulled out by vacuuming and (2) the gas flushing stage, flushing the chosen modified gas mix into the package. While this method achieves lower residual O_2 levels than the first one, it results in a lower packaging rate (Mullan and McDowell, 2011).

There is not a single effective MAP recipe. For instance, for inanimate fresh animal food, microbes can be controlled by increasing CO_2 and reducing O_2. However, under these same conditions, inefficient control of fungi and bacteria growth is observed in fruits and vegetables. This is explained by the continued respiration process after the harvest of these kinds of products (Qu et al., 2022). In addition to this, the choice of gas is also dependent upon the legal regulations of the region in which products are commercialized. For instance, the three main gases used in Europe are O_2, CO_2, and nitrogen (N_2), while carbon monoxide (CO), widely used in the USA, is currently not approved in the European Union (Mullan and McDowell, 2011).

The role and importance of each particular gas in MAP are related to its properties. O_2 promotes several deteriorative reactions in food including fat oxidation, browning reactions, and pigment oxidation. Moreover, most of the common spoilage bacteria and fungi require O_2 for growth. Therefore, to increase the shelf life, the pack atmosphere should contain a low concentration of this gas. Nevertheless, in some products, a low concentration of O_2 can result in undesirable color changes, senescence, and growth of anaerobic pathogenic bacteria. CO_2 is a colorless gas with a slightly pungent odor at very high concentrations. It has been reported that CO_2 is effective against psychrotrophs, and, in general, its activity is markedly greater at temperatures below 10°C than at higher ones (King and Nagel, 1975). N_2 is an inert and tasteless gas, without any antimicrobial activity. Nevertheless, the growth of aerobic spoilage organisms is inhibited or stopped by flushing N_2 to remove O_2. It is also effective in preventing package collapse. CO is a colorless, tasteless, and odorless gas with proved fungistatic action. Its ability to inhibit the browning of cut and physically damaged vegetable tissue has led to its use in preventing browning in leafy vegetables. Its fungistatic properties have also been employed to inhibit mold growth (Kader, 1983). Finally, noble gases, such as helium, argon, xenon, and neon, are sometimes included in MAP gas mixtures; however, literature on their application and benefits is limited (Mullan and McDowell, 2011).

3.5 Modified humidity packaging

MAP helps to preserve fresh fruits and vegetables by regulating the RH, reducing the amount of O_2 and increasing the concentration of CO_2 within the packaging (Bai and Plotto, 2011; Fonseca et al., 2002). This helps to prevent deterioration caused by water loss and the ongoing metabolic processes that occur after the produce has been harvested. However, disruptions in the cold chain and the ripening of climacteric fruit can result in a rapid increase in CO_2 and a decrease in O_2 levels within the MAP, leading to the development of anaerobic conditions. To address this, modified humidity packaging (MHP) maintains a high level of humidity while avoiding the potential for anaerobic conditions (Bai et al., 2019; Rodov et al., 2010). Commercial clamshells for small fruits are usually designed with large openings, but this design may result in the loss of benefits from high humidity (Sousa-Gallagher and Mahajan, 2013). In a study by Bai et al. (2019), the quality of various small fruits (sweet cherries, strawberries, blueberries) was compared in a commercial clamshell with an opening ratio of 2.83% to an MHC with an opening ratio of 0.44% (an 84% reduction in the opening ratio compared to commercial packaging) (Fig. 3). Authors found that MHP packaging maintained a RH of 98–100%, while the RH in commercial packaging was 93.6–96.8%. The differences significantly impacted fruit quality. Namely, vulnerable tissues such as cherry pedicels, strawberry calyxes, litchi pericarp, and cherry tomato skins rapidly deteriorated in commercial packaging. On the other hand, blueberries and strawberries packaged in MHP were firmer than fruit in conventional packaging. Nevertheless, decay incidence in strawberries was higher in MHP. These results suggest a complex relationship between high humidity and decay incidence in fruit that needs to be assessed when choosing low-vent packaging.

Some concerns have been raised regarding the reduced ventilation in MHP packaging. First, opening reduction causes a significant pressure drop and decreases cooling efficiency (Ngcobo et al., 2013). Also, condensation may occur in reduced openings (Linke and Geyer, 2013), particularly when packaged produce is transferred from cold to ambient temperature, which occurs quite often during marketing and in the consumer's kitchen (Rodov et al., 2010).

Figure 3 Comparison between a commercial clamshell (bottom, opening/surface ratio 2.83%) and a modified humidity clamshell (top, opening 0.44%) with litchi fruits.

4 Edible coatings and films

After being harvested, fruits and vegetables continuously experience vapor loss and exchange of O_2 and CO_2 between their interior and exterior, primarily through stomata or lenticels (Fig. 4). To address this, thin and biodegradable natural polymer sheets called edible coatings (ECs) and edible films (EFs) can act as selective barriers for gases like O_2, CO_2, and ethylene, in turn reducing water loss, oxidative processes, microbiological growth, and firmness loss (Dhall, 2013; Ghidelli and Pérez-Gago, 2018; Baldwin et al., 2011). While their goal is essentially the same, the difference between EC and EF is that the former is applied through immersion or spraying in liquid form, while the latter is a preformed solid sheet that is placed onto the food product (Sharma and Rao, 2015; Arredondo Ochoa, 2012).

There are several factors to consider for a successful application of ECs and EFs, including the choice of materials, their biodegradability, the physical and mechanical properties of the coatings and films, the addition of functional ingredients, and associated regulatory aspects. These will be discussed in further detail on the following pages.

4.1 Coating requisites and desirable properties

A coating must meet many performance and safety requirements. One of the most important requisites is the aforementioned gas exchange feature. Coatings must allow the diffusion of gases in order to avoid anaerobic conditions during postharvest storage, but they must also be restrictive enough to delay ripening and senescence. Moreover, it is useful to know the mechanical properties of

Figure 4 The emission of bubbles from the lenticels on the surface of a pear when air is forced through the seed cavity while the fruit is submerged in water.

coatings, which are generally evaluated in terms of thickness, tensile strength, elongation at break, and Young's modulus. The resulting values for these measurements depend on several factors, such as the type and amount of plasticizer added to the matrix, ambient temperature and RH, the inclusion of cross-linking agents, and film-forming process conditions, among others. Fruit coatings must not be tacky and they are sometimes applied to impart gloss and color. In addition, they must not develop off-flavors, ferment or crack, and they are expected to spread evenly, dry quickly, and remove easily from equipment (Baldwin et al., 2011).

4.2 Structural matrix materials

A wide range of structural materials are currently used in coating formulations. While the coating material should be dispersible or soluble in certain solvents, other components such as plasticizers and emulsifiers might be added during preparation to improve integrity, stability, and functionality. The resulting physical and mechanical properties of EC and EF depend on the type of film-forming materials and their structural cohesion. ECs are generally classified on the basis of the structural material of which they are composed. Hence, four groups can be differentiated: (1) polysaccharide-based, (2) protein-based, (3) lipid-based, and (4) composite coatings, which include a combination of the previously mentioned materials (Dea et al., 2011; Panahirad et al., 2021). Each category exhibits advantages and limitations. For instance, the molecular structure complexity of polysaccharide-based ECs allows for the production of films with excellent mechanical and structural properties, but they result in a poor humidity barrier. Similarly, proteins can form sheets of outstanding stiffness and thermal stability, but they are also deficient barriers to control water loss (Fernández et al., 2017). The hydrophilic nature of proteins and polysaccharides usually requires the addition of lipids to improve the moisture barrier of the coatings. Oppositely, lipid-based coatings are known for their protective properties against dehydration, as well as for providing shine and reducing senescence. However, lipid formulations form sheets with poor mechanical properties since they do not adhere efficiently to the surface of products, and they are fragile, frequently exhibiting cracks. Thus, they are usually amended with protein components or polysaccharides resulting in composite coatings (Ghidelli and Pérez-Gago, 2018).

4.2.1 Polysaccharide-based coatings

Polysaccharides as coating structural materials have been broadly applied in the past few years particularly because of their availability, low price, solubility, stability, safety, nontoxicity, and transparency (Panahirad et al., 2021). The most common film-forming polysaccharides are chitosan, alginate, cellulose, carrageenan, pectin, and starch.

Chitosan, a nontoxic polycationic compound available as waste of fisheries, is the most frequent polysaccharide used as coating base material due to its intrinsic edible and antimicrobial properties (Tkaczewska, 2020; Göksen et al., 2020). In the last decades, synergistic effects were reported in fruit and vegetable coated with chitosan and biological antimicrobial agents such as silver and zinc nanoparticles and essential oils (EOs) (Peralta-Ruiz et al., 2020; Zhang and Jiang, 2020; Romanazzi et al., 2009). Moreover, it has been shown that the combination of chitosan with other polysaccharides improves its functional properties. For instance, a chitosan-methyl cellulose film inhibited pathogenic bacteria growth on fresh-cut melon and pineapple (Sangsuwan et al., 2008).

Alginate and carrageenan are seaweed-derived polysaccharides that have been widely used in the development of EC and EF due to their renewability, sustainability, and gelling ability. EFs formed by these compounds exhibit low mechanical resistance; nevertheless, they can react with polyvalent cations producing water-insoluble polymers (Solano-Doblado et al., 2018). The addition of bioactive compounds to these films has been assessed for the preservation of different fresh products (Jayakody et al., 2022; Wani et al., 2021).

Cellulose is the main structural component in plants and the most abundant source of complex carbohydrates in the world. Its linear structure makes it tough, flexible, transparent, and resistant to fats and oils, and this results in a coating with excellent mechanical and structural properties (Arnon et al., 2015). Carboxymethyl cellulose (CMC) is the most popular cellulose derivative used for postharvest products (Jafarizadeh et al., 2011). CMC coatings containing bioactive additives exhibit excellent O_2, aroma, and oil barrier properties and significant anti-senescence effects in climacteric fruit (Saekow et al., 2019; Saxena et al., 2020).

Pectin is one of the most complex macromolecules in nature with a branched structure. Native pectin films are nontoxic, biodegradable, and exhibit high water vapor transmission rates (Panahirad et al., 2021). Also, they have been shown to restrict the loss of nutrients and volatile materials and to prevent microbial contamination during fruit storage (Mohamed et al., 2020). Pectin films have also been amended with various bioactive compounds to improve their mechanical, antioxidant, and antimicrobial properties (Shivangi et al., 2021; Bai et al., 2019).

Among carbohydrates, native and modified starches are considered among the most promising materials due to their wide sources, price, and good film-forming ability (Dai et al., 2019). However, they have the disadvantages of poor mechanical properties and easy absorption of water, which increases the risk of mildew and leads to a further significant decrease in mechanical properties. Several studies have shown that chitosan addition to starch-based EC increases antimicrobial activity and improves tensile strength and elongation at the break of films (Sánchez-Ortega et al., 2016). Moreover, it has been demonstrated that the chemical modification of starch by cross-linking allowed for the formation of films with better physical and chemical properties than the corresponding native starch, with a significant preservation effect on postharvest pears (Dai et al., 2020).

4.2.2 Protein-based coatings

Protein structural materials are derived from animal and plant sources. As long as food-grade proteins and additives are used, the resulting film can be considered safe and edible (Krochta and De Mulder-Johnston, 1997). Proteins, due to their molecular structure, exhibit a wide range of functional properties and potential for molecular bonds (Huff, 2013). This feature can be exploited to improve film properties through chemical, physical, and enzymatic methods (Carpiné et al., 2015; Montalvo et al., 2012). In the same way, barrier and mechanical properties might be affected and impaired by many factors. For instance, uncontrolled temperature and pH might denature proteins and modify the mentioned properties.

Whey protein isolate (WPI) is a traditional industrial product that can be self-assembled into WPI nanofibers, which can be also made into EC and EF with remarkable flexibility, antioxidant activity, and serve as an excellent barrier for gas, aroma compounds, and oils as compared to polysaccharides and lipids films (Feng et al., 2019). Soybean protein isolate (SPI) is suitable for EC and EF because of its low permeability to O_2 and CO_2 and reasonable cost. The application of coatings based on SPI was effective in controlling weight loss and firmness in fresh-cut apples (Alves et al., 2017), softening in apricots (Zhang et al., 2018) and browning of fresh-cut artichokes (Ghidelli et al., 2015).

Casein is a milk protein found in the form of micelles. It is commonly used to prepare emulsions due to its amphipathic nature (Momin et al., 2021). Casein's thermal stability, biodegradability, and ability to bind with small ions and molecules make it a highly desirable biomaterial for the preparation of EC (Głąb and Boratyński, 2017). It was reported that the application of casein-based EC helped to maintain firmness and color in guavas (Beulah et al., 2021), weight loss in grapes (Murmu and Mishra, 2018), and firmness in cherries and apples (Certel et al., 2004; Volpe et al., 2018).

Zein, a protein obtained from the endosperm of maize (*Zea mays* L), is soluble in 70–80% ethanol and has a high content of water-repelling amino acids. Compared to other protein-based EC and EF, zein films provide a better transmission of water vapor and volatiles. The prevention of color change, inhibition of pathogenic bacteria, and reduction of firmness and weight loss were reported for zein films in several crops, such as garlic cloves (Torun and Ozdemir, 2022), nectarines and apples (Mendes-Oliveira et al., 2022), and guavas (Santos et al., 2018).

4.2.3 Lipid-based coatings

Lipids are defined as small, hydrophobic, and natural molecules such as fats, waxes, sterols, and fat-soluble vitamins, among others. Due to their non-polar feature, they are generally used as a barrier against water-vapor transfer.

Lipids can be used as the main structural material of EC or as additives to protein or polysaccharide matrixes. Lipid-based coatings have high-barrier properties in comparison to polysaccharide and protein-based coatings. Nevertheless, they are deprived of structural integrity and mechanical strength and they sometimes cause undesirable organoleptic properties due to the greasy surface and rancidity (Sharma et al., 2019; Yousuf et al., 2021; Abdullah et al., 2009),

The addition of vegetable oils to EC and EF is a common practice to enhance film resistance against moisture and to prolong their functional and mechanical properties. For instance, the inclusion of sunflower oil in starch improved the barrier properties of the resulting film (García et al., 2000).

Fatty acids are also involved in the manufacture of hydrocolloid-based EC and EF. Several reports showed that the lipid part efficiently reduced the moisture permeability and enhanced the functioning of coatings. It is important to note that fatty acids vary in chain length and degree of saturation. Shorter chains and higher saturation result in more efficient control of moisture transfer (Sharma et al., 2019).

Waxes are considered the most efficient materials to decrease water-vapor permeability due to their hydrophobic nature. In addition to this feature, they are valued for their safety, the improvement of the product's visual appearance, and the reduction of phytopathogen development (Yousuf et al., 2021). The most frequently reported waxes for fruit and vegetable coatings are beeswax (Zambrano-Zaragoza et al., 2020), carnauba wax (Germano et al., 2019), and candelilla wax (Oregel-Zamudio et al., 2017).

4.3 Additives

The final performance of EC and EF depends not only on their mechanical and physical properties but also on their bioactive and functional characteristics. Various ingredients are usually added to EC to modify, improve or add desired properties or to enhance their efficiency and functionality (Baldwin et al., 2011; Sharma et al., 2019).

In order to improve tensile strength, elastic modulus, elongation at break, and other mechanical characteristics, the addition of binding agents, fillers, cross linkers, emulsifiers, and plasticizers is a common practice. The most widely used plasticizers are propylene glycol, glycerol, polyethylene glycol, sucrose, sorbitol, and water (Nicolau Lapeña et al., 2020; Myllärinen et al., 2002; Fadini et al., 2013). Among emulsifiers, Tween and oleic acid are the most popular (Sharma et al., 2019).

The addition of active compounds to coating matrixes helps to preserve the product's quality, delay senescence, and extend shelf life. For instance, antimicrobial agents slow down the growth of rotting and pathogenic microbes, preventing the development of molds and increasing food safety. The most commonly used antimicrobials are GRAS salts, organic acids, microbial bacteriocins, polypeptides, plant extracts and EOs, mineral salts, metallic ions nanoparticles, nitrites, sulfites, and parabens (Panahirad et al., 2021). Also, antioxidants such as ascorbic acid, citric acid, and tocopherol are added to coating matrices to prevent oxidative rancidity, degradation, and color loss (Youssef et al., 2012).

The development of films and coatings loaded with probiotics and prebiotic compounds is an emerging and promising technology. Functional coatings containing probiotic and prebiotic compounds have been successfully applied on fresh blueberries, melon, and apple pieces (Bambace et al., 2019; de Oliveira et al., 2021).

In addition to the aforementioned inclusion,, there has been significant progress in the development of a new generation of EC and EF over the past few years. These innovative coatings use nanotechnological solutions to create coatings that release in a controlled manner different active compounds. The application of nanoparticles and their nanoencapsulation has been addressed by several researchers, being the most frequent nanomaterials in these reports metal and metal oxides known by their antimicrobial properties. These nanoparticles have been successfully incorporated into EFs such as chitosan, alginate, carrageenan, and gellan (Xing et al., 2019; Zhang and Jiang, 2020; Ghidelli and Pérez-Gago, 2018)

4.4 Future perspectives and regulatory aspects

A new generation of EC based on the incorporation of active ingredients, such as antimicrobials, antioxidants, antibrowning agents, texture modifiers, and nutrients, is currently one of the approaches with the greatest interest in managing the quality of fresh fruits and vegetables (Panahirad et al., 2021). Nevertheless, an effective shelf life extension is not the only factor that determines coatings' success; consumer concerns and acceptance of the offered technology must be taken into account. In this regard, there are some key points that should be considered while producing an EC. Namely, (1) sensory attributes of integrated coated products such as appearance, taste, flavor, and texture; (2) proper labeling of the ingredients; (3) marketing approaches drawing attention toward the consumer benefits; and (4) other factors such as price, convenience, nutritional content, and hygienic standards, among others.

Moreover, the applicability of EC and EF is a current challenge for scientists and producers. Many practical and legal considerations must be fulfilled before producing and marketing an EC or EF. One of the most important facts is that the materials used for EC and EF manufacture will be subjected to the same legal evaluations as other food additives (Pirozzi et al., 2021; Valencia-Chamorro and Torres-Morales, 2016). Food administrations around the world including the Food and Drug Administration (FDA) and European Food Safety Authority (EFSA) have implemented various codes to register and food contact materials to protect consumers' health against potentially hazardous materials. When introducing a newly developed EC or EF to the FDA or EFSA, full information on structural materials, additives, and active ingredients must be provided. Also, an analysis of the potential migration of these compounds to food material is required in order to authenticate the safety of the treated product (Paidari et al., 2021).

5 Active packaging

Active packaging for fruit and vegetables is a type of food packaging that is designed to interact with the internal or external environment of the package in order to maintain or extend the shelf life of the produce. This is achieved by incorporating various substances into the packaging material, such as ethylene absorbers, moisture regulators, and antimicrobial agents, which help to control the growth of microorganisms, slow down ripening, and prevent spoilage.

One common type of active packaging for fruits and vegetables is ethylene scavenging packaging. Ethylene is a natural plant hormone that plays an important role in the ripening and senescence (aging) of many fruits and vegetables. It is produced by the fruit and vegetables themselves as they mature, and also as a response to stress, injury, or infection. Ethylene scavenging packaging contains materials that absorb ethylene gas and prevent it from reaching the produce, helping to prolong the shelf life of fruit and vegetables.

Another type of active packaging for fruit and vegetables is moisture-regulating packaging. These packages contain materials that can absorb or release moisture in response to changes in the environment, helping to maintain the optimal humidity levels for the produce and prevent moisture-related spoilage.

Antimicrobial packaging is also becoming increasingly popular for fruit and vegetables. These packages contain materials that release antimicrobial agents, helping to control the growth of microorganisms and prevent spoilage.

5.1 Ethylene scavenger and control

Ethylene is a phytohormone that is involved in the regulation of numerous physiological functions in plants. Climacteric fruits and vegetables exposed to high concentrations of ethylene can experience crumbling, accelerated maturation and quality deterioration, augmented susceptibility to pathogens, and physiological disorders during post-harvest processing, transportation, and storage (Mariah et al., 2022). There are two types of ethylene that affect ripening, as defined by their sources, endogenous (ethylene produced by the fruit itself) and exogenous (ethylene artificially introduced to the fruit) (Janjarasskul and Suppakul, 2018).

Ethylene scavengers are types of active packaging material that are used to scavenge ethylene from the headspace of packaging. These ethylene-capturing materials are usually deployed in the form of sachets that are placed inside the packaging, as a coating on the surface of another type of packaging material, or in the form of ethylene-scavenging active films (Wei et al., 2021). Ethylene scavengers can reduce the rate of ethylene generation by blocking ethylene binding sites, thus slowing the self-stimulating ethylene generation positive feedback loop (Ruperez et al., 2022).

5.1.1 1-Methylcyclopropene

1-Methylcyclopropene (1-MCP) is a commonly used endogenous ethylene inhibitor that blocks the hormonal action of ethylene by competing with ethylene for ethylene receptors (Janjarasskul and Suppakul, 2018; Sisler and

Serek, 1997), resulting in an inhibition of the synthesis of ethylene in fruit ethylene production system II (Wei et al., 2021). 1-MCP has been reported to effectively slow ethylene release, delay softening, reduce malondialdehyde content, and inhibit the activity of cell wall degrading enzymes of mangoes (Li et al., 2022a). The application of 1-MCP reduced softening by an average of 12% in apple fruit and delayed the ripening process by lowering the rate of sucrose hydrolysis (Falagan and Terry, 2020). In addition, 1-MCP was highly effective in reducing disease incidence and induced multiple changes in the bacterial and fungal microbiota in postharvest pear fruit (Zhang et al., 2021).

According to Satekge and Magwaza (2022), the apple and many other fruit industries have made extensive use of 1-MCP to prolong the shelf life of fresh produce. This has been achieved through pre-harvest spraying, short-term postharvest gas treatment, and by incorporating controlled-release 1-MCP into the packaging of fresh produce

To extend the shelf life of fresh and fresh-cut produce, Toivonen and Lu (2006) developed capsules of 1-MCP in a cyclodextrin matrix and a hygroscopic compound, which releases 1-MCP after absorbing high humidity. Meanwhile, Hotchkiss et al. (2007) developed packaging films that incorporate a 1-MCP/α-cyclodextrin complex into several common packaging materials by heat-pressing and solution-casting methods. The rate of release depends on RH and temperature, and LDPE was found to result in a higher and faster release than PS, PVC, and PP polymers. Sarker and Liu (2017) synthesized four boron derivatives of methylene cyclopropane that are stable at ambient conditions and gradually release 1-MCP upon contact with water, depending on their reaction rates. Huang et al. (2022) demonstrated the adsorption and release of 1-MCP using easily attainable microporous metal-organic frameworks (MOFs). The effectiveness of the encapsulation and release system has been validated by a fruit preservation test.

5.1.2 Ethylene scavengers

There are three types of ethylene scavengers classified based on their mechanisms, namely physical, chemical, and biological scavengers. Porous materials such as activated carbon and zeolite belong to the physical scavengers category, as they can physically trap or adsorb ethylene molecules. On the other hand, powerful oxidizing agents or catalysts such as potassium permanganate, palladium chloride, and titanium dioxide fall under the chemical scavengers category, as they can effectively remove ethylene from the headspace packaging. Earlier research conducted by Bai et al. (1989) involved the use of Kurthia spp., a type of bacteria isolated from soil, as a biological scavenger to extend the shelf life of apples (Fig. 5). These bacteria are expected to consume ethylene as a carbon source (Bai et al., 1989).

Figure 5 Apples in MAP with palladium chloride ethylene remover (top) and bacteria ethylene remover (*Kurthia* spp.) (bottom).

Packing ethylene scavengers, such as zeolite and titanium dioxide, in a small sachet has also been reported as effective (Wei et al., 2021). A biodegradable active packaging containing zeolite extended the shelf life of fresh broccoli florets, reduced the metabolism of broccoli stored at 12°C, and preserved the color and vitamin C content for 7 days (Marzano-Barreda et al., 2021). A chitosan coating incorporating nano-titanium dioxide particles significantly decreased the rate of shrinkage, decay, respiration, ethylene production, electrolyte leakage, superoxide anion production, and malondialdehyde accumulation in Ginkgo biloba seeds (Tian et al., 2019). MOFs in active packaging have also been widely applied as ethylene scavengers (Sultana et al., 2022). Ethylene scavengers which adsorb ethylene in order to remove it from packaging materials are generally porous structures which include activated alumina, activated carbon, and different types of clays (Rodríguez et al., 2014). Types of ethylene binding clays include silica, diatomaceous earth, anti-caking agent, China clay, quartz, and Oya stone in powder form (Rodríguez et al .2014). These clays possess a higher surface area and have a basic structural component of micro and/or nano size. Due to the highly specific surface area of porous materials, these ethylene adsorbents are capable of effectively absorbing ethylene released from fruits and vegetables (Janjarasskul and Suppakul, 2018).

5.2 Controlled release antimicrobials

5.2.1 Oxidants

Many oxidants such as ozone, hydrogen peroxide, and chlorine are used to maintain the quality of fresh fruits and vegetables and control post-harvest decay due to their antimicrobial effects (Cerioni et al., 2013). Encysting these antimicrobial gases into different solid matrices using encapsulation technologies, such as polymeric films, capsule powders, and electrospun nonwovens, has been widely applied for food preservation (Zaitoon et al., 2022)

Chlorine dioxide (ClO_2) is the preferential form of chlorine used for water treatment due to its strong oxidation capacity, which is 2.5 times higher than that of chlorine in the form of hypochlorous acid (Zhong et al., 2021). Gaseous ClO_2 is more effective than its aqueous counterpart in controlling foodborne pathogens on fresh produce due to its high diffusivity and penetrability (Sun et al., 2019). The controlled release of gaseous ClO_2 by means of specialized packaging is of growing interest as it allows for a continuous application of ClO_2 gas in order to maintain fresh produce safety (Sheng and Zhu, 2021). A controlled-release ClO_2 pad made using a crystalline form of ClO_2 was applied for blueberry preservation. ClO_2 gas was gradually and continuously released during storage, which significantly reduced the population of *E. coli* and *Colletotrichum acutatum* when compared to the control (Sun et al., 2014). A ClO_2 pouch containing a 0.5 g ClO_2 slurry was also applied by this group for grapefruit and tomato preservation (Sun et al., 2017b; Sun et al., 2017c; Sun et al., 2017a). The ClO_2 pouch reduced *E. coli* inoculated on grape tomatoes by 3.1 log CFU/g during 14-day storage at 20°C (Sun et al., 2017c). The continuous application of ClO_2 gas via packaging can also improve the quality and extend the shelf life of fresh produce. Polylactic acid films containing microencapsulated ClO_2 extended the shelf life of mangoes by maintaining peel and pulp color, reducing weight loss, and maintaining vitamin C and titratable acidity contents (Zhang et al., 2019). A corrugated box coated with ClO_2-generating agents also significantly reduced the weight and firmness loss of strawberries and maintained their internal concentration of titratable acid, total soluble solids, and ascorbic acid (Li et al., 2018). ClO_2 pouches containing a ClO_2 slurry or powder also improved the quality of tomatoes and blueberries, respectively (Sun et al., 2017c; Sun et al., 2014).

Nitric oxide (NO), as a ubiquitous gas signaling molecule with strong antimicrobial and antioxidant activities, modulates various physiological and biochemical processes and stress responses in plants (Wei et al., 2022). However, the gaseous state and short shelf life of NO requires a formulation that can control its storage and release. Treatment with a NO donor, sodium nitroprusside (SNP), delayed the decline of the physiological quality of grapes and reduced weight loss and respiration rate. The SNP treatment also increased the activities of superoxide dismutase and polyphenol oxidase and inhibited the activities of peroxidase and catalase of grape during the reaction to fungal pathogen infection (Shi et al., 2022). The NO treatment effectively inhibited lesion diameter and disease incidence on inoculated mango fruit and regulated hydrogen peroxide levels by reinforcing the activities of peroxidase, catalase, superoxide dismutase, and ascorbate peroxidase. It also increased the accumulation of lignin, total phenolics, anthocyanin, and flavonoids and the activities of chitinase and beta-1,3-glucanase (Ren et al., 2020). The effect of NO released from nitrate trimethylolethane cyclocarbonate on the quality of banana was investigated, and treatment with NO and light-emitting diode exposure effectively prolonged the yellowing time of banana, maintained a high level of phenolics, ascorbic acid, and glutathione, reduced chlorophyll degradation, inhibited the accumulation of reducing sugar and malondialdehyde, and enhanced superoxide dismutase activity (Zhou et al., 2020). NO-releasing chitosan nanoparticles reduced weight loss, glutathione, respiration rate, reactive oxygen species, and ethylene production and increased antioxidant enzyme activity and the content of soluble solids and ascorbic acid in sweet cherries (Ma et al., 2019).

Sulfur dioxide (SO_2) can be endogenously produced and rapidly transformed into sulfur-containing compounds (e.g. hydrogen sulfide, cysteine, and methionine) to maintain homeostasis in plant cells (Li et al., 2022b). Treatment with an SO_2-releasing paper insert in packaging showed brighter color associated with freshness, and higher flavonoid and chlorophyll contents of longan fruits (Han et al., 2020). Dual-release SO_2-generating pads were effective against *Botrytis cinerea*, resulting in slowed decay and spread of gray mold, while retaining the fresh appearance of stems of grapes (Mustonen, 1992). An SO_2 treatment also increased the concentration of cysteine, hydrogen sulfide, and glutathione and decreased the expression of VvSURTL, VvATPS1, VvATPS2, and VvAPR3 in grapes (Zhang et al., 2022). SO_2-emitting PE packaging effectively controlled blueberry decay and improved fruit quality during post-harvest storage (Saito et al., 2020; Cho et al., 2019).

5.2.2 Volatile compounds

EOs and their constituents are widely used as fumigants, due to their strong antimicrobial properties and high volatility, biodegradability, and safety (Namiota and Bonikowski, 2021). Carvacrol is a natural antimicrobial capable of inhibiting several microorganisms. The encapsulation of this compound can increase its stability and water solubility and facilitate its controlled release (Ayres Cacciatore et al., 2020). The use of a controlled-release carvacrol powder delayed the decay and maintained the safety of blueberries (Sun et al., 2021). A chitosan film with thymol and carvacrol-loaded MOF had significantly improved thermal stability, light barrier properties, tensile strength, and hydrophobicity over an unmodified film. More importantly, UiO-66-NH(2) restrained the release rate of thymol and carvacrol compared to a pure chitosan film and endowed the chitosan film with improved antimicrobial activity against *E. coli*, *Staphylococcus aureus*, and *Penicillium citrinum* (Xu et al., 2023). A nanocarrier-encapsulated EO enabled the controlled release of the oil, stabilizing the packaging contents for an increased duration (Sharma et al., 2022). A microdevice consisting of mesoporous silica particles loaded with thymol, carvacrol, or cinnamaldehyde, and then functionalized with zein, efficiently inhibited bacterial growth via the release of the loaded EO components (Poyatos-Racionero et al., 2021).

Ethanol has been widely used for the storage of fruit and vegetables. A novel ethanol gel with a controlled release rate was prepared to employ a gelatification reaction between ethanol and sodium stearate to overcome the disadvantage of conventional ethanol emitters. The application of a controlled-release ethanol emitter for the storage of Chinese bayberry fruit demonstrated its ability to reduce the decay rate, maintain firmness, and inhibit increased malondialdehyde concentration at 4°C (Mu et al., 2017). The effect of storage temperatures and ethanol vapor released by an active packaging system on the quality of freshly peeled shallots was investigated. The package tested was a solid PP tray incorporating an ethanol vapor-controlled release sachet. Microbial proliferation in terms of yeast and mold populations could be delayed through a combination of 10°C and ethanol vapor released from the sachet (Utto et al., 2018). A combination treatment of ethanol and hot water reduced the decay rate of peaches, nectarines, and grapes (Gabler et al., 2005; Margosan et al., 1997).

Trans-2-hexenal, being a natural compound with antimicrobial activity, is commonly used as a preservative of fresh fruit and vegetable (Cui et al., 2021). It was incorporated into an MAP cellulosic pad, which was then covered with a PA coating to control the release of the trans-2-hexenal. Fungal infection of strawberries was fully inhibited in the active packaging after 14 days at 4°C, although the strawberries were softer and darker than those in the control packages (Heras-Mozos et al., 2021). The inclusion of trans-2-hexenal in the storage atmosphere resulted in a significant extension of the shelf life of apple slices. The presence of trans-2-hexenal in the packaging atmosphere substantially prolonged the lag phases of inoculated yeast and reduced the growth potential of naturally occurring bacteria (Corbo et al., 2000). Several acetates, which were expected to be metabolites of trans-2-hexenal in fruit, were released from treated strawberry; however, these acetates did not inhibit fungal growth (Wakai et al., 2019).

6 Future trends in research

The future of fruit and vegetable packaging is likely to see the integration of new materials, technologies, and designs that aim to improve sustainability, food safety, and convenience (Iseppi et al., 2022; Ketelsen et al., 2020; Moustafa et al., 2019; Rangappa et al., 2020). Some possible trends include the following.

6.1 Packaging materials

Sustainable and environmentally friendly attributes are becoming more and more important (Iseppi et al., 2022; Ketelsen et al., 2020; Moustafa et al., 2019). Biodegradable packaging materials are materials that can break down into natural elements, such as water, CO_2, and organic matter, in the environment through processes like composting and biodegradation (Nur Hanani et al., 2014; Thulasisingh et al., 2021). Packaging materials made from biodegradable and compostable materials such as plant-based polymers and starch will become more prevalent, reducing waste, and reducing the impact on the environment (Thulasisingh et al., 2021; Moustafa et al., 2019). Some examples of biodegradable packaging materials are listed in Table 2.

Table 2 List of biodegradable packaging materials

Type of biodegradable packaging	Description	References
Paper and corrugated fiberboard	These materials are made from renewable resources and can easily biodegrade in composting conditions	Fadiji et al. (2018); Gaikwad et al. (2016)
Plant-based plastics	These are plastics made from renewable resources such as cornstarch, sugarcane, and potato starch. They are biodegradable and can break down in the environment	Basumatary et al. (2022); Santhosh et al. (2021); Zhang and Sablani (2021)
Biodegradable PE bags	These bags are made from a blend of traditional plastic and biodegradable materials, making them a more environmentally friendly alternative to traditional plastic bags	Shaikh et al. (2021); Thulasisingh et al. (2021)
Natural fibers	Natural fibers such as cotton, jute/hemp, bamboo, and burlap can be used to create biodegradable packaging materials	Sydow and Bieńczak (2019)
Mushroom packaging	This type of packaging is made from a mixture of agricultural waste and mushroom spores. It is biodegradable and compostable	Abhijith et al. (2018)
Algae-based packaging	Algae can be processed into a biodegradable material that can be used to make packaging	Mostafavi and Zaeim (2020)
Agricultural waste	Agricultural waste products like straw and maize husks can be used to make biodegradable packaging materials	Mostafavi and Zaeim (2020)
Edible packaging	Packaging made from edible materials that can replace traditional packaging and reduce waste	Janjarasskul and Krochta (2010); Petkoska et al. (2021)

Packaging that can be reused multiple times, such as containers made from durable materials, reduce waste and promote sustainability. Reusable packaging that is commonly used can range from large containers such as fruit harvest and storage bins, with a weight of one short ton (2000 pounds) or one metric ton (1000 kg), to small containers like berry baskets found in farmers' markets, which can weigh as little as one hundred grams or several ounces.

6.2 Smart packaging

Smart packaging is a term used to describe packaging that incorporates advanced technology and materials to enhance the functionality and performance of the packaging beyond simple containment and protection of the product. It can provide features such as extending the shelf life of the product, indicating freshness or authenticity, monitoring temperature and humidity levels, and providing product-specific information, such as instructions for use. Additionally, smart packaging can be designed to be environmentally friendly, reducing waste and promoting sustainability. Intelligent packaging, active packaging, and interactive packaging are some examples of smart packaging (Biji et al., 2015).

The most popular features provided by smart packaging include the following:

1. Time-temperature indicators: Packaging that changes color to indicate whether the produce inside has been exposed to temperature levels during certain periods that could make it unsafe for consumption (Wang et al., 2015).
2. Intelligent packaging/labels: Packaging that contains sensors and other technology that can monitor the freshness, temperature, and humidity levels of the produce, safety, and quality of the produce and provide information to consumers (Shinde et al., 2018; Ghaani et al., 2016; Yam et al., 2005).
3. QR codes: Packaging with QR codes that can provide consumers with information about the produce, such as the source, nutritional information, and expiration date (Rotsios et al., 2022; Almenar, 2020).
4. Radio-frequency identification tags: Packaging that contains a small tag that can be scanned to provide information on the product's journey from the farm to the store (Occhiuzzi et al., 2020).
5. Near-field communication (NFC) technology: Packaging with NFC chips that allow consumers to access information about the produce by tapping their smartphones on the package (Escobedo et al., 2021; Escobedo et al., 2017; El Matbouly et al., 2022).

6. MAP and active packaging: Packaging that controls the atmosphere inside the package to extend the shelf life of the produce and prevent spoilage.

6.3 Convenient and personalized packaging

Convenient and personalized packaging refers to packaging that is designed to meet the specific needs and preferences of individual consumers. This type of packaging is designed to be easy to use, carry, store, and dispose of and often includes features such as resealable or reusable closures, portion control, and ergonomic design (Ragaert et al., 2004; Vandichel and D'Alesio, 2015).

Personalized packaging is also designed to meet the specific needs of individual consumers and can include features such as custom sizes, shapes, colors, and branding. This type of packaging is often used to create a unique and memorable customer experience and can help build brand loyalty and customer engagement (Yu et al., 2015; Yan et al., 2020).

In today's fast-paced, on-the-go society, convenient and personalized packaging is becoming increasingly important as consumers seek products that can easily fit into their busy lifestyles. Companies that invest in convenient and personalized packaging are able to differentiate themselves from their competition, create a more memorable customer experience, and build stronger brand relationships with their customers.

6.4 Other packaging trends

The fresh fruit and vegetable packaging industry is constantly evolving, with new trends emerging all the time. Some of the current and emerging trends include the following:

1 Packaging materials with advanced printing technology will provide more detailed and accurate information about the produce, such as its origin, nutritional information, and other important details (Luo, 2022; Ouyang, 2015).
2 E-commerce: The growth of online grocery shopping is expected to drive the development of new packaging solutions to ensure the safe delivery of fresh produce to consumers (Spruit and Almenar, 2021; Gee et al., 2020).
3 Increased use of vacuum packaging: This method involves removing the air from the packaging to slow down the spoilage of the produce and extend its shelf life (Onyeaka et al., 2022).

7 Conclusion

In conclusion, the development of packaging materials and designs has made significant contributions to the physical protection of fruits and vegetables throughout the supply chain. Plastic bins have emerged as the preferred choice due to their lightweight nature, water and chemical resistance, and ease of cleaning. While wooden bins present sanitation challenges, they possess antimicrobial properties. The choice of packaging materials influences the transfer of microorganisms, with hydrophilic surfaces promoting biofilm formation. Various cushioning materials help prevent physical damage to fruits during handling. Additionally, individual paper packaging with additives has been effective in addressing postharvest decay and disorders. Achieving a balance between physical protection and hygienic considerations is essential for the safe and efficient handling of fresh produce.

Modified atmosphere packaging (MAP) has become widely used for preserving the freshness of fruits and vegetables. By altering the gas composition within the packaging, MAP reduces water loss, delays senescence, and prevents microbial contamination. Passive MAP occurs through product respiration and gas permeation, while active MAP involves gas flushing or vacuum-gas flushing. The selection of packaging materials is crucial, with flexible and semi-rigid plastics being common choices. The choice of gas depends on the desired atmosphere and legal regulations. MAP offers advantages such as extended shelf life and reduced reliance on chemical preservatives, but transportation and logistics present challenges. Modified humidity packaging (MHP) addresses the need to maintain high humidity while preventing anaerobic conditions. Overall, MAP continues to be a vital technology for improving the quality and safety of fresh produce.

Edible coatings (ECs) and edible films (EFs) hold promise for postharvest preservation of fruits and vegetables. These thin and biodegradable natural polymer sheets act as selective barriers for gases such as oxygen, carbon dioxide, and ethylene, reducing water loss, oxidative processes, microbiological growth, and firmness loss. The choice of materials, including polysaccharides, proteins, and lipids, plays a crucial role in determining the properties of the coatings and films. Additives such as plasticizers, emulsifiers, antimicrobials, and antioxidants can further enhance the performance and functionality of these coatings. By carefully considering these factors, ECs and EFs can contribute to extending the shelf life and maintaining the quality of fresh produce.

Active packaging plays a critical role in extending the shelf life of fruits and vegetables by interacting with their internal or external environment. Ethylene scavenging packaging, such as 1-methylcyclopropene (1-MCP), helps slow down the ripening process and inhibit the activity of cell wall degrading enzymes. Various materials, including zeolite, titanium dioxide, and metal-organic frameworks, are used as ethylene scavengers to absorb ethylene gas and prevent it from reaching the produce. Moisture-regulating packaging helps maintain optimal humidity levels, while antimicrobial packaging releases agents to control the growth of microorganisms. Controlled-release antimicrobials like chlorine dioxide, nitric oxide, and essential oils are incorporated into packaging to ensure produce safety. Overall, active packaging techniques offer effective solutions for enhancing the quality and extending the shelf life of fruits and vegetables.

The future of fruit and vegetable packaging holds promising advancements aimed at improving sustainability, food safety, and convenience. Integration of new materials, technologies, and designs will shape this evolution. Sustainable packaging materials, including biodegradable options such as plant-based polymers, paper, and mushroom packaging, will become more prevalent, reducing waste and minimizing environmental impact. Smart packaging will incorporate advanced features like time-temperature indicators, intelligent labels, QR codes, RFID tags, and NFC technology to enhance functionality, provide information to consumers, and ensure product quality. Convenient and personalized packaging will cater to the needs of busy consumers, offering resealable closures, portion control, and ergonomic design.

8 Where to look for further information

There are several resources you can consult for further information about fresh fruit and vegetable packaging. Here are some suggestions:

The Food and Agriculture Organization (FAO), governments and regulatory bodies, such as the FDA, and the United States Department of Agriculture (USDA) have guidelines and regulations on the packaging and labeling of fresh produce. These guidelines can provide information on food safety and packaging requirements for different types of fresh produce. For example:

- FAO Code of Practice for Packaging and Transport of Fresh Fruit and Vegetables (CXC 44-1995) (https://www.fao.org/fao-who-codexalimentarius/codex-texts/codes-of-practice/en/).
- FDA "Standards for the Growing, Harvesting, Packing, and Holding of Produce for Human Consumption: What You Need to Know About the FDA Regulation: Guidance for Industry small Entity Compliance Guide" (https://www.fda.gov/media/107298/download).
- USDA Agriculture Handbook 66 (AH-66): The Commercial Storage of Fruits, Vegetables, and Florist and Nursery Stock. (https://www.ars.usda.gov/ARSUserFiles/oc/np/CommercialStorage/CommercialStorage.pdf).

Fresh produce and packaging industry associations also provide industry specific research and information on packaging materials, design, and regulations. Here we list some of these associations and their websites:

- The Institute of Packaging Professionals (https://www.iopp.org/i4a/pages/index.cfm?pageID=1).
- The Flexible Packaging Association (https://www.flexpack.org/).

Attending industry events and trade shows, such as the International Fresh Produce Association (IFPA)'s events (https://www.freshproduce.com/events/?pageNumber=1), can also provide opportunities to learn more about fresh produce packaging and connect with industry professionals.

There are also several trade publications, such as *Packaging Digest*, *Produce Packaging*, *Fresh Produce Journal*, and *Packaging World*, that covers the latest trends, news, and developments in the fresh produce packaging industry. They can provide information not only on new packaging materials, but also insights from industry experts.

Finally, academic research papers and articles on the packaging and fresh produce can provide a more in-depth understanding of the scientific aspects such as the packaging design and the effect of different packaging materials on the shelf life and quality of fresh produce. Some examples of academic journals include *Food Packaging and Shelf Life*, *Postharvest Biology and Technology*, and *Journal of Food Science* magazines.

9 References

Abdullah, T. L., Shokrollah, H., Sijam, K. and Abdullah, S. (2009). Control of Huanglongbing (HLB) disease with reference to its occurrence in Malaysia. *African Journal of Biotechnology* 8(17): 4007–4015.

Abejón, R., Bala, A., Vazquez-Rowe, I., Aldaco, R. and Fullana-i-Palmer, P. (2020). When plastic packaging should be preferred: Life cycle analysis of packages for fruit and vegetable distribution in the Spanish peninsular market. *Resources, Conservation and Recycling* 155: 104666.

Abeysundara, P. D. A., Dhowlaghar, N., Nannapaneni, R., Schilling, M. W., Mahmoud, B., Sharma, C. S. and Ma, D. P. (2018). Salmonella enterica growth and biofilm formation in flesh and peel cantaloupe extracts on four food-contact surfaces. *International Journal of Food Microbiology* 280: 17–26.

Abhijith, R., Ashok, A. and Rejeesh, C. R. (2018). Sustainable packaging applications from mycelium to substitute polystyrene: A review. *Materials Today: Proceedings* 5(1): 2139–2145.

Admane, N., Genovese, F., Altieri, G., Tauriello, A., Trani, A., Gambacorta, G., Verrastro, V. and Di Renzo, G. C. (2018). Effect of ozone or carbon dioxide pre-treatment during long-term storage of organic table grapes with modified atmosphere packaging. *LWT* 98: 170–178.

Published by Burleigh Dodds Science Publishing Limited, 2023.

Alasalvar, C., Al-Farsi, M. and Quantick, P. (2005). Effect of chill storage and modified atmosphere packaging (MAP) on antioxidant activity, anthocyanins, carotenoids, phenolics and sensory quality of ready-to-eat shredded orange and purple carrots. *Food Chemistry* 89(1): 69–76.

Alden, K. M., Omid, M., Rajabipour, A., Tajeddin, B. and Soltani Firouz, M. (2019). Quality and shelf-life prediction of cauliflower under modified atmosphere packaging by using artificial neural networks and image processing. *Computers and Electronics in Agriculture* 163: 104861.

Almenar, E. (2020). Innovations in packaging technologies for produce. In: *Controlled and Modified Atmospheres for Fresh and Fresh-Cut Produce*. Elsevier, pp. 211-264.

Alves, M. M., Gonçalves, M. P. and Rocha, C. M. R. (2017). Effect of ferulic acid on the performance of soy protein isolate-based edible coatings applied to fresh-cut apples. *LWT* 80: 409–415.

An, J., Zhang, M., Lu, Q. and Zhang, Z. (2006). Effect of a prestorage treatment with 6-benzylaminopurine and modified atmosphere packaging storage on the respiration and quality of green asparagus spears. *Journal of Food Engineering* 77(4): 951-957.

Arnon, H., Granit, R., Porat, R. and Poverenov, E. (2015). Development of polysaccharides-based edible coatings for citrus fruits: A layer-by-layer approach. *Food Chemistry* 166: 465–472.

Arredondo Ochoa, T. (2012). *Diseño de Empaques Comestibles Activos a Base de Almidón Modificado para Su Posible Aplicación en Alimentos en Fresco*. Facultad de Quimica. Universidad Autonoma de Queretaro. Queretaro, Mexico.

Aviat, F., Gerhards, C., Rodriguez-Jerez, J. J., Michel, V., Bayon, I. L., Ismail, R. and Federighi, M. (2016). Microbial safety of wood in contact with food: A review. *Comprehensive Reviews in Food Science and Food Safety* 15(3): 491-505.

Aviat, F., Le Bayon, I., Federighi, M. and Montibus, M. (2020). Comparative study of microbiological transfer from four materials used in direct contact with apples. *International Journal of Food Microbiology* 333: 108780.

Ayhan, Z. (2019). Packaging and the shelf life of fruits and vegetables. In: *Reference Module in Food Science*. Elsevier.

Ayres Cacciatore, F., Dalmas, M., Maders, C., Ataíde Isaía, H., Brandelli, A. and da Silva Malheiros, P. (2020). Carvacrol encapsulation into nanostructures: Characterization and antimicrobial activity against foodborne pathogens adhered to stainless steel. *Food Research International* 133: 109143.

Bai, J., Abe, K., Iwata, T., Hashimoto, T. and Morita, A. (1989). Effect of a bacterial ethylene absorbent on keeping quality of apples removed from CA storage. *Journal of the Japanese Society for Cold Preservation of Food* 15(3): 119–124.

Bai, J., Baldwin, E., Tsantili, E., Plotto, A., Sun, X., Wang, L., Kafkaletou, M., Wang, Z., Narciso, J., Zhao, W., Xu, S., Seavert, C. and Yang, W. (2019). Modified humidity clamshells to reduce moisture loss and extend storage life of small fruits. *Food Packaging and Shelf Life* 22: 100376.

Bai, J. and Plotto, A. (2011). Coatings for fresh fruits and vegetables. In: Baldwin, E., Hagenmaier, R. and Bai, J. (Eds.), *Edible Coatings and Films to Improve Food Quality* (2nd edn.). Boca Raton: CRC Press, pp. 185–242.

Bai, J., Saftner, R. A. and Watada, A. E. (2003). Characteristics of fresh-cut honeydew (*Cucumis xmelo* L.) available to processors in winter and summer and its quality maintenance by modified atmosphere packaging. *Postharvest Biology and Technology* 28(3): 349–359.

Bai, J.-H., Saftner, R. A., Watada, A. E. and Lee, Y. S. (2001). Modified atmosphere maintains quality of fresh-cut cantaloupe (*Cucumis melo* L.). *Journal of Food Science* 66(8): 1207–1211.

Baker, K. F. and Heald, F. D. (1932). *Some Problems Concerning Blue Mold in Relation to Cleaning and Packing of Apples*. Phytopathology 22(8): 879-898.

Baldwin, E. A., Hagenmaier, R. and Bai, J. (2011). *Edible Coatings and Films to Improve Food Quality*. CRC Press.

Bambace, M. F., Alvarez, M. V. and Moreira, M. D. R. (2019). Novel functional blueberries: Fructo-oligosaccharides and probiotic lactobacilli incorporated into alginate edible coatings. *Food Research International* 122: 653–660.

Basumatary, I. B., Mukherjee, A., Katiyar, V. and Kumar, S. (2022). Biopolymer-based nanocomposite films and coatings: Recent advances in shelf-life improvement of fruits and vegetables. *Critical Reviews in Food Science and Nutrition* 62(7): 1912–1935.

Beaudry, R. (2007). MAP as a basis for active packaging. In: *Intelligent and Active Packaging for Fruits and Vegetables*. Boca Raton, FL: CRC Press, pp. 31–55.

Belay, Z. A., Caleb, O. J. and Opara, U. L. (2019). Influence of initial gas modification on physicochemical quality attributes and molecular changes in fresh and fresh-cut fruit during modified atmosphere packaging. *Food Packaging and Shelf Life* 21: 100359.

Berardinelli, A., Guarnieri, A., Phuntsho, J. and Ragni, L. (2001). Fruit damage assessment in peach packing lines. *Applied Engineering in Agriculture* 17(1): 57–62.

Beulah, A., Sucharitha, K., Dheeraj, S. and Prameela, P. (2021). Effect of casein edible coating on the postharvest quality of fresh guava fruits during ambient storage conditions. *Carpathian Journal of Food Science and Technology* 13: 5–20.

Biji, K. B., Ravishankar, C. N., Mohan, C. O. and Srinivasa Gopal, T. K. (2015). Smart packaging systems for food applications: A review. *Journal of Food Science and Technology* 52(10): 6125–6135.

Blakistone, B. (1999). Introduction. In: Blakistone, B. (eds) Principles and applications of modified atmosphere packaging of foods. . Springer, Boston, MA. pp. 1-13.

Brody, A. L. (2005). Commercial uses of active food packaging and modified atmosphere packaging systems. In: *Innovations in Food Packaging*. Elsevier, pp.457–474.

Brooks, C. (1920). *Diseases of Apples in Storage*. United States Department of Agriculture.

Carpiné, D., Dagostin, J. L. A., Bertan, L. C. and Mafra, M. R. (2015). Development and characterization of soy protein isolate emulsion-based edible films with added coconut oil for olive oil packaging: Barrier, mechanical, and thermal properties. *Food and Bioprocess Technology* 8(8): 1811–1823.

Cerioni, L., Lazarte Mde, L., Villegas, J. M., Rodríguez-Montelongo, L. and Volentini, S. I. (2013). Inhibition of Penicillium expansum by an oxidative treatment. *Food Microbiology* 33(2): 298–301.

Certel, M., Uslu, M. K. and Ozdemir, F. (2004). Effects of sodium caseinate- and milk protein concentrate-based edible coatings on the postharvest quality of Bing cherries. *Journal of the Science of Food and Agriculture* 84(10): 1229–1234.

Cho, H. S., Jeong, H. J., Park, M. Y., Lee, S. M. and Hwang, I. T. (2019). Fruit quality affected by ethylene-absorbent and sulfur dioxide generating pad during cold storage in blueberry (*Vaccinium corymbosum*). *Hortscience* 54(9): S262-S262.

Conesa, A., Verlinden, B. E., Artés-Hernández, F., Nicolaï, B. and Artés, F. (2007). Respiration rates of fresh-cut bell peppers under supertamospheric and low oxygen with or without high carbon dioxide. *Postharvest Biology and Technology* 45(1): 81–88.

Cooley, J. S. and Crenshaw, J. H. (1931). *Control of Botrytis Rot of Pears With Chemically Treated Wrappers*. United States Department of Agriculture.

Corbo, M. R., Lanciotti, R., Gardini, F., Sinigaglia, M. and Guerzoni, M. E. (2000). Effects of hexanal, trans-2-hexenal, and storage temperature on shelf life of fresh sliced apples. *Journal of Agricultural and Food Chemistry* 48(6): 2401–2408.

Cozzolino, R., De Giulio, B., Pellicano, M. P., Pace, B., Capotorto, I., Martignetti, A., D'Agresti, M., Laurino, C. and Cefola, M. (2021). Volatile, quality and olfactory profiles of fresh-cut polignano carrots stored in air or in passive modified atmospheres. *LWT* 137: 110408.

Cozzolino, R., Martignetti, A., Cefola, M., Pace, B., Capotorto, I., De Giulio, B., Montemurro, N. and Pellicano, M. P. (2019). Volatile metabolites, quality and sensory parameters of "Ferrovia" sweet cherry cold stored in air or packed in high CO_2 modified atmospheres. *Food Chemistry* 286: 659–668.

Cui, K., He, L., Cui, G., Zhang, T., Chen, Y., Zhang, T., Mu, W. and Liu, F. (2021). Biological activity of trans-2-hexenal against the storage insect pest *Tribolium castaneum* (Coleoptera: Tenebrionidae) and mycotoxigenic storage fungi. *Journal of Economic Entomology* 114(2): 979–987.

Dai, L., Zhang, J. and Cheng, F. (2019). Effects of starches from different botanical sources and modification methods on physicochemical properties of starch-based edible films. *International Journal of Biological Macromolecules* 132: 897–905.

Dai, L., Zhang, J. and Cheng, F. (2020). Cross-linked starch-based edible coating reinforced by starch nanocrystals and its preservation effect on graded Huangguan pears. *Food Chemistry* 311: 125891.

de Oliveira, K. Á. R., Fernandes, K. F. D. and de Souza, E. L. (2021). Current advances on the development and application of probiotic-loaded edible films and coatings for the bioprotection of fresh and minimally processed fruit and vegetables. *Foods* 10(9): 2207.

Dea, S., Ghidelli, C., Pérez-Gago, M. and Plotto, A. (2011). Coatings for minimally processed fruits and vegetables. In: *Edible Coatings and Films to Improve Food Quality*. Boca Raton, FL: CRC Press, pp. 243–289.

Dhall, R. K. (2013). Advances in edible coatings for fresh fruits and vegetables: A review. *Critical Reviews in Food Science and Nutrition* 53(5): 435–450.

Drake, S. R., Elfving, D. C. and Braden, L. (2004). 'd' Anjou' Pear fruit quality as influenced by paper wraps infused with inorganic and organic materials. *Journal of the American Pomological Society* 58(3): 129–134.

El Matbouly, H., Nikbakhtnasrabadi, F. and Dahiya, R. (2022). RFID near-field communication (NFC)-based sensing technology in food quality control. In: *Biosensing and Micro-Nano Devices*. Springer, pp. 219–241.

Escalona, V. H., Aguayo, E. and Artés, F. (2007a). Modified atmosphere packaging improved quality of kohlrabi stems. *LWT - Food Science and Technology* 40(3): 397–403.

Escalona, V. H., Aguayo, E. and Artés, F. (2007b). Quality changes of fresh-cut kohlrabi sticks under modified atmosphere packaging. *Journal of Food Science* 72(5): S303–S307.

Escobedo, P., Bhattacharjee, M., Nikbakhtnasrabadi, F. and Dahiya, R. (2021). Flexible strain and temperature sensing NFC tag for smart food packaging applications. *IEEE Sensors Journal* 21(23): 26406–26414.

Escobedo, P., Erenas, M. M., Lopez-Ruiz, N., Carvajal, M. A., Gonzalez-Chocano, S., de Orbe-Payá, I., Capitán-Valley, L. F., Palma, A. J. and Martínez-Olmos, A. (2017). Flexible passive near field communication tag for multigas sensing. *Analytical Chemistry* 89(3): 1697–1703.

Fadiji, T., Berry, T. M., Coetzee, C. J. and Opara, U. L. (2018). Mechanical design and performance testing of corrugated paperboard packaging for the postharvest handling of horticultural produce. *Biosystems Engineering* 171: 220–244.

Fadini, A. L., Rocha, F. S., Alvim, I. D., Sadahira, M. S., Queiroz, M. B., Alves, R. M. V. and Silva, L. B. (2013). Mechanical properties and water vapour permeability of hydrolysed collagen–cocoa butter edible films plasticised with sucrose. *Food Hydrocolloids* 30(2): 625–631.

Falagan, N. and Terry, L. A. (2020). 1-Methylcyclopropene maintains postharvest quality in Norwegian apple fruit. *Food Science and Technology International* 26(5): 420–429.

Feng, Z., Li, L., Wang, Q., Wu, G., Liu, C., Jiang, B. and Xu, J. (2019). Effect of antioxidant and antimicrobial coating based on whey protein nanofibrils with TiO_2 nanotubes on the quality and shelf life of chilled meat. *International Journal of Molecular Sciences* 20(5): 1184.

Floros, J. D. and Matsos, K. I. (2005). Introduction to modified atmosphere packaging. In: Han, J. H. (Ed.), *Innovations in Food Packaging* (Vol. 10). London: Academic Press, pp. 159–172.

Fonseca, S. C., Oliveira, F. A. R. and Brecht, J. K. (2002). Modelling respiration rate of fresh fruits and vegetables for modified atmosphere packages: A review. *Journal of Food Engineering* 52(2): 99–119.

Gabler, F. M., Smilanick, J. L., Ghosoph, J. M. and Margosan, D. A. (2005). Impact of postharvest hot water or ethanol treatment of table grapes on gray mold incidence, quality, and ethanol content. *Plant Disease* 89(3): 309–316.

Gaikwad, K. K., Singh, S. and Lee, Y. S. (2016). Functional corrugated board with organic and inorganic materials in food packaging applications: A review. *Korean Journal of Packaging Science & Technology* 22(3): 49–58.

García, M. A., Martino, M. N. and Zaritzky, N. E. (2000). Lipid addition to improve barrier properties of edible starch-based films and coatings. *Journal of Food Science* 65(6): 941–944.

Gee, I. M., Heard, B. R., Webber, M. E. and Miller, S. A. (2020). The future of food: Environmental lessons from e-commerce. *Environmental Science and Technology* 54(23): 14776–14784.

Germano, T. A., Aguiar, R. P., Bastos, M. S. R., Moreira, R. A., Ayala-Zavala, J. F. and de Miranda, M. R. A. (2019). Galactomannan-carnauba wax coating improves the antioxidant status and reduces chilling injury of 'Paluma' guava. *Postharvest Biology and Technology* 149: 9–17.

Ghaani, M., Cozzolino, C. A., Castelli, G. and Farris, S. (2016). An overview of the intelligent packaging technologies in the food sector. *Trends in Food Science and Technology* 51: 1–11.

Ghidelli, C., Mateos, M., Rojas-Argudo, C. and Pérez-Gago, M. B. (2015). Novel approaches to control browning of fresh-cut artichoke: Effect of a soy protein-based coating and modified atmosphere packaging. *Postharvest Biology and Technology* 99: 105–113.

Ghidelli, C. and Pérez-Gago, M. B. (2018). Recent advances in modified atmosphere packaging and edible coatings to maintain quality of fresh-cut fruits and vegetables. *Critical Reviews in Food Science and Nutrition* 58(4): 662–679.

Głąb, T. K. and Boratyński, J. (2017). Potential of casein as a carrier for biologically active agents. *Topics in Current Chemistry* 375(4): 71.

Göksen, G., Fabra, M. J., Ekiz, H. I. and López-Rubio, A. (2020). Phytochemical-loaded electrospun nanofibers as novel active edible films: Characterization and antibacterial efficiency in cheese slices. *Food Control* 112: 107133.

Gómez, P. A. and Artés, F. (2005). Improved keeping quality of minimally fresh processed celery sticks by modified atmosphere packaging. *LWT: Food Science and Technology* 38(4): 323–329.

Gonzalez-Aguilar, G. A., Buta, J. G. and Wang, C. Y. (2003). Methyl jasmonate and modified atmosphere packaging (MAP) reduce decay and maintain postharvest quality of papaya 'Sunrise'. *Postharvest Biology and Technology* 28(3): 361–370.

Han, D., Luo, T., Zhang, L., Wu, J., Wu, H., Wu, Z., Li, J., Wang, J. and Pan, X. (2020). Optimized precooling combined with SO_2-released paper treatment improves the storability of longan (*Dimocarpus longan* Lour.) fruits stored at room temperature. *Food Science and Nutrition* 8(6): 2827–2838.

Heald, F. D. and Ruehle, G. D. (1931). The rots of Washington apples in cold storage. *Washington Agricultural Expert Statistics Bulletin* 253: 48.

Heras-Mozos, R., Garcia-Moreno, A., Monedero-Prieto, M., Tone, A. M., Higueras, L., Hernández-Muñoz, P. and Gavara, R. (2021). Trans-2-hexenal-based antifungal packaging to extend the shelf life of strawberries. *Foods* 10(9): 2166.

Hotchkiss, J. H., Watkins, C. B. and Sanchez, D. G. (2007). Release of 1-methylcyclopropene from heat-pressed polymer films. *Journal of Food Science* 72(5): E330–E334.

Huang, L.-L., Yu, L., Li, B., Li, B., Wang, H. and Li, J. (2022). Adsorption and release of 1-methylcyclopropene by metal-organic frameworks for fruit preservation. *ACS Materials Letters* 4(6): 1053–1057.

Huff, K. (2013). *Active and Intelligent Packaging: Innovations for the Future*. Department of Food Science and Technology, Virginia Polytechnic Institute and State University. Blacksburg, Virginia, USA.

Iseppi, R., Zurlini, C., Cigognini, I. M., Cannavacciuolo, M., Sabia, C. and Messi, P. (2022). Eco-friendly edible packaging systems based on live-Lactobacillus kefiri MM5 for the control of Listeria monocytogenes in fresh vegetables. *Foods* 11(17): 2632.

Jafarizadeh, H., Osman, A., Tan, C. and Abdul, R. (2011). Development of an edible coating based on chitosan-glycerol to delay 'Berangan' banana (*Musa sapientum* cv. Berangan) ripening process. *International Food Research Journal* 18.

Janjarasskul, T. and Krochta, J. M. (2010). Edible packaging materials. *Annual Review of Food Science and Technology* 1: 415–448.

Janjarasskul, T. and Suppakul, P. (2018). Active and intelligent packaging: The indication of quality and safety. *Critical Reviews in Food Science and Nutrition* 58(5): 808–831.

Jayakody, M. M., Vanniarachchy, M. P. G. and Wijesekara, I. (2022). Seaweed derived alginate, agar, and carrageenan based edible coatings and films for the food industry: A review. *Journal of Food Measurement and Characterization* 16(2): 1195–1227.

Jiang, T., Luo, S., Chen, Q., Shen, L. and Ying, T. (2010). Effect of integrated application of gamma irradiation and modified atmosphere packaging on physicochemical and microbiological properties of shiitake mushroom (*Lentinus edodes*). *Food Chemistry* 122(3): 761–767.

Kader, A. A. (1983). Physiological and biochemical effects of carbon monoxide added to controlled atmospheres on fruits. *Acta Horticulturae* (138): 221–226.

Ketelsen, M., Janssen, M. and Hamm, U. (2020). Consumers' response to environmentally-friendly food packaging: A systematic review. *Journal of Cleaner Production* 254: 120123.

King, A. D. and Nagel, C. W. (1975). Influence of carbon dioxide upon the metabolism of *Pseudomonas aeruginosa*. *Journal of Food Science* 40(2): 362–366.

Krochta, J. and De Mulder-Johnston, C. (1997). Edible and biodegradable polymer films: Challenges and opportunities. *Food Technology* 51: 61–72.

Kumar, P. and Sethi, S. (2021). Influence of modified atmospheres on shelf life and quality of fresh-cut apples. *Journal of Packaging Technology and Research* 5(3): 209–216.

Li, R., Ma, J., Gu, H., Jia, W., Shao, Y. and Li, W. (2022a). 1-Methylcyclopropene counteracts ethylene promotion of fruit softening and roles of MiERF2/8 and MiPG in postharvest mangoes. *Frontiers in Plant Science* 13: 971050.

Li, Z. G., Li, X. E. and Chen, H. Y. (2022b). Sulfur dioxide: An emerging signaling molecule in plants. *Frontiers in Plant Science* 13: 891626.

Li, Y., Pi, Q. M., You, H. H., Li, J. Q., Wang, P. C., Yang, X. and Wu, Y. (2018). A smart multi-functional coating based on anti-pathogen micelles tethered with copper nanoparticles via a biosynthesis method using l-vitamin C. *RSC Advances* 8(33): 18272–18283.

Liamnimitr, N., Thammawong, M., Techavuthiporn, C., Fahmy, K., Suzuki, T. and Nakano, K. (2018). Optimization of bulk modified atmosphere packaging for long-term storage of 'Fuyu'persimmon fruit. *Postharvest Biology and Technology* 135: 1–7.

Linke, M. and Geyer, M. (2013). Condensation dynamics in plastic film packaging of fruit and vegetables. *Journal of Food Engineering* 116(1): 144–154.

López-Gálvez, F., Gómez, P. A., Artés, F., Artés-Hernández, F. and Aguayo, E. (2021). Interactions between microbial food safety and environmental sustainability in the fresh produce supply chain. *Foods* 10(7): 1655.

Lu, D., Zhang, M., Wang, S., Cai, J., Zhou, X. and Zhu, C. (2010). Nutritional characterization and changes in quality of *Salicornia bigelovii* Torr. during storage. *LWT: Food Science and Technology* 43(3): 519–524.

Luo, X. (2022). Application of inkjet-printing technology in developing indicators/sensors for intelligent packaging systems. *Current Opinion in Food Science* 46: 100868.

Ma, Y. Y., Fu, L., Hussain, Z., Huang, D. and Zhu, S. (2019). Enhancement of storability and antioxidant systems of sweet cherry fruit by nitric oxide-releasing chitosan nanoparticles (GSNO-CS NPs). *Food Chemistry* 285: 10–21.

Maraveas, C. (2020). Production of sustainable and biodegradable polymers from agricultural waste. *Polymers* 12(5): 1127.

Marcela Fernandez, N., Echeverria, D. C., Andres Mosquera, S. A. and Paz, S. P. (2017). Estado actual del uso de recubrimientos comestibles en frutas y hortalizas. *Biotecnoloía en el Sector Agropecuario y Agroindustrial* 15(2): 134–141.

Margosan, D. A., Smilanick, J. L., Simmons, G. F. and Henson, D. J. (1997). Combination of hot water and ethanol to control postharvest decay of peaches and nectarines. *Plant Disease* 81(12): 1405–1409.

Mariah, M. A. A., Vonnie, J. M., Erna, K. H., Nur'Aqilah, N. M., Huda, N., Abdul Wahab, R. and Rovina, K. (2022). The emergence and impact of ethylene scavengers techniques in delaying the ripening of fruits and vegetables. *Membranes (Basel)* 12(2): 117.

Marzano-Barreda, L. A., Yamashita, F. and Bilck, A. P. (2021). Effect of biodegradable active packaging with zeolites on fresh broccoli florets. *Journal of Food Science and Technology* 58(1): 197–204.

Mendes-Oliveira, G., Gu, G., Luo, Y., Zografos, A., Minas, I. and Nou, X. (2022). Edible and water-soluble corn zein coating impregnated with nisin for Listeria monocytogenes reduction on nectarines and apples. *Postharvest Biology and Technology* 185: 111811.

Mohamed, S. A. A., El-Sakhawy, M. and El-Sakhawy, M. A.-M. (2020). Polysaccharides, protein and lipid-based natural edible films in food packaging: A review. *Carbohydrate Polymers* 238: 116178.

Momin, M. C., Jamir, A. R., Ankalagi, N., Henny, T. and Devi, O. B. (2021). Edible coatings in fruits and vegetables: A brief review. Pharmacology Innovation *Journal* 10: 71–78.

Montalvo, C., López-Malo, A. and Palou, E. (2012). Películas comestibles de proteína: características, propiedades y aplicaciones. *Emas Selectos de Ingeniería de Alimentos* 6: 32–46.

Mostafavi, F. S. and Zaeim, D. (2020). Agar-based edible films for food packaging applications: A review. *International Journal of Biological Macromolecules* 159: 1165–1176.

Moustafa, H., Youssef, A. M., Darwish, N. A. and Abou-Kandil, A. I. (2019). Eco-friendly polymer composites for green packaging: Future vision and challenges. *Composites Part B: Engineering* 172: 16–25.

Mu, H., Gao, H., Chen, H., Fang, X. and Han, Q. (2017). A novel controlled release ethanol emitter: Preparation and effect on some postharvest quality parameters of Chinese bayberry during storage. *Journal of the Science of Food and Agriculture* 97(14): 4929–4936.

Muftuoğlu, F., Ayhan, Z. and Esturk, O. (2012). Modified atmosphere packaging of Kabaaşı apricot (*Prunus armeniaca* L.'Kabaaşı'): Effect of atmosphere, packaging material type and coating on the physicochemical properties and sensory quality. *Food and Bioprocess Technology* 5(5): 1601–1611.

Mullan, M. and McDowell, D. (2011). Modified atmosphere packaging. In: Kirwan, R. (Ed.), *Food and Beverage Packaging Technology* (2nd edn.). Blackwell Publishing Ltd., pp. 263–294.

Murmu, S. B. and Mishra, H. N. (2018). The effect of edible coating based on Arabic gum, sodium caseinate and essential oil of cinnamon and lemon grass on guava. *Food Chemistry* 245: 820–828.

Mustonen, H. M. (1992). The efficacy of a range of sulfur-dioxide generating pads against Botrytis-cinerea infection and on out-turn quality of calmeria table grapes. *Australian Journal of Experimental Agriculture* 32(3): 389–393.

Myllärinen, P., Partanen, R., Seppälä, J. and Forssell, P. (2002). Effect of glycerol on behaviour of amylose and amylopectin films. *Carbohydrate Polymers* 50(4): 355–361.

Namiota, M. and Bonikowski, R. (2021). The current state of knowledge about essential oil fumigation for quality of crops during postharvest. *International Journal of Molecular Sciences* 22(24): 13351.

Newman, K. L., Bartz, F. E., Johnston, L., Moe, C. L., Jaykus, L. A. and Leon, J. S. (2017). Microbial load of fresh produce and paired equipment surfaces in packing facilities near the US and Mexico border. *Journal of Food Protection* 80(4): 582–589.

Ngcobo, M. E. K., Delele, M. A., Opara, U. L. and Meyer, C. J. (2013). Performance of multi-packaging for table grapes based on airflow, cooling rates and fruit quality. *Journal of Food Engineering* 116(2): 613–621.

Nicolau Lapeña, I., Colás-Medà, P., Alegre, I., Aguiló-Aguayo, I., Muranyi, P. and Viñas, I. (2020). Aloe vera gel: An update on its use as a functional edible coating to preserve fruits and vegetables. *Progress in Organic Coatings* 151: 106007.

Nur Hanani, Z. A., Roos, Y. H. and Kerry, J. P. (2014). Use and application of gelatin as potential biodegradable packaging materials for food products. *International Journal of Biological Macromolecules* 71: 94–102.

Occhiuzzi, C., D'uva, N., Nappi, S., Amendola, S., Giallucca, C., Chiabrando, V., Garavaglia, L., Giacalone, G. and Marrocco, G. (2020). Radio-frequency-identification-based intelligent packaging: Electromagnetic classification of tropical fruit ripening. *IEEE Antennas and Propagation Magazine* 62(5): 64–75.

Oliveira, A., Coelho, M., Alexandre, E. M. C., Gomes, M. H., Almeida, D. P. F. and Pintado, M. (2015). Effect of modified atmosphere on phytochemical profile of pasteurized peach purées. *LWT: Food Science and Technology* 64(2): 520–527.

Oms-Oliu, G., Raybaudi-Massilia Martínez, R. M., Soliva-Fortuny, R. and Martín-Belloso, O. (2008a). Effect of superatmospheric and low oxygen modified atmospheres on shelf-life extension of fresh-cut melon. *Food Control* 19(2): 191–199.

Oms-Oliu, G., Soliva-Fortuny, R. and Martín-Belloso, O. (2008b). Physiological and microbiological changes in fresh-cut pears stored in high oxygen active packages compared with low oxygen active and passive modified atmosphere packaging. *Postharvest Biology and Technology* 48(2): 295–301.

Onyeaka, D. H., Nwaizu, C.-C. and Ekaette, I. (2022). Mathematical modeling for thermally treated vacuum-packaged foods: A review on sous vide processing. *Trends in Food Science and Technology* 126: 73–85.

Oregel-Zamudio, E., Angoa-Pérez, M. V., Oyoque-Salcedo, G., Aguilar-González, C. N. and Mena-Violante, H. G. (2017). Effect of candelilla wax edible coatings combined with biocontrol bacteria on strawberry quality during the shelf-life. *Scientia Horticulturae* 214: 273–279.

Ouyang, Y. (2015). *Advanced Printing and Packaging Materials and Technologies*. Trans. Tech. Publishers.

Paidari, S., Zamindar, N., Tahergorabi, R., Kargar, M., Ezzati, S., Shirani, N. and Musavi, S. H. (2021). Edible coating and films as promising packaging: A mini review. *Journal of Food Measurement and Characterization* 15(5): 4205–4214.

Panahirad, S., Dadpour, M., Peighambardoust, S. H., Soltanzadeh, M., Gullón, B., Alirezalu, K. and Lorenzo, J. M. (2021). Applications of carboxymethyl cellulose- and pectin-based active edible coatings in preservation of fruits and vegetables: A review. *Trends in Food Science and Technology* 110: 663–673.

Panea, B. and Ripoll, G. (2019). Plant-derived extracts feed-addition and packaging type influence consumer sensory perception of pork. *Nutrients* 11(11): 2652.

Patil, S. B., Raut, V., Gadekar, A. and Patil, S. M. (2011). Effect of packaging and cushioning materials on shelf life of custard apple fruits under different storage conditions. *Asian Journal of Horticulture* 6(1): 264–268.

Patrignani, F., Siroli, L., Gardini, F. and Lanciotti, R. (2016). Contribution of two different packaging material to microbial contamination of peaches: Implications in their microbiological quality. *Frontiers in Microbiology* 7: 938.

Peralta-Ruiz, Y., Tovar, C. D. G., Sinning-Mangonez, A., Coronell, E. A., Marino, M. F. and Chaves-Lopez, C. (2020). Reduction of postharvest quality loss and microbiological decay of tomato "Chonto" (*Solanum lycopersicum* L.) using chitosan-E essential oil-based edible coatings under low-temperature storage. *Polymers* 12(8): 1822.

Petkoska, A. T., Daniloski, D., D'Cunha, N. M., Naumovski, N. and Broach, A. T. (2021). Edible packaging: Sustainable solutions and novel trends in food packaging. *Food Research International* 140: 109981.

Pinto, L., Palma, A., Cefola, M., Pace, B., D'Aquino, S., Carboni, C. and Baruzzi, F. (2020). Effect of modified atmosphere packaging (MAP) and gaseous ozone pre-packaging treatment on the physico-chemical, microbiological and sensory quality of small berry fruit. *Food Packaging and Shelf Life* 26: 100573.

Pirozzi, A., Pataro, G., Donsì, F. and Ferrari, G. (2021). Edible coating and pulsed light to increase the shelf life of food products. *Food Engineering Reviews* 13(3): 544–569.

Poyatos-Racionero, E., Guari-Borras, G., Ruiz-Rico, M., Morellá-Aucejo, Á., Aznar, E., Barat, J. M., Martínez-Máñez, R., Marcos, M. D. and Bernardos, A. (2021). Towards the enhancement of essential oil components' antimicrobial activity using new Zein protein-gated mesoporous silica microdevices. *International Journal of Molecular Sciences* 22(7): 3795.

Qu, P., Zhang, M., Fan, K. and Guo, Z. (2022). Microporous modified atmosphere packaging to extend shelf life of fresh foods: A review. *Critical Reviews in Food Science and Nutrition* 62(1): 51–65.

Ragaert, P., Verbeke, W., Devlieghere, F. and Debevere, J. (2004). Consumer perception and choice of minimally processed vegetables and packaged fruits. *Food Quality and Preference* 15(3): 259–270.

Rangappa, S. M., Parameswaranpillai, J., Thiagamani, S. M. K., Krishnasamy, S. and Siengchin, S. (2020). *Food Packaging: Advanced Materials, Technologies, and Innovations*. Boca Raton: CRC Press.

Reche, J., García-Pastor, M. E., Valero, D., Hernández, F., Almansa, M. S., Legua, P. and Amorós, A. (2019). Effect of modified atmosphere packaging on the physiological and functional characteristics of Spanish jujube (*Ziziphus jujuba* Mill.) cv. Phoenix. *Scientia Horticulturae* 258: 108743.

Ren, Y., Xue, Y., Tian, D., Zhang, L., Xiao, G. and He, J. (2020). Improvement of postharvest anthracnose resistance in mango fruit by nitric oxide and the possible mechanisms involved. *Journal of Agricultural and Food Chemistry* 68(52): 15460–15467.

Rivera, C. S., Blanco, D., Salvador, M. L. and Venturini, M. E. (2010). Shelf-life extension of fresh tuber aestivum and tuber melanosporum truffles by modified atmosphere packaging with microperforated films. *Journal of Food Science* 75(4): E225–E233.

Robertson, G. (2005). *Food Packaging: Principles and Practice* (2nd edn.). Boca Raton: CRC Press (Vol. 550).

Rodov, V., Ben-Yehoshua, S., Aharoni, N. and Cohen, S. (2010). Modified humidity packaging of fresh produce. *Horticultural Reviews* 37: 281–329.

Rodriguez, J. and Zoffoli, J. P. (2016). Effect of sulfur dioxide and modified atmosphere packaging on blueberry postharvest quality. *Postharvest Biology and Technology* 117: 230–238.

Romanazzi, G., Mlikota Gabler, F. M., Margosan, D., Mackey, B. E. and Smilanick, J. L. (2009). Effect of chitosan dissolved in different acids on its ability to control postharvest gray mold of table grapes. *Phytopathology* 99(9): 1028–1036.

Rotsios, K., Konstantoglou, A., Folinas, D., Fotiadis, T., Hatzithomas, L. and Boutsouki, C. (2022). Evaluating the use of QR codes on food products. *Sustainability* 14(8): 4437.

Rudra, S. G., Singh, V., Jyoti, S. D. and Shivhare, U. S. (2013). Mechanical properties and antimicrobial efficacy of active wrapping paper for primary packaging of fruits. *Food Bioscience* 3: 49–58.

Ruperez, D., Gracia-Valles, N., Clavero, E., Silva, F. and Nerín, C. (2022). Mechanochemically scaled-up alpha cyclodextrin nanosponges: Their safety and effectiveness as ethylene scavenger. *Nanomaterials (Basel)* 12(17): 2900.

Saekow, M., Naradisorn, M., Tongdeesoontorn, W. and Hamauzu, Y. (2019). Effect of carboxymethyl cellulose coating containing ZnO-nanoparticles for prolonging shelf life of persimmon and tomato fruit. *Journal of Food Science and Agricultural Technology (JFAT)* 5: 41–48.

Saito, S., Obenland, D. and Xiao, C. L. (2020). Influence of sulfur dioxide-emitting polyethylene packaging on blueberry decay and quality during extended storage. *Postharvest Biology and Technology* 160: 111045.

Sánchez-Ortega, I., García-Almendárez, B. E., Santos-López, E. M., Reyes-González, L. R. and Regalado, C. (2016). Characterization and antimicrobial effect of starch-based edible coating suspensions. *Food Hydrocolloids* 52: 906–913.

Sangsuwan, J., Rattanapanone, N. and Rachtanapun, P. (2008). Effect of chitosan/methyl cellulose films on microbial and quality characteristics of fresh-cut cantaloupe and pineapple. *Postharvest Biology and Technology* 49(3): 403–410.

Santhosh, R., Nath, D. and Sarkar, P. (2021). Novel food packaging materials including plant-based byproducts: A review. *Trends in Food Science and Technology* 118: 471–489.

Santos, T. M., Men de Sá Filho, M. S. M., Silva, E. O., Silveira, M. R. S. D., Miranda, M. R. A., Lopes, M. M. A. and Azeredo, H. M. C. (2018). Enhancing storage stability of guava with tannic acid-crosslinked zein coatings. *Food Chemistry* 257: 252–258.

Sarker, M. I. and Liu, L. S. (2017). Boron complexes with gradual 1-methylcyclopropene releasing capability. U.S. Patent 9,622,479.

Satekge, T. K. and Magwaza, L. S. (2022). Postharvest application of 1-methylcyclopropene (1-MCP) on climacteric fruits: Factors affecting efficacy. *International Journal of Fruit Science* 22(1): 595–607.

Saxena, A., Saxena, T. M., Raju, P. S. and Bawa, A. S. (2013). Effect of controlled atmosphere storage and chitosan coating on quality of fresh-cut jackfruit bulbs. *Food and Bioprocess Technology* 6(8): 2182–2189.

Saxena, A., Sharma, L. and Maity, T. (2020). Chapter 34: Enrichment of edible coatings and films with plant extracts or essential oils for the preservation of fruits and vegetables. In: Pal, K., Banerjee, I. and Sarkar, P. (Eds.), *Biopolymer-Based Formulations*. Elsevier, pp. 859–880.

Selcuk, N. and Erkan, M. (2014). Changes in antioxidant activity and postharvest quality of sweet pomegranates cv. Hicrannar under modified atmosphere packaging. *Postharvest Biology and Technology* 92: 29–36.

Shaikh, S., Yaqoob, M. and Aggarwal, P. (2021). An overview of biodegradable packaging in food industry. *Current Research in Food Science* 4: 503–520.

Sharma, P., Shehin, V. P., Kaur, N. and Vyas, P. (2019). Application of edible coatings on fresh and minimally processed vegetables: A review. *International Journal of Vegetable Science* 25(3): 295–314.

Published by Burleigh Dodds Science Publishing Limited, 2023.

Sharma, S., Mulrey, L., Byrne, M., Jaiswal, A. K. and Jaiswal, S. (2022). Encapsulation of essential oils in nanocarriers for active food packaging. *Foods* 11(15): 2337.

Sharma, S. and Rao, T. V. R. (2015). Xanthan gum based edible coating enriched with cinnamic acid prevents browning and extends the shelf-life of fresh-cut pears. *LWT: Food Science and Technology* 62(1): 791–800.

Sheng, L. and Zhu, M. J. (2021). Practical in-storage interventions to control foodborne pathogens on fresh produce. *Comprehensive Reviews in Food Science and Food Safety* 20(5): 4584–4611.

Shi, J., Huang, D., Du, Y., et al. (2022). Effects of exogenous nitric oxide treatment on grape berries against *Botrytis cinerea* and *Alternaria alternata* related enzymes and metabolites. *Plant Disease*. Epub ahead of print. DOI: 10.1094/PDIS-04-22-0928-RE.

Shinde, R., Rodov, V., Krishnakumar, S. and Subramanian, J. (2018). Active and intelligent packaging for reducing postharvest losses of fruits and vegetables. *Postharvest Biology and Nanotechnology*: 171–189.

Shivangi, S., Dorairaj, D., Negi, P. S. and Shetty, N. P. (2021). Development and characterisation of a pectin-based edible film that contains mulberry leaf extract and its bio-active components. *Food Hydrocolloids* 121: 107046.

Siro, I., Devlieghere, F., Jacxsens, L., Uyttendaele, M. and Debevere, J. (2006). The microbial safety of strawberry and raspberry fruits packaged in high-oxygen and equilibrium-modified atmospheres compared to air storage. *International Journal of Food Science and Technology* 41(1): 93–103.

Siroli, L., Patrignani, F., Serrazanetti, D. I., Chiavari, C., Benevelli, M., Grazia, L. and Lanciotti, R. (2017). Survival of spoilage and pathogenic microorganisms on cardboard and plastic packaging materials. *Frontiers in Microbiology* 8: 2606.

Sisler, E. C. and Serek, M. (1997). Inhibitors of ethylene responses in plants at the receptor level: Recent developments. *Physiologia Plantarum* 100(3): 577–582.

Smock, R. (1957). A comparison of treatments for control of the apple scald disease. *Proceedings of the American Society for Horticultural Science* 69(1): 91-100.

Solano-Doblado, L. G., Alamilla-Beltrán, L. and Jiménez-Martínez, C. (2018). Películas y recubrimientos comestibles funcionalizados. *TIP Revista Especializada en Ciencias Químico-Biológicas* 21.

Sousa-Gallagher, M. J. and Mahajan, P. V. (2013). Integrative mathematical modelling for MAP design of fresh-produce: Theoretical analysis and experimental validation. *Food Control* 29(2): 444–450.

Spotts, R. A. and Cervantes, L. A. (1994). Contamination of harvest bins with pear decay fungi and evaluation of disinfestants on plastic and wood bin material. In: International Society for Horticultural Science (ISHS) (Ed.), *Acta Horticulturae*. Belgium: Leuven, Vol. 367, pp. 419–425.

Spruit, D. and Almenar, E. (2021). First market study in e-commerce food packaging: Resources, performance, and trends. *Food Packaging and Shelf Life* 29: 100698.

Sultana, A., Kathuria, A. and Gaikwad, K. K. (2022). Metal-organic frameworks for active food packaging: A review. *Environmental Chemistry Letters* 20(2): 1479–1495.

Sun, X., Bai, J., Ference, C., Wang, Z., Zhang, Y., Narciso, J. and Zhou, K. (2014). Antimicrobial activity of controlled-release chlorine dioxide gas on fresh blueberries. *Journal of Food Protection* 77(7): 1127–1132.

Sun, X., Baldwin, E. and Bai, J. (2019). Applications of gaseous chlorine dioxide on postharvest handling and storage of fruits and vegetables - A review. *Food Control* 95: 18–26.

Sun, X., Baldwin, E., Ference, C., Narciso, J., Plotto, A., Ritenour, M., Harrison, K., Gangemi, D. and Bai, J. (2017a). The effect of controlled-release chlorine dioxide on the preservation of grapefruit. *Hortscience* 52(1): 122–126.

Sun, X., Baldwin, E., Plotto, A., Narciso, J., Ference, C., Ritenour, M., Harrison, K., Gangemi, J. and Bai, J. (2017b). Controlled-release of chlorine dioxide in a perforated packaging system to extend the storage life and improve the safety of grape tomatoes. *Journal of Visualized Experiments: JoVE* 122. DOI: 10.3791/55400.

Sun, X., Zhou, B., Luo, Y., Ference, C., Baldwin, E., Harrison, K. and Bai, J. (2017c). Effect of controlled-release chlorine dioxide on the quality and safety of cherry/grape tomatoes. *Food Control* 82: 26–30.

Sun, X., Cameron, R. G., Plotto, A., Zhong, T., Ference, C. M. and Bai, J. (2021). The effect of controlled-release carvacrol on safety and quality of blueberries stored in perforated packaging. *Foods* 10(7): 1487.

Sydow, Z. and Bieńczak, K. (2019). The overview on the use of natural fibers reinforced composites for food packaging. *Journal of Natural Fibers* 16(8): 1189–1200.

Tano, K., Oulé, M. K., Doyon, G., Lencki, R. W. and Arul, J. (2007). Comparative evaluation of the effect of storage temperature fluctuation on modified atmosphere packages of selected fruit and vegetables. *Postharvest Biology and Technology* 46(3): 212–221.

Tao, F., Zhang, M., Hangqing, Y. and Jincai, S. (2006). *RETRACTED: Effects of Different Storage Conditions on Chemical and Physical Properties of White Mushrooms after Vacuum Cooling*. Elsevier.

Thulasisingh, A., Kumar, K., Yamunadevi, B., Poojitha, N., SuhailMadharHanif, S. and Kannaiyan, S. (2021). Biodegradable packaging materials. *Polymer Bulletin*: 1–30.

Tian, F., Chen, W., Wu, C., Kou, X., Fan, G., Li, T. and Wu, Z. (2019). Preservation of Ginkgo biloba seeds by coating with chitosan/Nano-TiO_2 and chitosan/Nano-SiO_2 films. *International Journal of Biological Macromolecules* 126: 917–925.

Timm, E., Bollen, A., Rue, B. D. and Woodhead, I. M. (1998). Apple damage and compressive forces in bulk bins during orchard transport. *Applied Engineering in Agriculture* 14(2): 165–172.

Tkaczewska, J. (2020). Peptides and protein hydrolysates as food preservatives and bioactive components of edible films and coatings - A review. *Trends in Food Science and Technology* 106: 298–311.

Toivonen, P. and Lu, C. (2006). *Compositions and Methods to Improve the Storage Quality of Packaged Plants*. Google Patents.

Torun, M. and Ozdemir, F. (2022). Milk protein and zein coatings over peeled garlic cloves to extend their shelf life. *Scientia Horticulturae* 291: 110571.

Toscano, S., Rizzo, V., Licciardello, F., Romano, D. and Muratore, G. (2021). Packaging solutions to extend the shelf life of green asparagus (*Asparagus officinalis* L.) 'Vegalim'. *Foods* 10(2): 478.

Tudela, J. A., Marín, A., Garrido, Y., Cantwell, M., Medina-Martínez, M. S. and Gil, M. I. (2013). Off-odour development in modified atmosphere packaged baby spinach is an unresolved problem. *Postharvest Biology and Technology* 75: 75–85.

Utto, W., Preutikul, R., Malila, P., Noomhorm, A. and Bronlund, J. E. (2018). Delaying microbial proliferation in freshly peeled shallots by active packaging incorporating ethanol vapour-controlled release sachets and low storage temperature. *Food Science and Technology International* 24(2): 132–144.

Valencia-Chamorro, S. and Torres-Morales, J. (2016). Recubrimientos comestibles aplicados en productos de IV y V gamma. *Revista Iberoamericana de Tecnología Postcosecha* 17(2): 162–174.

Vandichel, R. and D'Alesio, S. (2015). A secondary packaging solution development suitable for E-groceries. *Cities* 2(37): 37.

Villegas, B. M., Hall, N. O., Ryser, E. T. and Marks, B. P. (2020). Influence of physical variables on the transfer of Salmonella Typhimurium LT2 between potato (*Solanum tuberosum*) and stainless steel via static and dynamic contact. *Food Microbiology* 92: 103607.

Volpe, S., Torrieri, E. and Cavella, S. (2018). Use of chitosan and chitosan-caseinate coating to prolong shelf life of minimally processed apples. *Italian Journal of Food Science* 30(5SI): 30–37.

Waghmare, R. B. and Annapure, U. S. (2013). Combined effect of chemical treatment and/or modified atmosphere packaging (MAP) on quality of fresh-cut papaya. *Postharvest Biology and Technology* 85: 147–153.

Wakai, J., Kusama, S., Nakajima, K., Kawai, S., Okumura, Y. and Shiojiri, K. (2019). Effects of trans-2-hexenal and cis-3-hexenal on post-harvest strawberry. *Scientific Reports* 9(1): 10112.

Wang, L., Hong, K., Xu, R., Zhao, Z. and Cao, J. (2021). The alleviation of cold-stimulated flesh reddening in 'Friar' plum fruit by the elevated CO2 with polyvinyl chloride (PVC) packaging. *Scientia Horticulturae* 281: 109997.

Wang, L., Wang, F., Zhang, Y., Ma, Y., Guo, Y. and Zhang, X. (2020). Enhancing the ascorbate–glutathione cycle reduced fermentation by increasing NAD^+ levels during broccoli head storage under controlled atmosphere. *Postharvest Biology and Technology* 165: 111169.

Wang, S., Liu, X., Yang, M., Zhang, Y., Xiang, K. and Tang, R. (2015). Review of time temperature indicators as quality monitors in food packaging. *Packaging Technology and Science* 28(10): 839–867.

Wani, S. M., Gull, A., Ahad, T., Malik, A. R., Ganaie, T. A., Masoodi, F. A. and Gani, A. (2021). Effect of gum arabic, xanthan and carrageenan coatings containing antimicrobial agent on postharvest quality of strawberry: Assessing the physicochemical, enzyme activity and bioactive properties. *International Journal of Biological Macromolecules* 183: 2100–2108.

Wei, H., Seidi, F., Zhang, T., Jin, Y. and Xiao, H. (2021). Ethylene scavengers for the preservation of fruits and vegetables: A review. *Food Chemistry* 337: 127750.

Wei, L., Zhang, J., Wei, S., Hu, D., Liu, Y., Feng, L., Li, C., Qi, N., Wang, C. and Liao, W. (2022). Nitric oxide enhanced salt stress tolerance in tomato seedlings, involving phytohormone equilibrium and photosynthesis. *International Journal of Molecular Sciences* 23(9): 4539.

Wellman, R. F. and Heald, F. D. (1938). Steam sterilization of apple boxes for blue mold. *Washington Agricultural Expert Statistics Bulletin* 357: 16.

Xing, Y., Li, W., Wang, Q., Li, X., Xu, Q., Guo, X., Bi, X., Liu, X., Shui, Y., Lin, H. and Yang, H. (2019). Antimicrobial nanoparticles incorporated in edible coatings and films for the preservation of fruits and vegetables. *Molecules* 24(9): 1695.

Xu, Y., Chen, L., Zhang, Y., Huang, Y., Cao, J. and Jiang, W. (2023). Antimicrobial and controlled release properties of nanocomposite film containing thymol and carvacrol loaded UiO-66-NH($_2$) for active food packaging. *Food Chemistry* 404(A): 134427.

Published by Burleigh Dodds Science Publishing Limited, 2023.

Yam, K. L., Takhistov, P. T. and Miltz, J. (2005). Intelligent packaging: Concepts and applications. *Journal of Food Science* 70(1): R1–R10.

Yan, Y., Tang, C. and Zhang, L. (2020). Personalized design of food packaging driven by user preferences. In: *Advances in Usability and User Experience. Proceedings of the AHFE 2019 International Conferences on Usability & User Experience, and Human Factors and Assistive Technology*, Washington DC, USA. Springer, pp. 514–523.

Yang, Q., Zhang, X., Wang, F. and Zhao, Q. (2019). Effect of pressurized argon combined with controlled atmosphere on the postharvest quality and browning of sweet cherries. *Postharvest Biology and Technology* 147: 59–67.

Ye, J.-J., Li, J.-R., Han, X.-X., Zhang, L., Jiang, T. and Xia, M. (2012). Effects of active modified atmosphere packaging on postharvest quality of shiitake mushrooms (*Lentinula edodes*) stored at cold storage. *Journal of Integrative Agriculture* 11(3): 474–482.

Youssef, K., Ligorio, A., Sanzani, S. M., Nigro, F. and Ippolito, A. (2012). Control of storage diseases of citrus by pre- and postharvest application of salts. *Postharvest Biology and Technology* 72: 57–63.

Yousuf, B., Sun, Y. and Wu, S. (2021). Lipid and lipid-containing composite edible coatings and films. *Food Reviews International* 38: 574–597. DOI: 10.1080/87559129.2021.1876084.

Yu, Z., Xu, H., Yang, Z. and Guo, B. (2015). Personalized travel package with multi-point-of-interest recommendation based on crowdsourced user footprints. *IEEE Transactions on Human-Machine Systems* 46(1): 151–158.

Yun, X., Wang, Y., Li, M., Jin, Y., Han, Y. and Dong, T. (2017). Application of permselective poly(ε-caprolactone) film for equilibrium-modified atmosphere packaging of strawberry in cold storage. *Journal of Food Processing and Preservation* 41(6): e13247.

Zaitoon, A., Luo, X. and Lim, L. T. (2022). Triggered and controlled release of active gaseous/volatile compounds for active packaging applications of agri-food products: A review. *Comprehensive Reviews in Food Science and Food Safety* 21(1): 541–579.

Zambrano-Zaragoza, M. L., Quintanar-Guerrero, D., Del Real, A., González-Reza, R. M., Cornejo-Villegas, M. A. and Gutiérrez-Cortez, E. (2020). Effect of nano-edible coating based on beeswax solid lipid nanoparticles on Strawberry's preservation. *Coatings* 10(3): 253.

Zhang, B., Huang, C., Zhang, L., Wang, J., Huang, X., Zhao, Y., Liu, Y. and Li, C. (2019). Application of chlorine dioxide microcapsule sustained-release antibacterial films for preservation of mangos. *Journal of Food Science and Technology* 56(3): 1095–1103.

Zhang, H. and Sablani, S. (2021). Biodegradable packaging reinforced with plant-based food waste and by-products. *Current Opinion in Food Science* 42: 61–68.

Zhang, L., Chen, F., Lai, S., Wang, H. and Yang, H. (2018). Impact of soybean protein isolate-chitosan edible coating on the softening of apricot fruit during storage. *LWT* 96: 604–611.

Zhang, M., Meng, X., Bhandari, B., Fang, Z. and Chen, H. (2015). Recent application of modified atmosphere packaging (MAP) in fresh and fresh-cut foods. *Food Reviews International* 31(2): 172–193.

Zhang, W. and Jiang, W. (2020). Antioxidant and antibacterial chitosan film with tea polyphenols-mediated green synthesis silver nanoparticle via a novel one-pot method. *International Journal of Biological Macromolecules* 155: 1252–1261.

Published by Burleigh Dodds Science Publishing Limited, 2023.

Zhang, Y., Gao, C., Masum, M. M. I., Cheng, Y., Wei, C., Guan, Y. and Guan, J. (2021). Dynamic microbiome changes reveal the effect of 1-methylcyclopropene treatment on reducing post-harvest fruit decay in "doyenne du comice" pear. *Frontiers in Microbiology* 12: 729014.

Zhang, Z., Wei, J., Wang, M., Zhang, J. and Wu, B. (2022). Induced sulfur metabolism by sulfur dioxide maintains postharvest quality of 'Thompson Seedless' grape through increasing sulfite content. *Journal of the Science of Food and Agriculture* 102(3): 1174–1184.

Zhao, X., Xia, M., Wei, X., Xu, C., Luo, Z. and Mao, L. (2019). Consolidated cold and modified atmosphere package system for fresh strawberry supply chains. *LWT* 109: 207–215.

Zhong, T., Zhang, J., Sun, X., Kou, J., Zhang, Z., Bai, J. and Ritenour, M. A. (2021). The potential of gaseous chlorine dioxide for the control of citrus postharvest stem-end rot caused by lasiodiplodia theobromae. *Plant Disease* 105(11): 3426–3432.

Zhou, H. W., Yuan, B., Qiao, H. S., Leilei, G., Bingbing, Z., Wei, C. and Chongjiang, C. (2020). LED light-triggered release of nitric oxide from NTC to delay the ripening of banana. *LWT: Food Science and Technology* 134: 110129.

Zhu, Y., Wu, F., Trmcic, A., Wang, S. and Warriner, K. (2020). Microbiological status of reusable plastic containers in commercial grower/packer operations and risk of Salmonella cross-contamination between containers and cucumbers. *Food Control* 110: 107021.

Chapter 11

The role of good manufacturing practice and hazard analysis and critical control point systems in maintaining the safety of minimally processed fresh produce

Carol A. Wallace, Jan M. Soon and Shingai P. Nyarugwe, University of Central Lancashire, UK

1 Introduction

Hazard analysis and critical control point (HACCP) systems, supported by good manufacturing practices (GMPs), have been the backbone of food safety management systems since the latter part of the twentieth century. First developed as part of the US space program to ensure safe food for the early manned space flights (Ross-Nazzal, 2007), HACCP was pioneered in the wider food industry by the Pillsbury Company which launched the approach to the industry in the early 1970s. International acceptance of HACCP quickly followed, and it was promoted by the World Health Organisation (WHO) as the most cost-effective way to assure safe food production (WHO, 1997). The internationally accepted HACCP principles were defined in 1993 (CAC, 1993) by the joint United Nations Food and Agriculture Organisation (FAO)/WHO Codex Committee on Food Hygiene. At the same time and with members in common, the *National Advisory Committee on Microbiological Criteria for Foods* (NACMCF) published its similar HACCP principles document (NACMCF, 1992).

http://dx.doi.org/10.19103/AS.2023.0121.18

GMPs pre-date the HACCP principles and are defined as actions applied to the production of food that help ensure food products are consistently produced and controlled according to safety and quality standards (de Oliveira et al., 2016). With the development of HACCP and need for good hygienic control of operational standards to support HACCP and management of critical control points (CCPs), GMPs became more formally applied and the term 'prerequisite programs' was developed to show their importance as a prerequisite to HACCP application in food businesses. This chapter will outline current best practices in the application of GMP and HACCP systems (CCFH, 2020) in maintaining the safety of minimally processed fresh produce.

Definitions of different levels of processing of foods are not universally agreed and may differ when provided from different perspectives. Several definitions regarding 'minimally processed foods' come from the perspective of nutritional science, rather than food science or food safety, where minimally processed foods may be regarded as natural foods altered by methods that include removal of inedible or unwanted parts, and also processes that include drying, crushing, grinding, powdering, fractioning, filtering, roasting, boiling, non-alcoholic fermentation, pasteurisation, chilling, freezing, placing in containers and vacuum packaging (Monteiro et al., 2019). According to Monteiro et al. (2019), the main aims of minimal processing are to extend the life of products, enabling their storage for longer use, and often to make their preparation easier or more diverse. With such a wide group of processes being applied to foods that are considered minimally processed, the range of foods meeting the definition is also wide and varied, e.g. grains, legumes, vegetables, fruits, nuts, herbs, spices, milk, meat and other foods, and, importantly for food safety, encompasses a multiplicity of potential food safety hazards.

In the context of the microbiological safety of fresh produce, minimally processed products will include ready-to-eat produce categories with no lethal kill step or equivalent, often described as 'fresh-cut produce' (FDA, 2008). This includes fruit and vegetables that have received some preparation steps, such as washing, other surface treatments, trimming, shredding, coring, peeling, slicing and chopping. Product examples include washed and bagged leafy greens, cut melon and other fruits, broccoli/cauliflower florets, peeled carrots, cut celery sticks, fruit salads, etc. Minimal processing may include hurdle technology, a combination of techniques and factors to reduce or control pathogens to create safe foods (Leistner and Gorris, 1995; Wallace et al., 2018). Mukhopadhyay and Gorris (2014) state that the concept of hurdle technology fits well with the consumer trend for minimally processed foods.

Studies on implementation of GMPs and HACCP-based food safety management programs among fresh produce and minimally processed fruits and vegetables producers indicated that establishments with GMPs and HACCP produced products with better microbiological quality. In a recent study, Maldonade et al. (2019) reported that food industries and retailers that adopted GMPs showed reduced contamination rates in minimally processed fresh produce. Regarding HACCP, in Greece, Kokkinakis and Fragkiadakis (2007) compared the microbiological quality of tomatoes in establishments with and without HACCP. The majority of tomato samples from HACCP sites contained < 20 CFU/g *E. coli* whilst sites operating without or lacking in HACCP mostly contained > 100 CFU/g *E. coli*; in addition, *Listeria monocytogenes* was not detected in tomatoes from sites producing under GAP and HACCP but was detected in between 26% and 38% of samples from sites without these hygiene and safety controls. Similarly, a study of fruit and vegetable juices in Korea found samples without HACCP processes showed higher levels and proportion of microbial contamination than samples with HACCP processes and a correlation between the microbial quality of fruit and vegetable juices and the application of sterilisation and HACCP (Lee et al., 2021). Frankish et al. (2022) investigated food safety management systems in Australian Apple Packhouses and found higher performance in packhouses with better-designed controls and/or where there was higher employee food safety knowledge. Conversely, low requirements for food safety knowledge amongst workers, limited validation of preventive controls and limited use of expert knowledge for risk assessment were found to be common food safety system weaknesses (Frankish et al., 2022).

Systems and practices to assure the safety of food products, such as GMP and HACCP, apply in the same way to minimally processed fresh produce as they do for all other food products. The challenges of assuring safety relate to the range of microbiological hazards that may be present in raw produce and the lack of a lethal kill step that can be applied within a HACCP system. This chapter considers key hazards of interest in fresh produce, lessons from outbreaks in the minimally processed fresh produce sector and outlines current best practices in the application of GMP and HACCP to assure the safety of fresh produce for consumption.

2 Key hazards in minimally processed fresh produce and lessons from previous outbreaks and incidents

2.1 Key hazards in minimally processed foods

Outbreaks emanating from minimally processed foods have been mainly associated with *Salmonella* spp., *E. coli*, *L. monocytogenes*, *Bacillus cereus*, *Campylobacter* spp., *Staphylococcus* spp., *Yersinia enterocolitica*, *Shigella* spp., *Clostridium botulinum*, *Clostridium perfringens*, *Aeromonas* spp., hepatitis A and norovirus (Carstens et al., 2019; EFSA Panel on Biological Hazards, 2013). The FDA defines foodborne illness outbreaks as an event where at least two or more people are identified to have the same illness emanating from the same contaminated food (FDA, 2021a). Food products that have been implicated in these outbreaks include minimally processed fresh produce such as melons, cantaloupe, mangoes and leafy vegetables including cabbages, lettuce and spinach, root vegetables, cauliflower, broccoli, asparagus and others such as onions and radishes (Hussain and Gooneratne, 2017). EFSA Panel on Biological Hazards (2013) suggested that most food/pathogen outbreak combinations involving fresh produce were: (i) *Salmonella* spp. and leafy greens, bulb and stem vegetables, tomatoes, melons, sprouted seeds or raspberries; (ii) pathogenic *E. coli* and fresh pods or legumes; (iii) norovirus and leafy greens or bulb and stem vegetables, tomatoes, carrots or raspberries; (iv) *Shigella* spp. and fresh pods, fresh herbs, legumes or carrots.

2.2 Sources of contamination

Minimally processed fresh produce are slightly altered from their natural state using processes that are unable to completely prevent, eradicate or reduce the microorganisms to sufficiently safe levels (Siroli et al., 2015). Carstens et al. (2019) reported that pathogenic microorganisms are found in these types of produce, as they are not subjected to interventions to reduce or eliminate the occurrence of such (EFSA Panel on Biological Hazards, 2014e). The unit operations involved such as peeling, slicing and shredding commonly used in minimal processing introduce and stimulate growth of any surviving microorganisms in the inner tissues (Siroli et al., 2015). The most common routes of contamination for minimally processed fruits and vegetables include pre-harvest contaminants from the soil, irrigation water, spraying before harvesting, insufficiently treated organic amendments and contamination/cross-contamination during harvesting and post-harvest from tools, water, equipment, people, packaging, transport, processing and storage (Carstens et al., 2019; EFSA Panel on Biological Hazards, 2014e). For example, *Salmonella*, *Shigella*, *Yersinia* and norovirus are mostly found in minimally processed bulb and stem vegetables as well as carrots as these have prolonged periods of contact with the soil during their growth (EFSA Panel on Biological Hazards, 2014e). The microorganisms can be found on the surface of plants (epiphytes) and can become internalised through damaged plant tissue or internalised via root colonisation (endophytes) (Warriner et al., 2003). Epiphytic bacteria are typically gram negatives such as pseudomonads and the leaf surfaces are a suitable habitat of these pathogens as they tend to aggregate in epidermal cells. Through damaged plants, the stomata provide entry into the sub-surface structures, e.g. substomatal cavity and other internal parts of the leaves where pathogens can reside. However, pathogens can also be found in undamaged produce (endophytes) by uptake through the roots (Hirneisen et al., 2012). Examples of internalisation of pathogens include *Salmonella* in sprouted seeds; *E. coli* O157:H7 in lettuce, cabbages and apples;

Salmonella in tomatoes; *Bacillus subtilis* L-forms in strawberries, and norovirus in spinach and onions (Hirneisen et al., 2012; Warriner et al., 2003).

Contamination can also be amplified during processing if appropriate preventive controls are not in place (Marler, 2022). Moreover, pathogens are likely to survive on the surfaces of these fruits and vegetables at both ambient and refrigeration temperatures for several days/weeks (EFSA Panel on Biological Hazards, 2014e). Webb et al. (2019) highlighted psychrotrophic *B. cereus* as a food safety hazard to chilled minimally processed foods as spores could survive the mild treatments and the minimal processing involved. Being a facultative anaerobic bacterium, it is capable of multiplying at 8 °C and below and ubiquitous in the environment presenting a hazard to the consumption of minimally processed chilled foods (Webb et al., 2019). Moreover, when pathogens are internalised, minimal processing, including washing, surface sterilisation, use of antimicrobial agents, and UV radiation have proven ineffective (Warriner et al., 2003).

2.3 Examples of foodborne disease outbreaks in minimally processed food products

2.3.1 Leafy greens

Leafy greens have been categorised as the highest concern to food safety from a fresh produce perspective (FAO/WHO, 2008) and food safety issues relating to leafy greens is a long-standing concern. *E. coli, Salmonella* and norovirus have been reported and implicated in several foodborne outbreaks concerning leafy green vegetables (EFSA Panel on Biological Hazards, 2014a; EFSA Panel on Biological Hazards, 2014b; FAO/WHO, 2008). In November 2021, the US Centers for Disease Control and Prevention (CDC) reported ten people in seven states infected with *E. coli* O157:H7 owing from consumption of spinach (FDA, 2021b). A study by Marshall et al. (2020) on *Shiga* toxin-producing *E. coli* outbreaks during 2009 to 2018 in Canada and the USA revealed STEC O157 linked to leafy green vegetables as the most common source of foodborne STEC O157. In this study, most outbreaks were linked to romaine lettuce (54%). Environmental assessments suggest a possible link between agricultural and manufacturing practices, including the growing, harvesting and post-harvest practices, the intrinsic factors and STEC contaminations (Marshall et al., 2020). A similar strain was reported in December 2021 where traceback evidence also indicated romaine lettuce to be the likely outbreak source. Testing of canal water in the region where the lettuce is grown identified the outbreak strain (*E. coli* O157:H7), suggesting the water could have contaminated the lettuce (Marler, 2022).

A multistate outbreak of listeriosis associated with packaged leafy green salads was reported between 2015 and 2016. Self et al. (2019) suggested that technologies that enable increased shelf life of fresh produce, the ability of *L. monocytogenes* to grow in cold environments and form biofilms, and manufacturing practices could have contributed to the presence of *L. monocytogenes* in the identified processing facility. The US FDA also reported a *Salmonella* Typhimurium outbreak that resulted in 31 reported illnesses in 2021. In their report, water management practices and sanitation practices were indicated to be insufficient, and the study highlighted the importance of analysing all microbial hazards, including those of nearby and adjacent land uses (Food Safety News, 2022).

2.3.2 Sprouted seeds, onions and tomatoes

Sprouted seeds, onions and tomatoes were identified as the second highest level of concern together with berries and melons (FAO/WHO, 2008). The Shiga toxin-producing *E. coli* (STEC) O104:H4 outbreak, which occurred in Germany and France in 2011, was one of the largest outbreaks of this kind. The outbreak involved at least 4075 cases and 50 deaths in 16 countries in Europe and North America (CDC, 2013; European Food Safety Authority, 2011), and it was linked to different sprouted seed mixtures with fenugreek seeds imported from Egypt as the most probable cause (European Food Safety Authority, 2011). For onions, *Salmonella* Oranienburg was reported by the FDA as of November 2021. In this outbreak, at least 892 illnesses and 183 hospitalisations were reported, and the source was identified as fresh whole onions imported from Mexico with at least 38 states and Puerto Rico reporting such cases (FDA, 2021b). Valadez et al. (2012) listed outbreaks related with tomatoes from 1990 to 2009. *Salmonella* was the most implicated in the outbreaks. In 2019, an outbreak in Sweden was also linked to *Salmonella* in tomatoes. *Salmonella* can be transferred to the flesh through further handling and can also persist on the surface of intact tomatoes. Growth can also be promoted on sliced, diced, cut tomatoes and some tomato products especially when stored at temperatures which will allow growth. There is also evidence for the survival of *Salmonella* in tomato juice (EFSA Panel on Biological Hazards, 2014c).

2.3.3 Fruits

Of the 841 foodborne disease outbreaks reported in the USA in 2017, which resulted in 14481 illnesses, fruits accounted for at least 521 illnesses, which were the most illnesses in the produce sector and second single food category (to turkey) causing outbreak-related illnesses (CDC, 2019). The FAO/WHO reported berries and melons to be the second highest level of concern together with onions and tomatoes (FAO/WHO, 2008). Berries (strawberries, raspberries, blueberries and blackberries) have been implicated in outbreaks

(EFSA Panel on Biological Hazards, 2014a). Examples include the 2019 outbreak of hepatitis A virus linked to fresh blackberries, 2016 norovirus outbreak linked to strawberries and grapes, 2011 outbreak of *E. coli* O157:H4 linked to strawberries and the 2009 multistate *Salmonella* outbreak linked to blueberries. Minimally processed berries are usually not subjected to physical interventions that would result in substantial elimination or reduction of the occurrence of *Salmonella* and norovirus (EFSA Panel on Biological Hazards, 2014a). Regarding melons, multistate/multicountry outbreaks have been reported. Between 15 March and 6 July 2021, a multicountry outbreak of *Salmonella* Braenderup sequence type 22 (ST22) linked to imported melons was confirmed in 348 cases. The cases were reported in 12 EU (European Union)/European Economic Area (EEA) countries and the United Kingdom. Melons and watermelons are a good matrix for growth of bacteria and pathogens including *Salmonella,* owing to the considerable amounts of protein (0.8%) and sugar levels, which depend on the cultivar. Damage to the fruit surface, e.g. due to bruising or cracking during harvest and cutting the melons prior to consumption, introduces the bacteria in the flesh of the fruit (EFSA Panel on Biological Hazards, 2014d). The pH (5.1-6-7) of the flesh presents a good substrate for the growth of *Salmonella*. Risk factors also include ring fungi rot, yeasts and spoilage moulds (e.g. *Cladosporium cladosporioides, Geotrichum candidum*) (EFSA Panel on Biological Hazards, 2014d).

To prevent and/or eliminate these hazards and ultimately the foodborne outbreaks from occurring, several key measures are required. These include removing or discouraging animals near areas where fresh produce is cultivated and in agricultural areas, avoiding the use of manure-based agricultural amendments, the disinfection of irrigation water, reducing the use of agricultural chemicals (Beretti and Stuart, 2008; Olimpi et al., 2019). Soil health and quality and practices that safeguard the same should also be considered and implemented and drainage also considered (Alegbeleye et al., 2018). Moreover, portable water should be used in all agricultural processes, reducing splashing, avoiding use of sprayers and decontamination of seeds (Alegbeleye et al., 2018). The FDA established standards for growing, harvesting, packing and holding of produce for human consumption which specify processes, procedures, practices and requirements for training, worker health and hygiene, quality of agricultural water, biological soil amendments of animal origin and human waste, domesticated and wild animals, and equipment, tools, buildings and sanitation to prevent the occurrence of hazards and minimise the risk contamination and adverse health effects (FDA, 2015). Many of these points form part of good agricultural practices and GMPs are considered in the following section.

3 Developments in good manufacturing practices for minimally processed fresh produce

GMPs are a prerequisite program for HACCP implementation as they lay the foundation for a safe and hygienic environment to protect food against potential food safety hazards, ensure proper building and equipment design and control of food manufacturing processes, help to protect consumer health and environment and reduce food waste. The Codex Alimentarius Code of Practice on the *General Principles of Food Hygiene* (CCFH, 2020) and *Code of Hygienic Practice for fresh fruits and vegetables* (CAC/RCP 53 - 2003) provide guidance on design and facilities, process control, maintenance and sanitation and personal hygiene. The US FDA established the current good manufacturing practices (CGMPs) in manufacturing, packing or holding human food in 21 CFR Part 110 (US FDA, 2008) and in September 2015, US FDA modernised the CGMPs and established them in 21 CFR Part 117 with the requirements for hazard analysis and risk-based preventive controls (US FDA, 2020).

The elements of GMP include personnel, building, facilities and surroundings, equipment, cleaning and sanitation operations, production and process controls, documentation, and recordkeeping and traceback and recall.

3.1 Personnel

Personnel hygiene practices have been identified as effective strategies to minimise cross-contamination of foodborne pathogens and prevent foodborne outbreaks. Personnel hygiene practices include wearing hairnets, beard covers, overalls and gloves to prevent contamination. Although single-use gloves are used as a physical barrier to ensure food hygiene, however the ability of gloves as bacterial and viral vectors has been demonstrated (Qi et al., 2020; Sobolik et al., 2021). It is highly important that handwashing be carried out appropriately and consistently even if a gloving policy is in place. Effective and high handwashing compliance reduced risks of foodborne illnesses (Sobolik et al., 2021). Employees should be trained to understand the consequences of illness and open lesions and should be trained not to enter or work in the production area if they exhibit symptoms of a contagious illness. Studies have reported that furloughing ill workers could potentially reduce downstream infection amongst consumers (Duret et al., 2017; Sobolik et al., 2021; Yang et al., 2019). Yang et al. (2019) revealed up to 40% reduction in norovirus cases per year by excluding ill workers from work whilst symptomatic and for 2 days post-symptoms.

Food safety and hygiene training must be conducted for all staff. New staff could be trained during induction, followed by regular food safety training for all staff. Handwashing techniques could be taught in person via practical demonstration, videos or infographics. Some training utilised GloGerm®

handwashing training kits to demonstrate cross-contamination pathways. Employee training should cover hazards such as biological, physical, chemical and allergens, risks and measures to prevent and control contamination. Although food safety training was found to increase employees' knowledge, however this may not necessarily translate into food safety behaviour (Insfran-Rivnarola et al., 2020; Zanin et al., 2017). It is important that staff understand why specific controls are in place as this helps to build trust and that by adhering to safe food handling practices, they help to improve food safety (Frankish et al., 2021). A positive food safety culture environment where supervisors or site-managers acted as role models and demonstrated fair methods of enforcing rules using courteous communication were found to significantly improve workers' hand hygiene practices before work and after breaks (Nieto-Montenegro et al., 2008). Hence, effective training should be followed by involvement of the supervisory personnel as role models to workers.

3.2 Premises

Food plant and buildings should be of adequate size for processing, packaging and storage of product that does not result in contamination, product adulteration or creation of insanitary conditions. The grounds and building layout should facilitate movement of materials through the plant in a hygienic manner. The grounds surrounding the food plant are kept in good condition, grass and weeds are mowed and kept short to reduce pests. Waste materials are removed, stored and disposed adequately without attracting pests and do not pose as a source of contamination. It is important to avoid large standing water bodies by adequately draining the area to reduce pests and potential contamination via seepage and foot-borne filth into the food plant. Floors, walls and ceilings should be durable, easily cleanable and able to withstand moisture, chemical and mechanical abuse. The floor should not be slippery and should have no chips or cracks in floor coverings. The floor is bonded to the wall by semi-circular joints (rounded) to facilitate easy cleaning and draining. Food plants can be zoned to prevent cross-contamination of food products. Zones can be applied to manage facilities and are divided into the following groups, i.e. Zone 1 (areas with the strictest hygiene requirements and where surfaces are in direct contact with food, e.g. conveyor belts); Zone 2 (includes food processing equipment but excludes food contact surfaces); Zone 3 (areas not in direct contact with food but can harbour microorganisms and interact with Zones 1 and 2, e.g. floor, wall, drain cover); and Zone 4 (external areas such as break rooms, office) (Gu et al., 2019; Simmons and Wiedmann, 2018). GMP failures associated with premises have been identified as possible routes of contamination in a fresh produce processing facility. For example, stagnant water on the packing facility floor and inadequate GMPs were identified as factors contributing to a multistate outbreak of listeriosis associated with cantaloupe in 2011. A truck was kept near the melon processing line and was used to transport melons to a cattle farm; this may have introduced Listeria into the processing line (McCollum et al., 2013).

Equipment should be designed and constructed of materials that can be easily accessed and cleaned and should be located, maintained and calibrated according to its intended use (US FDA, 2020). Detailed information on hygienic design principles is also available in the European Hygienic Engineering and Design Group (EHEDG) guidelines (https://www.ehedg.org/) (EHEDG, 2018). The formation of biofilm on equipment is highly problematic in the minimally processed fruits and vegetables industry (Brooks and Flint, 2008). Unhygienic equipment design and harbourage areas were found to play a role in the persistence of foodborne pathogens as cleaning and sanitising agents are unable to reach hard to clean areas (Carpentier and Cerf, 2011). Studies on hygienic equipment design and its impact on biofilm formation have been carried out. By improving the flow pattern in industrial equipment (Cunault et al., 2018) and reducing poor hygienic features such as horizontal surfaces, right angles, poor-quality welds and closed corners in the bottom of vats, this helps to reduce surface contamination (Cunault et al., 2019). Bouvier et al. (2021) reported how the influence of vat design in the vegetable processing industry would affect biofilm contamination surfaces. Hygienic design features such as no horizontal surfaces and open angles exceeding 100° were proposed by the authors.

3.3 Products and process control

Specifications and controls should be developed for raw fruits and vegetables, processing aids, packaging materials and gases, e.g. in Technical Data Sheets or product descriptions (see section 4). The information should include information such as product name, composition of the material, organoleptic information, physical, chemical and microbiological criteria, presence of food allergens, labelling, shipping and storage information, shelf life, handling instructions, condition for and responsibilities in case of inspection failure (Amsbary, 2013). Setting specifications and inspection of incoming raw materials help to reduce potential food safety hazards. Visually inspect trucks, containers and carriers of raw materials to ensure vehicles are clean, and the fresh produce meets visual indicators of cleanliness. Monitor incoming temperatures and if raw materials are supposed to be chilled, verify the internal temperature. Similarly, finished product specifications should be agreed with customers. The processing of raw materials should be monitored to ensure the safety and quality of the end products.

As in receiving and checking specifications of raw materials, finished goods should be stored under appropriate storage temperatures and transported in trucks that are in good condition, clean and pest-free. For example, US FDA (2008) recommends a storage temperature of 41°F (5°C) or lower for cut melons, cut leafy greens and other fresh-cut product. If the finished goods being loaded are refrigerated or frozen product, the trailer should be pre-cooled, and temperature verified. Maintaining cold storage conditions during processing, shipping and storage helps to reduce/prevent the growth of foodborne pathogens and maintain the organoleptic and quality properties (Luciano et al., 2022; US FDA, 2008). Fresh produce is often washed to remove soil, debris, dust and plant tissue exudates. Washing could also help to reduce the microbial load on the product (Goodburn and Wallace, 2013). The washing procedures depend on type of product (leafy greens, tomatoes) and format (fresh-cut, shredded, whole product) and facilities available including type of disinfectant and washing practices (e.g. overhead wash, immersion, single-wash, double-pass) (Lopez-Galvez et al., 2021). However, delicate fruits, e.g. raspberries, are generally not washed as washing may damage the structure; hence, good agricultural practices (GAP) during primary production are essential (Julien-Javaux et al., 2019).

3.4 Procedures

Procedures should be developed, maintained and updated routinely and accordingly and implemented effectively. Standard Operating Procedures (SOP) are a set of written instructions and refer to the overall operational procedures of the site. For example, the operating procedure to operate a flume tank involves what, when, how, where and other operational parameters. Meanwhile, Sanitation Standard Operating Procedures (SSOPs) relate to all sanitation-related processes or activities. For example, what is the procedure for cleaning the flume tank, how is it cleaned, how frequent or when will it be cleaned, location of flume tank, who will be cleaning it, verification of cleaning activity and other operational parameters. Hence, SOPs and SSOPs, when written clearly, can help to minimise errors and miscommunication between staff within and between food businesses, suppliers and customers. In addition to developing and implementing the written procedures, records should be kept documenting the implementation and monitoring of the SSOPs and if any corrective actions were taken (Hui, 2015).

Failure to document and/or keep accurate records can jeopardise public health as evidenced by the salmonellosis outbreak associated with fresh produce in 2008 in the USA. One of the factors contributing to the delay in the investigation of the outbreak was due to poor traceability, including inconsistent and/or poor record keeping (Smith and Hoorfar, 2012). Tomatoes were implicated as the source of contamination, but further epidemiologic and microbial swabs identified the outbreak strain from jalapeño pepper at a Texas importer facility and Serrano peppers at a farm in Mexico. Thus, documentation and accurate recordkeeping are essential to GMP and key to effective traceability. In general, procedures should be clearly written to minimise errors and facilitate effective communication following staff turnover or for training. Accurate recordkeeping is essential to trace activities, especially who performed or operated the procedure and when it occurred. Documents must be approved, signed and dated with the latest version number, and obsolete documents should be removed to prevent inadvertent use. Written records must be auditable, and any correction made to the document or record must be signed or initialled and dated whilst the correction must permit the reading of the original information (Patel and Chotai, 2011).

Good and accurate recordkeeping will facilitate effective traceability and product recall. Tracing fresh produce is often challenging due to issues of combining and/or repacking of produce from multiple suppliers, fresh fruits and vegetables may be sold loose or without packaging and are highly perishable (Smith and Hoorfar, 2012). Due to the complexity and large amount of data that needs to be gathered across the supply chain, today firms rely on technologies such as barcodes, Radio Frequency Identification (RFID), QR codes and wireless sensor networks for real-time tracking and remote monitoring (Tagarakis et al., 2021). However, smaller firms are still able to rely on 'one-step-forward, one-step-back' to trace the raw materials or ingredients back to the immediate supplier and identify the businesses to which their products have been supplied. If each fresh produce business operator complies with the approach, the chain of information will be intact and improve the likelihood of tracing the products back to the origins and locate finished products. Similar to a HACCP team when developing HACCP plan, there should be a product recall team with members from various departments involved in procurement, processing, quality assurance and warehousing. Simulated or mock recalls should be conducted to ensure effective traceability and recall.

The implementation of GMP is a collective effort from the management and staff to ensure personnel, premises, products, processes and procedures are in place to ensure the safety of minimally processed fresh produce. The ineffective implementation and/or non-compliance of GMP can result in serious consequences such as foodborne illnesses, food recalls, litigation and fines, loss of public trust, profit and reputation. The consequences are not limited to the individual site or company and may affect other similar fresh produce industry. It is evident that GMP plays an essential role in food manufacturing environment and lays the solid foundation for the effective implementation of HACCP plan.

4 The hazard analysis and critical control point system and minimally processed fresh produce

The internationally accepted approach to control of food safety hazards through the HACCP system is defined by the Codex Alimentarius Commission Committee on Food Hygiene in the *General Principles of Food Hygiene CXC 1-1969* (CCFH, 2020). The Codex HACCP principles (Table 1) describe how to analyse hazards, determine CCPs and manage them on a day-to-day basis through applying CCP monitoring procedures and taking appropriate corrective actions following a deviation from a critical limit.

The General Principles of Food Hygiene CXC 1-1969 document, hereafter referred to as Codex, outlines the general principles that are intended to be followed by food businesses at all stages of the food chain, including primary producers and importers as well as processors/manufacturers, warehousing and logistics operators, retailers and traders, and food service (CCFH, 2020). Key responsibilities for food businesses are to understand the consequences of food safety hazards that are relevant to their operations regarding consumer health and to make sure that these hazards are appropriately identified and managed. From a foundational point of view, this is by the application of good hygienic practices (GHPs) (see section 3 on GMP), and where there are significant hazards that cannot be controlled solely by GHPs, then they need to be addressed through the development and implementation of a HACCP plan (CCFH, 2020).

Codex recognises that it is not always feasible to apply HACCP at primary production and that implementation of HACCP may be challenging for some businesses (CCFH, 2020). This may be particularly relevant in areas where it is difficult to apply a traditional CCP such as a cook process, as might be the case in many fresh produce items, and this has led to criticisms of HACCP not being suitable for primary agriculture due to the lack of lethal control measures

Table 1 The Codex HACCP Principles (CCFH, 2020)

Number	Codex HACCP principle
1	Conduct a hazard analysis and identify control measures.
2	Determine the critical control points (CCPs).
3	Establish validated critical limits.
4	Establish a system to monitor control of CCPs.
5	Establish the corrective actions to be taken when monitoring indicates a deviation from a critical limit at a CCP has occurred.
6	Validate the HACCP plan and then establish procedures for verification to confirm that the HACCP system is working as intended.
7	Establish documentation concerning all procedures and records appropriate to these principles and their application.

(Sperber, 2005; Tapia et al., 2009). In a minimally processed produce setting, it is likely that emphasis on prerequisite programs/GMP/GHP (section 3) will be crucial, but it is also possible that combinations of control measures that provide a hurdle technology effect might be employed and formulation (Fig. 1) or treatment controls may be suitable to be managed as CCPs. For example, non-thermal hurdle technology such as ultrasound assisted washing combined with disinfectants was found to reduce foodborne pathogens during fresh produce washing (Wang and Wu, 2022). The use of simple, cheap and existing hurdle techniques such as mechanical peeling, washing in chlorinated water, use of micro-perforated packaging and cold storage was also found to produce safe and nutritious fruits after 8 days of storage (Diaz-Delgado et al., 2022).

Indeed, HACCP principles can be applied flexibly and an understanding of likely hazards, their significance and options for their control allows the principles of HACCP to be applied in all food business types. Further, the discipline encompassed by a HACCP approach to identify and analyse hazards, determine appropriate control options and ensure that these are rigorously applied and monitored gives additional benefits to any food operation beyond the food safety control system, including improvements in food quality and enhancement of organisational culture, and the ability to meet legal obligations and private standards requirements (Mortimore and Wallace, 2013).

Legal requirements for HACCP vary around the world making the HACCP governance situation quite complex. In some countries or groups of countries, there is legislation requiring application of HACCP for all food products; in other countries, there may be legislation for certain types of food products but not all. Terminology also differs in legislation, so in some countries, there may

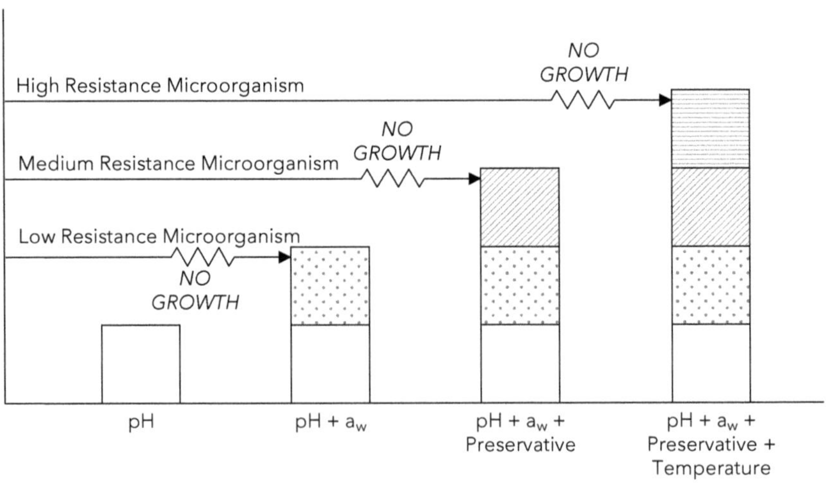

Figure 1 Example of hurdle technology as a formulation control measure: additive effects of combined factors (Adapted from Wallace et al., 2018)

be requirements to apply HACCP or HACCP principles; in others, there may be a requirement to create a food safety plan or to perform hazard analysis and implement appropriate controls. Therefore, food businesses need to satisfy themselves of the legal requirements for the countries in which they operate. Similarly, private standards requirements may vary but most food chain standards, particularly those benchmarked to the requirements of the Global Food Safety Initiative (GFSI, 2020), require HACCP application in line with Codex HACCP principles. This means that both food safety regulatory inspections and third-party certification audits will be assessing HACCP systems and it is important for businesses preparing and handling minimally processed fresh produce to be able to apply HACCP effectively to their operations.

4.1 Applying hazard analysis and critical control point principles in minimally processed fresh produce

In applying HACCP principles, it is important to understand the definitions of specific terms used within the principles and application guidance. Key definitions for HACCP principle application are listed in Table 2.

In any food operation, it is necessary that management and personnel have the necessary knowledge and training to apply HACCP effectively and successful HACCP programs rely on commitment and a supportive culture. A multidisciplinary team-based approach to application of HACCP principles is recommended to ensure appropriate knowledge, expertise and experience to take vital food safety decisions regarding hazard significance and necessary controls. Preliminary work is done by the HACCP team to work though the Codex preliminary steps which make up steps 1-5 inclusive of the Codex Logic Sequence for Application of HACCP (CCFH, 2020) and provide the foundation for HACCP principle application.

- Step 1: Assemble a HACCP team. HACCP principles are applied by a multidisciplinary team of individuals with the expertise and skills to analyse hazards and take decisions about food safety controls.
- Step 2: Describe product. This provides background information as a basis for hazard analysis and helps the HACCP team members to familiarise themselves with the process and product characteristics.
- Step 3: Identify intended use. Since different consumer groups may have varying susceptibility to hazards, it is important to consider the consumer target group and, in some cases where foods are intended for particularly vulnerable customers, additional enhanced controls may be needed. Also important is how the product might be used, including abuse potential such as inappropriate storage temperatures, and ways in which consumers are known to use the product other than those intended by the food business. For example, McLauchlin et al. (2021) reported on a European outbreak

Table 2 Selected HACCP Definitions (CCFH, 2020)

Term	Definition
Acceptable level	A level of hazard in a food at or below which the food is considered to be safe according to its intended use
Control measure	Any action or activity that can be used to prevent or eliminate a hazard or reduce it to an acceptable level
Corrective action	Any action taken when a deviation occurs in order to re-establish control, segregate and determine the disposition of the affected product if any and prevent or minimise reoccurrence of the deviation
Critical control point (CCP)	A step at which a control measure or control measures, essential to control a significant hazard, is/are applied in a HACCP system
Critical limit	A criterion, observable or measurable, relating to a control measure at a CCP which separates acceptability from unacceptability of the food
Deviation	Failure to meet a critical limit or to follow a GHP procedure
HACCP plan	Documentation or set of documents, prepared in accordance with the principles of HACCP to ensure control of significant hazards in the food business
HACCP system	The development of a HACCP plan and the implementation of the procedures in accordance with that plan
Hazard	A biological, chemical or physical agent in food with the potential to cause an adverse health effect
Hazard analysis	The process of collecting and evaluating information on hazards identified in raw materials and other ingredients, the environment, in the process or in the food, and conditions leading to their presence to decide whether or not these are significant hazards
Monitor	The act of conducting a planned sequence of observations or measurements of control parameters to assess whether a control measure is under control
Prerequisite program	Programs including good hygiene practices, good agricultural practices and GMPs, as well as other practices and procedures such as training and traceability, that establish the basic environmental and operating conditions that set the foundation for implementation of a HACCP system
Significant hazard	A hazard identified by a hazard analysis, as reasonably likely to occur at an unacceptable level in the absence of control, and for which control is essential given the intended use of the food
Validation of control measures	Obtaining evidence that a control measure or combination of control measures, if properly implemented, is capable of controlling the hazard to a specified outcome
Verification	The application of methods, procedures, tests and other evaluations, in addition to monitoring, to determine whether a control measure is or has been operating as intended

of human listeriosis associated with frozen sweet corn consumption; the frozen sweet corn producers expected it to be cooked by the consumers but there was evidence of misuse of this food as a ready-to-eat product. For minimally processed produce, most products will be used directly by the consumer without further processes such as cooking so products need to be safe at the end of production and remain safe for consumption throughout shelf life.

- Step 4: Construct Flow Diagram. The process flow diagram is constructed to show all the activities in a production process in a stepwise manner and to show the interactions between the different activities. This is normally achieved with a block and arrow diagram, and it is important that sufficient detail of the operations is included to allow understanding of the process, e.g. temperatures, disinfectant strengths and wash times, etc.
- Step 5: On-site Confirmation of Flow Diagram. The process flow diagram is used by the HACCP team to perform hazard analysis by working through all process steps. Therefore, it is vital that the diagram is accurate, and this is achieved by on-site confirmation by HACCP team members or independent employees who can check the details and establish that nothing has been missed.

The remaining steps of the Codex Logic Sequence (CCFH, 2020) follow through the application of the seven Codex HACCP principles.

4.1.1 HACCP Principle 1: Conduct a hazard analysis and identify control measures (Logic sequence Step 6)

This principle requires the HACCP team to list all potential hazards that are likely to occur associated with each step in the process flow diagram, conduct a hazard analysis to identify the significant hazards and consider any measures to control identified hazards. The list of hazards should be specific, e.g. a microorganism of concern as a contaminant in a raw material or metal fragments due to damaged chopping and cutting equipment, and should denote the source of cause of the hazard being present at the process step. The HACCP team should then analyse the listed hazards to determine the significant hazards, i.e. those hazards that are reasonably likely to occur at an unacceptable level in the absence of control, and for which control is essential given the intended use of the food.

Hazard analysis is a critical part of the HACCP application process requiring knowledge, experience and expertise along with appropriate information to take decisions. Weaknesses in hazard analysis are likely to lead to weaknesses in HACCP plans and may result in too many CCPs to manage effectively or, worse, too few CCPs and possibly uncontrolled hazards (Wallace et al., 2014).

Guidance from Codex (CCFH, 2020) suggests that the following should be considered:

a) *hazards associated with producing or processing the type of food, including its ingredients and process steps (e.g. from surveys or sampling and testing of hazards in the food chain, from recalls, from information in the scientific literature or from epidemiological data).*

   When applying principle 1 for minimally processed produce, the hazards discussed in Section 2 of this chapter should be considered but the analysis should not be limited to a given list and should depend on the likely contaminants and their potential impact on consumer health in the specific product and process.

b) *the likelihood of occurrence of hazards, taking into consideration prerequisite programs, in the absence of additional control.*

   It is important to consider the likelihood of occurrence in the given product and its ingredients under normal growing handling and processing conditions. Whilst the potential role of prerequisite programs is considered, identification of likely occurrence in the absence of control, e.g. if prerequisites and other controls are insufficient, is an important part of the hazard evaluation.

c) *the likelihood and severity of adverse health effects associated with the hazards in the food in the absence of control.*

   This requires knowledge and understanding of the hazard in question, such as the infective dose for microbiological hazards and the potential for causing illness in the absence of control.

d) *identified acceptable levels of the hazards in the food, e.g. based on regulation, intended use, and scientific information.*

   An acceptable level is defined as 'A level of hazard in a food at or below which the food is considered to be safe according to its intended use'. Again, this requires appropriate technical knowledge and understanding within the HACCP team to make this judgement.

e) *the nature of the facility and the equipment used in making the food product.*

   This relates to the potential for cross-contamination with hazards and the capability to control hazards.

f) *survival or multiplication of pathogenic microorganisms.*

   Pathogenic microorganisms are not only a concern due to their nature as a contaminant but because they may also be able multiply to higher levels making them more harmful or they may be able to survive processes intended to control them, e.g. spore-forming microorganisms.

g) production or persistence in foods of toxins (e.g. mycotoxins), chemicals (e.g. pesticides, drug residues, allergens) or physical agents (e.g. glass, metal).

h) the intended use and/or probability of product mishandling by potential consumers that could render the food unsafe, and
i) conditions leading to the above.

Following the analysis of hazards, consideration needs to be given to what control measures can be applied to control each hazard. Control measures are actions or activities that can be used to prevent or eliminate a hazard or reduce it to an acceptable level (CCFH, 2020). This can be through application of GHP targeting general groups of hazards, e.g. a general reduction in microbiological levels, or specific hazards such as specific cleaning of equipment to prevent contamination of ready-to-eat foods with *Listeria monocytogenes,* the hazard involved in a 2011 multistate outbreak of cut melons (McCollum et al., 2013). Depending on the product and inherent hazards, control measures for significant hazards may also be required within the process, some of which will be at CCPs.

4.1.2 HACCP Principle 2: Determine the Critical Control Points (Logic Sequence Step 7)

This step involves deciding which of the identified control measures need to be applied at a CCP; these will be control measures for significant hazards since CCPs need to be established at steps where control is essential and where deviation could result in the production of potentially unsafe food. Determination of CCPs can be achieved using a CCP decision tree (CCFH, 2020, 2022) or via judgement within the HACCP team. The decision involves assessment of whether a specific control measure can be used at the process step being analysed, whether it can also be used later in the process, or whether there is a more appropriate control measure for the hazard at another step. In addition, it is important to determine whether the control measure is used in combination with a control measure at another step to control the same hazard, where both steps should be considered as CCPs.

4.1.3 HACCP Principle 3: Establish Validated Critical Limits (Logic Sequence Step 8)

Critical limits are the operational limits within which the CCP is under control and the product being produced is considered to be acceptable, i.e. safe. The limit denotes the boundary between safe and potentially unsafe product being produced and must be capable of being measured or observed so that monitoring is able to detect loss of control, i.e. deviation, at a CCP. Critical limits are always associated with the control measure(s) for a given hazard and are often maximum or minimum values for measurable parameters such as temperature, time, pH, Aw, moisture levels, available chlorine, contact time, conductance, or they can be observable parameters such as belt speed, flow rate, pump settings, etc.

The critical limits need to be scientifically validated regarding evidence that they are capable of controlling hazards to an acceptable level if properly implemented; and they need to be practically validated to demonstrate that they are always achieved within the normal operational variation of the production process. Often critical limits can be based on published literature values, regulatory guidance or third-party studies, e.g. from equipment manufacturers; however, where no existing evidence is available, the food business is responsible for validation of the chosen critical limits, e.g. through microbiological inactivation studies. See also Principle 6 in a later section and Codex Alimentarius Commission (CAC), 2008.

4.1.4 HACCP Principle 4: Establish a system to monitor control of CCPs (Logic Sequence Step 9)

Monitoring is done to show on an ongoing basis that a CCP is operating within its critical limit(s). Therefore, it is important that the chosen monitoring procedures are capable of detecting loss of control (CCP deviation) and also that the frequency of monitoring is set such that deviation can be detected in a timely manner. This enables rapid identification and isolation of potentially unsafe product following a CCP deviation. In some cases, usually where operational action limits that are tighter than critical limits are set, it will be possible to detect a trend towards loss of control and to instigate actions, e.g. process alterations, which prevent a deviation from occurring. This is due to the safety 'buffer zone' that operational action limits can allow. Many monitoring procedures are offline or not continuous, meaning that a check has to be done at the required interval, e.g. an operator checking a gauge or pump setting or a label check regarding allergens. Some monitoring procedures can be continuously monitored online, e.g. process times and temperatures. These should be linked to alarms or automatic divert functions to prevent potentially unsafe product going forward if the CCP deviates.

In addition to the monitoring procedure, the system should include record keeping and appropriate training for the personnel who will perform the CCP monitoring. Records of the monitoring results should be signed by the person who has done the monitoring and reviewed by a more senior member of staff. Training should include all the steps to take when monitoring indicates a deviation or drift towards a deviation, and the CCP monitor or other staff member close to the process should have authority to take immediate action where results indicate a deviation has occurred.

4.1.5 HACCP Principle 5: Establish the corrective actions to be taken when monitoring indicates a deviation from a critical limit at a CCP has occurred (Logic Sequence Step 10)

A corrective action is any action taken when a deviation occurs in order to re-establish control, segregate and determine the disposition of the affected product if any and prevent or minimise reoccurrence of the deviation (CCFH, 2020). A corrective action or actions need to be specified and documented for each CCP so that any deviation at a CCP can be dealt with effectively. The corrective action will include both actions to bring the CCP back under control and to make sure that any potentially unsafe product is handled appropriately so that consumers are protected. Often monitoring is periodic and not continuous, meaning that when a deviation occurs the corrective action needs to consider all product produced since the last 'good' or 'within critical limit' monitoring check. Depending on the frequency of monitoring, this may be a substantial amount of product, which will need segregating and expert evaluation regarding the need for product disposition or potential safe use. For example, in the case of minimally processed fresh produce, it may be possible to divert suspect material to a product where a more intensive process, e.g. with heat processing CCPs, could control the hazard in question. Corrective actions for all likely deviations should be discussed and documented during HACCP plan development so that the necessary steps can be followed when a deviation occurs.

4.1.6 HACCP Principle 6: Validate the HACCP plan and then establish procedures for verification to confirm that the HACCP system is working as intended (Logic Sequence Step 11)

Validation of the HACCP plan makes sure that it is capable of ensuring control of all relevant significant hazards for the business regarding the product and process being considered. HACCP plan validation looks across the HACCP plan to make sure that all elements are suitable for controlling these hazards, i.e. that hazards have been identified and analysed appropriately and suitable control measures determined, critical control points have been correctly identified, critical limits set at the correct levels for food safety, monitoring is suitable to detect loss of control of CCPs and corrective action would handle potentially unsafe product and bring the CCP back under control. Validation of control measures and their critical limits is particularly important since technical information is needed to establish that significant hazards will be controlled effectively within the specified parameters. This may include use of scientific literature, mathematical models, validation studies or use of expert guidance documents, e.g. the Codex Alimentarius Guidelines for the Validation of

Food Safety Control Measures (CAC, 2008). An example relevant to minimally processed produce is the validation of post-harvest dehydration of tree nuts to prevent aflatoxin contamination. The technical criteria for validation are listed in the Table 3 and the validation process in Fig. 2.

4.1.7 HACCP Principle 7: Establish documentation concerning all procedures and records appropriate to these principles and their application (Logic Sequence Step 12)

Documentation and record keeping is essential for food businesses to verify that their HACCP controls are in place, working in practice and being maintained (CCFH, 2020). Amounts of documentation needed will vary depending on the nature and size of the business and sector specific HACCP models and guidelines can be used as long as these have been developed by experts and reflect the operations of the business. For example, the European Food Safety Authority (EFSA) produced hazard analysis approaches (including flow

Table 3 Validation of post-harvest dehydration to prevent aflatoxin contamination of tree nuts - technical and decision criteria (adapted from CAC, 2008)

Control measure to be validated: post-harvest dehydration of tree nuts.			
Hazard	Food Safety Outcome required	Parameters	Decision Criteria
Aflatoxin contamination has been identified as a hazard that is reasonably likely to occur in tree nuts. Its control requires application of measures both pre-harvest and post-harvest. Post-harvest measures are focused on rendering the tree nuts incapable of supporting continued aflatoxin production by *Aspergillus* spp.	The recognised international standard for aflatoxin B1 is 20 μg/kg. However, to consider process and analytical uncertainties, the food safety outcome is set at 10 μg/kg	Aflatoxin-producing *Aspergillus* spp. cannot grow and synthesise the toxins when the water activity of the product falls below 0.70. The amount of aflatoxin that is produced post-harvest depends on the speed of dehydration and the rate of mould growth. Scientific literature suggests that germination of the spores and initiation of toxin synthesis can occur with 24 to 48 hours of exposure of post-harvest tree nuts to a moist environment. The level of aflatoxin B1 present in post-harvest tree nuts will also depend on the levels present prior starting dehydration.	The water activity in lots of tree nuts being treated can be consistently reduced to <0.70 within 24 hours. After dehydration "wet spots" that have a water activity ≥0.70 are absent from the lot. The level of aflatoxin B1 in the tree nuts after a water activity <0.70 has been attained does not exceed 10 μg/kg. The treatment includes appropriate packaging/storage of the dried tree nuts.

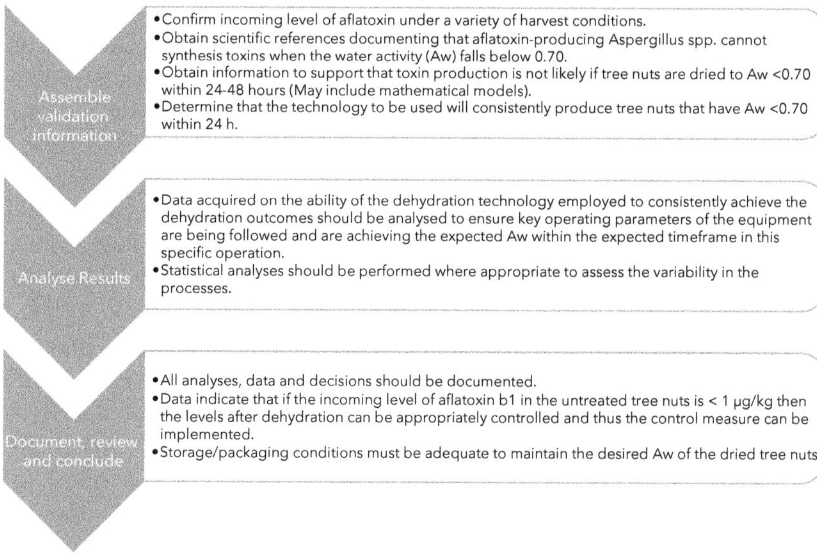

Figure 2 Validation of post-harvest dehydration to prevent aflatoxin contamination of tree nuts - validation process.

diagrams, hazard analysis and prerequisite program control activities) for certain small retail businesses including a grocery business handling vegetables, herbs, fruits, fresh-cut prepacked and dried fruits (EFSA Panel on Biological Hazards et al., 2017). A further example generic HACCP food safety plan with general guidance for crops was produced by GLOBAL G.A.P (GLOBAL G.A.P., 2020).

5 Conclusion and future trends

Fresh and minimally processed produce are subject to contamination with a range of microbiological hazards, often from the growing stage but also through cross-contamination during processing and handling. Outbreaks in this sector are well-documented and highlight the importance of implementing effective control strategies to protect consumers. The Codex general principles of food hygiene, including GMP guidance and HACCP principles, were updated in 2020 (CCFH, 2020), and these principles are likely to remain unchanged for some time. Similarly, the requirements for HACCP and GMP are highlighted in legislation and industry/private standards and there is much well-developed guidance on how to apply the necessary GMP in different settings. The key to food safety is for all food businesses to understand the likely hazards in their operations and to apply GMP throughout their operations to control the general operational conditions and to apply HACCP to identify and control significant food safety hazards to an acceptable level that manages risk to consumers. Future research trends are likely to include case studies in minimally processed operations where the experiences of applying HACCP to control the relevant significant hazards can be useful to the industry as a whole. Within the Codex Committee on Food Hygiene, further deliberations on the use of decision trees to identify CCPs resulted in encouragement to use these as flexible tools (CCFH, 2022) and a new decision tree that can be used alongside the latest version of the Codex HACCP documents (CCFH,2020) was adopted by Codex Alimentarius Commission in 2023 (CAC, 2023), although CCFH accepts that other proven decision trees may also be beneficial. Further research into new technologies which allow more effective control of hazards, e.g. through alternative control mechanisms to traditional heat processing and additional hurdle technology options, may offer benefits to the minimally processed produce sector. In addition, new technologies and approaches to monitoring can help to ensure that CCPs are under control within their critical limits. The use of non-thermal technologies like irradiation (Bisht et al., 2021), pulsed electric fields (Arshad et al., 2021), electrostatic sprays (Goodburn and Wallace, 2013; Almasoud et al., 2015), ultrasound washing (Wang and Wu, 2022) and high-pressure processing (Denoya et al., 2022) could be incorporated within HACCP to assure the safety of minimally processed fresh produce. The application of Internet of Things (IoT) such as blockchain technology (Wallace and Manning, 2020), integrated traceability systems (Gallo et al., 2021), digital twins (Defraeye et al., 2021) and use of sensors and artificial intelligence (Arya and Gangwar, 2021) offers promising research areas within the fresh produce industry.

6 Where to look for further information

Readers will find it helpful to obtain some of the guidelines cited in this chapter to review the detailed information provided. In particular, the General Principles of Food Hygiene document provides helpful guidance on hygienic practices (GMP/GHP) and HACCP application, and the Guidelines for the Validation of Food Safety Control Measures. The European Hygienic Engineering Design Group (EHEDG) provides a range of guidelines for hygienic design of different processing operations. Further authoritative guidance can be found through national food safety authorities, e.g. FDA's Guide to minimise microbial food safety hazards of fresh-cut fruits and vegetables. See Section 6.1 for further reading suggestions.

Research on the presence, behaviour and control of hazards in different food matrices is ongoing and much work is published in scientific journals and presented at food safety conferences. Some research funded by national or regional authorities, e.g. EU Horizon 2020 funded projects, has online reporting during the project so that interested parties can follow the progress and keep in touch with findings. Engagement with these sources of information is helpful for anyone wishing to challenge and strengthen food safety controls.

See Section 6.2 for suggestions of key journals, conferences and an example of a research project with ongoing reporting during the project timeline.

6.1 Further reading

- Codex Committee on Food Hygiene (2020). General Principles of Food Hygiene CXC 1-1069, Adopted in 1969, Amended in 1999. Revised in 1997, 2003, 2020. http://www.fao.org/fao-who-codexalimentarius/codex-texts/codes-of-practice/en/.
- Codex Alimentarius Commission (2008). Guidelines for the Validation of Food Safety Control Measures. CAC/GL 69 2008.
- EHEDG (2018). Hygienic design principles guidelines (https://www.ehedg.org/guidelines-working-groups/guidelines/guidelines).
- US FDA (2008). Guidance for industry: Guide to minimise microbial food safety hazards of fresh-cut fruits and vegetables.

6.2 Key journals, conferences and projects

- International Association for Food Protection Annual meeting plus relevant Professional Development Groups, e.g. the Fruit and Vegetable Safety and Quality PDG.
- Food Control.
- Trends in Food Science and Technology.
- Food Research International.
- Journal of Food Safety.
- EFSA Journal - outbreaks.
- Horizon (2020). Non-Thermal physical technologies to preserve healthiness of fresh and minimally processed fruit and vegetables (SHEALTHY). Grant ID: 817936. DOI 10.3030/817936.

7 References

Alegbeleye, O. O., Singleton, I. and Sant'Ana, A. S. (2018). Sources and contamination routes of microbial pathogens to fresh produce during field cultivation: A review. *Food Microbiology* 73, 177–208.

Almasoud, A., Hettiarachchy, N., Rayaprolu, S., Horax, R. and Eswaranandam, S. (2015). Electrostatic spraying of organic acids on biofilms formed by *E. coli* O157:H7 and *Salmonella typhimurium* on fresh produce. *Food Research International* 78, 27–33. http://doi.org/10.1016/j.foodres.2015.11.012.

Amsbary, R. (2013). Raw materials: Selection, specifications and certificate of analysis. *Quality Assurance & Food Safety*. Retrieved June 2013. Available at: https://www.qualityassurancemag.com/article/aib0613-raw-materials-requirements/.

Arshad, R. N., Abdul-Malek, Z., Roobab, U., Munir, M. A., Naderipour, A., Qureshi, M. I., Bekhit, A. E., Liu, Z.. and Aadil, R. M. (2021). Pulsed electric field: A potential alternative towards a sustainable food processing. *Trends in Food Science and Technology* 111, 43–54. https://doi.org/10.1016/j.tifs.2021.02.041.

Arya, P. S. and Gangwar, M. (2021). A proposed architecture: Detecting freshness of vegetables using Internet of Things (IoT) & deep learning prediction algorithm. In: *IEEE 3rd International Conference on Advances in Computing, Communication Control and Networking (ICACCCN)*, 17–18 December.

Beretti, M. and Stuart, D. (2008). Food safety and environmental quality impose conflicting demands on central coast growers. *California Agriculture* 62(2), 68–73.

Bisht, B., Bhatnagar, P., Gururani, P., Kumar, V., Tomarf, M. S., Sinhmar, R., Rathi, N. and Kumar, S. (2021). Food irradiation: Effect of ionizing and non-ionizing radiations on preservation of fruits and vegetables – A review. *Trends in Food Science and Technology* 114, 372–385. https://doi.org/10.1016/j.tifs.2021.06.002.

Bouvier, L., Cunault, C., Faille, C., Dallagi, H., Wauquier, L. and Benezech, T. (2021). Influence of the design of fresh-cut food washing tanks on the growth kinetics of *Pseudomonas fluorescens* biofilms. *iScience* 24(6), 102506.

Brooks, J. D. and Flint, S. H. (2008). Biofilms in the food industry: Problems and potential solutions. *International Journal of Food Science and Technology* 43(12), 2163–2176.

CAC (Codex Alimentarius Commission) (1993). Codex guidelines for the application of the hazard analysis critical control point system. In: *Training Considerations for the Application of the HACCP System to Food Processing and Manufacturing, WHO/FNU/FOS/93.3*. Geneva: World Health Organisation, Annex II.

CAC (Codex Alimentarius Commission) (2003). *Code of Hygienic Practice for Fresh Fruits and Vegetables (CAC/RCP 53-2003).*

CAC (Codex Alimentarius Commission) (2008). *Guidelines for the Validation of Food Safety Control Measures*. CAC/GL 69.

CAC (Codex Alimentarius Commission) (2023). Joint FAO/WHO food standards programme codex Alimentarius Commission Forty-Fifth Session FAO headquarters, Rome, Italy 21–25 November 2022, 12–13 December 2022, and 19 December 2022-6 February 2022. Report p8. https://www.fao.org/fao-who-codexalimentarius/committees/cac/meetings/de/Accessed May 2023.

Carpentier, B. and Cerf, O. (2011). Review – Persistence of *Listeria monocytogenes* in food industry equipment and premises. *International Journal of Food Microbiology* 145(1), 1–8.

Carstens, C. K., Salazar, J. K. and Darkoh, C. (2019). Multistate outbreaks of foodborne illness in the United States associated with fresh produce from 2010 to 2017. *Frontiers in Microbiology* 10, 2667.

CDC (2013). Outbreak of *Escherichia coli* O104:H4 infections associated with sprout consumption – Europe and North America, May–July 2011. *Morbidity and Mortality Weekly Report (MMWR)* 62(50), 1029–1031.

CDC (2019). *Surveillance for Foodborne Disease Outbreaks United States, 2017, Annual Report*. Atlanta, GA: United States Department of Health and Human Services.

Codex Committee on Food Hygiene (CCFH) (2020). *General Principles of Food Hygiene CXC 1069, Adopted in 1969, Amended in 1999. Revised in 1997, 2003.* Available at: http://www.fao.org/fao-who-codexalimentarius/codex-texts/codes-of-practice/en/.

Codex Committee on Food Hygiene (CCFH) (2022). *Report of the 52nd Session of the Codex Committee on Food Hygiene*. Available at: https://www.fao.org/fao-who-codexalimentarius/meetings/detail/en/?meeting=CCFH&session=52.

Cunault, C., Faille, C., Briandet, R., Postollec, F., Desriac, N. and Benezech, T. (2018). Pseudomonas sp. Biofilm development on fresh-cut food equipment surfaces – A growth curve – Fitting approach to building a comprehensive tool for studying surface contamination dynamics. *Food and Bioproducts Processing* 107, 70–87.

Cunault, C., Faille, C., Calabozo-Delgado, A. and Benezech, T. (2019). Structure and resistance to mechanical stress and enzymatic cleaning of *Pseudomonas fluorescens* biofilms formed in fresh-cut ready to eat washing tanks. *Journal of Food Engineering* 262, 154–161.

De Oliveira, C. A. F., da Cruz, A. G., Tavolaro, P. and Corassion, C. H. (2016). Chapter 10: Food safety: Good manufacturing practices (GMP), sanitation standard operating procedures (SSOP), hazard analysis and critical control point (HACCP). In: Barros-Velazques, J. (Ed.), *Antimicrobial Food Packaging*. Elsevier Inc., pp. 129–139.

Defraeye, T., Shrivastava, C., Berry, T., Verboven, P., Onwude, D., Schudel, S., Bühlmann, A., Cronje, P. and Rossi, R. M. (2021). Digital twins are coming: Will we need them in supply chains of fresh horticultural produce? *Trends in Food Science and Technology* 109, 245–258. https://doi.org/10.1016/j.tifs.2021.01.025.

Denoya, G. I., Polenta, G. A., Apostolo, N. M., Budde, C. O., Cristiani, M. and Vaudagna, S. R. (2022). High-pressure processing applied for enhancing the antioxidant content of minimally processed peaches. *International Journal of Food Science and Technology* 57(1), 684–694.

Diaz-Delgado, G. L., Rodriguez-Rodriguez, E. M., Dorta, E. and Lodo, M. G. (2022). Effects of peeling, film packaging and cold storage on the quality of minimally processed prickly pears (*Opuntia ficus-indica* L. Mill.). *Agriculture* 12(2), 281.

Duret, S., Pouillot, R., Fanaselle, W., Papafragkou, E., Liggans, G., Williams, L. and Van Doren, J. M. (2017). Quantitative risk assessment of norovirus transmission in food establishments: Evaluating the impact of intervention strategies and food employee behavior on the risk associated with Norovirus in foods. *Risk Analysis* 37(11), 2080–2106.

EFSA Panel on Biological Hazards (2013). Scientific opinion on the risk posed by pathogens in food of non-animal origin. Part 1 (outbreak data analysis and risk ranking of food/pathogen combinations). *EFSA Journal* 11(1), 3025.

EFSA Panel on Biological Hazards (2014a). Scientific opinion on the risk posed by pathogens in food of non-animal origin. Part 2 (*Salmonella* and Norovirus in berries). *EFSA Journal* 12(6), 3706.

EFSA Panel on Biological Hazards (2014b). Scientific opinion on the risk posed by pathogens in food of non-animal origin. Part 2 (*Salmonella* and Norovirus in leafy greens eaten raw as salads). *EFSA Journal* 12(3), 3600.

EFSA Panel on Biological Hazards (2014c). Scientific opinion on the risk posed by pathogens in food of non-animal origin. Part 2 (*Salmonella* and Norovirus in tomatoes). *EFSA Journal* 12(10), 3832.

EFSA Panel on Biological Hazards (2014d). Scientific opinion on the risk posed by pathogens in food of non-animal origin. Part 2 (*Salmonella* in melons). *EFSA Journal* 12(10), 3831.

EFSA Panel on Biological Hazards (2014e). Scientific opinion on the risk posed by pathogens in food of non-animal origin. Part 2 (*Salmonella*, *Yersinia*, *Shigella* and Norovirus in bulb and stem vegetables, and carrots). *EFSA Journal* 12(12), 3937.

EHEDG (2018). *Hygienic Design Principles* (3rd ed.). Available at: https://www.ehedg.org/.

European Food Safety Authority (2011). Shiga toxin-producing *E. coli* (STEC) O104: H4 2011 outbreaks in Europe: Taking Stock. *EFSA Journal* 9(10), 2390.

FAO/WHO (2008). *Microbiological Hazards in Fresh Fruits and Vegetables*. Geneva, Switzerland.

FDA (2008). *Guidance for Industry: Guide to Minimize Microbial Food Safety Hazards of Fresh-Cut Fruits and Vegetables*, FDA-2008-D-0108.

FDA (2021a). Outbreaks of foodborne illness. Available at: https://www.fda.gov/food/recalls-outbreaks-emergencies/outbreaks-foodborne-illness.

FDA (2021b). Public health advisories from investigations of foodborne illness outbreaks. Available at: https://www.fda.gov/food/outbreaks-foodborne-illness/public-health-advisories-investigations-foodborne-illness-outbreaks. Retrieved 14 December 2021.

Food and Drug Administration (2015). Standards for the growing, harvesting, packing, and holding of produce for human. *Consumption* 80, pp. 74353–74568.

Frankish, E. J., McAlpine, G., Mahoney, D., Oladele, B., Luning, P. A., Ross, T., Bowman, J. P. and Bozkurt, H. (2021). Review article: Food safety culture from the perspective of the Australian horticulture industry. *Trends in Food Science and Technology* 116, 63–74.

Frankish, E. J., Phan-Thien, K., Ross, T., McConchie, R., Luning, P. A. and Bozkurt, H. (2022). Performance assessment of food safety management systems in Australian apple packhouses in view of microbial control. *Food Control* 133, 108642. https://doi.org/10.1016/j.foodcont.2021.108642.

Gallo, A., Accorsi, R., Goh, A., Hsiao, H. and Manzini, R. (2021). A traceability-support system to control safety and sustainability indicators in food distribution. *Food Control* 124, 107866. https://doi.org/10.1016/j.foodcont.2021.107866.

GFSI (2020). *The Global Food Safety Initiative Benchmarking Requirements*. The Consumer Press Goods Forum. Available at: www.mygfsi.com.

Global Gap (2020). *Generic HACCP Plan for Crops Version 5.4*. Cologne, Germany: GLOBALG.A.P. Available at: https://www.globalgap.org/uk_en/documents/.

Goodburn, C. and Wallace, C. A. (2013). The microbiological efficacy of decontamination methodologies for fresh produce: A review. *Food Control* 32(2), 418–427.

Gu, G., Ottesen, A., Bolten, S., Wang, L., Luo, Y., Rideout, S., Lyu, S. and Nou, X. (2019). Impact of routine sanitation on the microbiomes in a fresh produce processing facility. *International Journal of Food Microbiology* 294, 31–41.

Hirneisen, K. A., Sharma, M. and Kniel, K. E. (2012). Human enteric pathogen internalization by root uptake into food crops. *Foodborne Pathogens and Disease* 9(5), 396–405.

Hui, Y. H. (2015). Chapter 15: Sanitation standard operating procedures. In: *Plant Sanitation for Food Processing and Food Service* (2nd edn.). Boca Raton: CRC Press, pp. 303–320.

Hussain, M. A. and Gooneratne, R. (2017). Understanding the fresh produce safety challenges. *Foods (Basel, Switzerland)* 6(3), 23. https://doi.org/10.3390/foods6030023.

Insfran-Rivanrola, A., Tlapa, D., Limon-Romero, J., Baez-Lopez, Y., Miranda-Ackerman, M., Arredondo-Soto, K. and Ontiveros, S. (2020). A systematic review and meta-analysis of the effects of food safety and hygiene training on food handlers. *Foods* 9(9), 1169.

Julien-Javaux, F., Gerard, C., Campagnoli, M. and Zuber, S. (2019). Strategies for the safety management of fresh produce from farm to fork. *Current Opinion in Food Science* 27, 145–152.

Kokkinakis, E. N. and Fragkiadakis, G. A. (2007). HACCP effect on microbiological quality of minimally processed vegetables: A survey in six mass-catering establishments. *International Journal of Food Science and Technology* 42(1), 18–23.

Lee, S., Han, A., Jo, S., Cheon, H., Song, H., Jang, A., Kim, D. and Lee, S.-Y. (2021). Microbiological quality and safety of commercial fresh fruit and vegetable juices in Korea. *LWT* 152, 112432.

Leistner, L. and Gorris, L. G. M. (1995). Food preservation by hurdle technology. *Trends in Food Science and Technology* 6(2), 41–46.

Lopez-Galvez, F., Allende, A. and Gil, M. I. (2021). Recent progress on the management of the industrial washing of fresh produce with a focus on microbiological risks. *Current Opinion in Food Science* 38, 46–51.

Luciano, W. A., Griffin, S., de Souza Pedrosa, G. T., Alvarenga, V., Valdramidis, V. and Magnani, M. (2022). Growth behavior of low populations of Listeria monocytogenes on fresh-cut mango, melon and papaya under different storage temperatures. *Food Microbiology* 102, 103930.

Maldonade, I. R., Ginani, V. C., Riquette, R. F. R., Gurgel-Goncalves, R., Mendes, V. S. and Machado, E. R. (2019). Good manufacturing practices of minimally processed vegetables reduce contamination with pathogenic microorganisms. *Revista do Instituto de Medicina Tropical de Sao Paulo* 61, e14, https://doi.org/10.1590/S1678-9946201961014.

Marler, B. (2022). Publisher's Platform: another E. coli outbreak in leafy greens – given the time of the year likely linked again to Yuma. Food Safety News, Opinion. Available at: https://www.foodsafetynews.com/2022/01/publishers-platform-another-e-coli-outbreak-in-leafy-greens-given-the-time-of-the-year-likely-linked-again-to-yuma/.

Marshall, K. E., Hexemer, A., Seelman, S. L., Fatica, M. K., Blessington, T., Hajmeer, M., Kisselburgh, H., Atkinson, R., Hill, K., Sharma, D., Needham, M., Peralta, V., Higa, J., Blickenstaff, K., Williams, I. T., Jhung, M. A., Wise, M. and Gieraltowski, L. (2020). Lessons learned from a decade of investigations of Shiga toxin-producing Escherichia coli outbreaks linked to leafy greens, United States and Canada. *Emerging Infectious Diseases* 26(10), 2319.

McCollum, J. T., Cronquist, A. B., Silk, B. J., Jackson, K. A., O'Connor, K. A., Cosgrove, S., Gossack, J. P., Parachini, S. S., Jain, N. S., Ettestad, P., Ibraheem, M., Cantu, V., Joshi, M., DuVernoy, T., Fogg, N. W., Gorny, J. R., Mogen, K. M., Spires, C., Teitell, P., Joseph, L. A., Tarr, C. L., Imanishi, M., Neil, K. P., Tauxe, R. V. and Mahon, B. E. (2013). Multistate outbreak of listeriosis associated with cantaloupe. *New England Journal of Medicine* 369(10), 944–953.

McLauchlin, J., Aird, H., Amar, C., Barker, C., Dallman, T., Lai, S., Painset, A. and Willis, C. (2021). An outbreak of human listeriosis associated with frozen sweet corn consumption: Investigations in the UK. *International Journal of Food Microbiology* 338, 108994. https://doi.org/10.1016/j.ijfoodmicro.2020.108994.

Monteiro, C. A., Cannon, G., Lawrence, M., Costa Louzada, M. L. and Pereira Machado, P. (2019). *Ultra-Processed Foods, Diet Quality, and Health Using the NOVA Classification System*. Rome: Food and Agriculture Organization.

Monteiro, C. A., Cannon, G., Levy, R. B., Moubarac, J. C., Louzada, M. L. C., Rauber, F., Khandpur, N., Cediel, G., Neri, D., Martinez-Steele, E., Baraldi, L. G. and Jaime, P. C. (2019). Ultra-processed foods: What they are and how to identify them. *Public Health Nutrition* 22(5), 936–941.

Mortimore, S. E. and Wallace, C. A. (2013). *HACCP: A Practical Approach* (3rd edn.). New York: Springer.

Mukhopadhyay, S. and Gorris, L. G. M. (2014). Hurdle technology. In: Batt, C. A. and Tortorellow, M. L. (Eds.), Encyclopedia of Food Microbiology (2nd edn.), pp. 221-227. Academic Press. ISBN 978-0-12-384733-1.

National Advisory Committee on Microbiological Criteria for Foods (NACMCF) (1992). Hazard analysis and critical control point system (adopted 20 March 1992). *International Journal of Food Microbiology* 16(1), 1-23.

Nieto-Montenegro, S., Brown, J. L. and LaBorde, L. F. (2008). Development and assessment of pilot food safety educational materials and training strategies for Hispanic workers in the mushroom industry using the Health Action Model. *Food Control* 19(6), 616-633.

Olimpi, E. M., Baur, P., Echeverri, A., Gonthier, D., Karp, D. S., Kremen, C., Sciligo, A. and De Master, K. T. (2019). Evolving food safety pressures in California's central coast region. *Frontiers in Sustainable Food Systems* 3.

Patel, K. T. and Chotai, N. P. (2011). Documentation and records: Harmonized GMP requirements. *Journal of Young Pharmacists* 3(2), 138-150.

Qi, Y., He, Y., Beuchat, L. R., Zhang, W. and Deng, X. (2020). Glove-mediated transfer of Listeria monocytogenes on fresh-cut cantaloupe. *Food Microbiology* 88, 103396.

Ricci, A., Chemaly, M., Davies, R., Esamez, P. S. F., Girones, R., Herman, L., Lindqvist, R., Nørrung, B., Robertson, L., Ru, G., Simmons, M., Skandamis, P., Snary, E., Speybroeck, N., Kuile, B. T., Threlfall, J., Wahlstrom, H., Allende, A., Barregard, L., Jacxsens, L., Koutsoumanis, K., Sanaa, M., Varzakas, T., Baert, K., Hempen, M., Rizzi, V., Van der Stede, Y. and Bolton, D. (2017). EFSA Panel on Biological Hazards (BIOHAZ). Hazard analysis approaches for certain small retail establishments in view of the application of their food safety management systems. *EFSA Journal* 15(3), 4697. https://doi.org/10.2903/j.efsa.2017.4697.

Ross-Nazzal, J. (2007). From farm to fork: How space food standards impacted the food industry and changed food safety standards. In: Dick, S. J. and Launius, R. D. (Eds.), *Societal Impact of Spaceflight*. Washington: National aeronautics and Space administration, Office of External Relations-History Division (NaSa Sp-2007-4801). Available at: http://history.nasa.gov/sp4801-part1.pdf.

Self, J. L., Conrad, A., Stroika, S., Jackson, A., Whitlock, L., Jackson, K. A., Beal, J., Wellman, A., Fatica, M. K., Bidol, S., Huth, P. P., Hamel, M., Franklin, K., Tschetter, L., Kopko, C., Kirsch, P., Wise, M. E. and Basler, C. (2019). Multistate outbreak of listeriosis associated with packaged leafy green salads, United States and Canada, 2015-2016. *Emerging Infectious Diseases* 25(8), 1461.

Simmons, C. K. and Wiedmann, M. (2018). Identification and classification of sampling sites for pathogen environmental monitoring programs for Listeria monocytogenes: Results from an expert elicitation. *Food Microbiology* 75, 2-17.

Siroli, L., Patrignani, F., Serrazanetti, D. I., Tabanelli, G., Montanari, C., Gardini, F. and Lanciotti, R. (2015). Lactic acid bacteria and natural antimicrobials to improve the safety and shelf-life of minimally processed sliced apples and lamb's lettuce. *Food Microbiology* 47, 74-84. https://doi.org/10.1016/j.fm.2014.11.008.

Smith, M. A. and Hoorfar, J. (2012). Investigating a large outbreak of salmonellosis in the UK involving fresh produce. In: Hoorfar, J. (Ed.), *Case Studies in Food Safety and Authenticity: Lessons from Real-Life Situations* (Vol. 4). Oxford: Woodhead Publishing, pp. 33-42.

Sobolik, J. S., Newman, K. L., Jaykus, L. A., Bihn, E. A. and Leon, J. S. (2021). Norovirus transmission mitigation strategies during simulated produce harvest and packing. *International Journal of Food Microbiology* 357, 109365.

Sperber, W. H. (2005). HACCP does not work from farm to table. *Food Control* 16(6), 511–514.

Tagarakis, A. C., Benos, L., Kateris, D., Tsotsolas, N. and Bochtis, D. (2021). Bridging the gaps in traceability systems for fresh produce supply chains: Overview and development of an integrated IoT-based system. *Applied Sciences* 11(16), 7596.

Tapia, M. S., Gómez-López, V. M. and Olaizola, C. (2009). HACCP implementation in the production of fresh-cut fruits and vegetables. *Stewart Postharvest Review* 4, 6.

US FDA (2008). Guidance for industry: Guide to minimize microbial food safety hazards of fresh-cut fruits and vegetables. Available at: https://www.fda.gov/regulatory-information/search-fda-guidance-documents/guidance-industry-guide-minimize-microbial-food-safety-hazards-fresh-cut-fruits-and-vegetables#ftn4.

US FDA (2020). Current good manufacturing practices (CGMPs) for food and dietary supplements. Available at: https://www.fda.gov/food/guidance-regulation-food-and-dietary-supplements/current-good-manufacturing-practices-cgmps-food-and-dietary-supplements.

Valadez, A. M., Schneider, K. R. and Danyluk, M. D. (2012). Outbreaks of foodborne diseases associated with tomatoes. *EDIS* 2012(6).

Wallace, C. A., Holyoak, L., Powell, S. C. and Dykes, F. C. (2014). HACCP – The difficulty with hazard analysis. *Food Control* 35(1), 233–240.

Wallace, C. A. and Manning, L. (2020). *Food Provenance: Assuring Product Integrity and Identity, CAB Reviews, ISSN 1749-8848*.

Wallace, C. A., Sperber, W. H. and Mortimore, S. E. (2018). *Food Safety for the 21st Century* (2nd edn.). Oxford, UK: Wiley-Blackwell.

Wang, J. and Wu, Z. (2022). Combined use of ultrasound-assisted washing with in-package atmospheric cold plasma processing as a novel non-thermal hurdle technology for ready-to-eat blueberry disinfection. *Ultrasonics Sonochemistry* 84, 105960.

Warriner, K., Ibrahim, F., Dickinson, M., Wright, C. and Waites, W. M. (2003). Internalization of human pathogens within growing salad vegetables. *Biotechnology and Genetic Engineering Reviews* 20(1), 117–134.

Webb, M. D., Barker, G. C., Goodburn, K. E. and Peck, M. W. (2019). Risk presented to minimally processed chilled foods by psychrotrophic *Bacillus cereus*. *Trends in Food Science and Technology* 93, 94–105. https://doi.org/10.1016/j.tifs.2019.08.024.

World Health Organization (1997). *Food Safety Issues, HACCP: Introducing the Hazard Analysis and Critical Control Point System*. Food Safety Unit, Programme of Food Safety and Food Aid, World Health Organisation. World Health Organization/Funeral Service Foundation/FOS/97.2.

Yang, W., Steele, M., Lopman, B., Leon, J. S. and Hall, A. J. (2019). The population-level impacts of excluding norovirus-infected food workers from the workplace: A mathematical modelling study. *American Journal of Epidemiology* 188(1), 177–187.

Zanin, L. M., da Cunha, D. T., de Rosso, V. V., Capriles, V. D. and Stedefledt, E. (2017). Knowledge, attitudes and practices of food handlers in food safety: An integrative review. *Food Research International* 100(1), 53–62.

Chapter 12

Improving safe consumer handling of fresh produce

Jennifer Quinlan and Melissa Kavanaugh, Drexel University, USA

1 Introduction

While consumption of fruits and vegetables remains steady and below the recommended levels, consumption of potatoes and juice has decreased and consumption of leafy greens and ready-to-eat (RTE) produce has increased. This is all occurring at the same time that foodborne illnesses and outbreaks associated with produce, particularly leafy greens and RTE salads and fresh cut fruit, continue to rise. There is an increasing interest in organic and sustainably grown produce. Additionally, consumers are increasingly choosing to 'eat locally'. That is, interest in produce from farmer's markets, and local organic produce has grown. Data indicate that organically grown and locally grown (farmer's market) produce are no safer, and may be less safe, than traditionally grown produce. Data also indicate that many consumers may not be aware

http://dx.doi.org/10.19103/AS.2023.0121.19

of the potential risks associated with both unprocessed and RTE produce. While safe produce handling education materials are available, they are often embedded in more extensive general safe food handling materials. There is a need to provide more direct produce safety education, especially to vulnerable populations including older adults, the immunocompromised and pregnant women. There is also an opportunity to develop such produce safety education materials using formative research and behaviour change models to optimize the efficacy of produce safety education materials for consumers.

2 Consumer produce consumption

For most Americans, overall intake of fruits and vegetables remains below recommended levels (USDA and HHS, 2020; Rehm et al., 2016); however, increasing fruit and vegetable consumption is a stated public health goal in the United States, as well as throughout many other Westernized nations (Government of Canada, 2019; USDA and HHS, 2020; NHS, 2019; Sante Publique France, 2019). The majority of Canadians, as well as Europeans, reported eating at least one fruit and one vegetable per day in 2015 and 2014, respectively (Eurostat, 2018; Polsky and Garriguet, 2020). Conversely, approximately 42% and 60% of American adults reported consuming at least one fruit or one vegetable per day, respectively (Hoy et al., 2021a,b). Population-based data seem to indicate that overall intake of fruits and vegetables among adults in the United States has not changed significantly between 2003 and 2018 (Bowman et al., 2021a). However, while total intake of fruits and vegetables may have remained constant, it does appear that consumption patterns have shifted modestly (Rehm et al., 2016). Among US adults, whole fruit consumption increased while fruit juice consumption decreased (Rehm et al., 2016). A similar trend is also apparent among US children over roughly the same time period (Bowman et al., 2021b; Kim et al., 2014). In contrast, most Canadian consumers decreased consumption of total fruits and vegetables between 2004 and 2015 (Polsky and Garriguet, 2020). The most commonly consumed whole fruits in the United States and Canada are bananas and apples (Hoy et al., 2021a; Polsky and Garriguet, 2020). Melons, which have been a source of foodborne outbreaks in both the United States and Canada, are also among the top five sources of fruit consumed in Canada (Polsky and Garriguet, 2020; CDC, 2022b; Government of Canada, 2018).

Similar to fruit intake, vegetable consumption has remained relatively constant among US consumers; however, American adults increased their consumption of dark leafy greens (i.e. kale or broccoli) and red/orange vegetables (i.e. carrots or red peppers) modestly (Bowman et al., 2021a; Rehm et al., 2016). Again, most Canadians reported consuming fewer vegetables overall in 2015 when compared to 2004, and decreases in both potato and lettuce consumption were largely responsible for the decrease (Polsky and Garriguet, 2020). Even with the decrease, the most frequently consumed vegetables in both the United States and Canada are still potatoes and lettuce (Canada) or leafy green salads (US) (Hoy et al., 2021b; Polsky and Garriguet, 2020; Sebastian et al., 2018).

In the most recent FDA Food Safety and Nutrition survey, 37% of consumers reported eating vegetables 1-2 times/day with only 8% reporting eating vegetables 3 or more times per day (Lando et al., 2021). Also according to the FDA survey, consumers knew that produce items such as apples and fresh vegetables are good sources of fibre and similarly knew that fruit juice is not a source of fibre. With respect to frozen vegetables, 5-12% of consumers report eating them without heating, depending upon the vegetable (Lando et al., 2021). Research indicates that those who consume frozen fruits and vegetables have higher intakes of produce compared to those who do not consume frozen produce (Storey and Anderson, 2018), so encouraging consumption of frozen fruits and vegetables may be a way to meet public health objectives. There is a need to continue to encourage consumers to include more fruits and vegetables in their diet, and yet at the same time, there is a need to increase consumers' knowledge about how to safely prepare, store and consume raw, RTE and frozen produce.

2.1 Consumption of high-risk produce items

Sprouts were historically a major source of foodborne illness outbreaks in the United States and Canada (CDC, 2022b; Health Canada, 2011). Within the past 5 years, sprouts were identified as the source of one *Salmonella* Montevideo and one *Escherichia coli* O103 outbreak in the United States; the most recent outbreak in Canada related to sprouts and *Salmonella* occurred in 2020 (CDC, 2022b; Government of Canada, 2020b). Produce safety research and educational initiatives have tended to mention safe food handling practices related to raw sprouts specifically, whereas if other sources of produce are mentioned, the recommendations or questions are typically more general (i.e. wash fruits and vegetables) (Cates et al., 2009; Godwin and Stone, 2012; Kosa et al., 2011; Medeiros et al., 2008). However, few US consumers (22%) reported eating raw sprouts over the past 12 months, and only 20% reported hearing or reading anything about possible health problems related to raw sprouts (Lando et al., 2021). This indicates that the low level of sprout consumption is more likely due to lack of access or taste preference rather than out of food safety concerns. In the future, it may be advisable to focus educational efforts on hazardous produce items that are routinely consumed, such as leafy greens (Hoy et al., 2021b; Polsky and Garriguet, 2020; Sebastian et al., 2018).

Tooby et al. (2021) examined consumption of high-risk foods among Canadians. Three of the four most commonly reported high-risk foods consumed were produce and included romaine lettuce (48.8%), prepackaged salad greens (46.1%) and melon (39.7%). Knowing the risks associated with a food didn't affect consumption of the food, and older adults who are at greater risk for complications from foodborne illness reported similar consumption patterns of high-risk foods as consumers at lower risk (Tooby et al., 2021). Additional research is needed because few food safety research studies have focused on determining how often high-risk foods are consumed (Hoelzer et al., 2012; Jacxsens et al., 2015; Tooby et al., 2021). This is a gap in the literature that could be filled through incorporating food frequency type questionnaires as part of future food safety research studies.

3 Pathogens isolated from produce samples

Fortunately, bacterial pathogenic contamination of produce at the point of sale to consumers is small, relative to the amount of fresh produce consumed. Studies that have attempted to find pathogens in produce at the point of purchase with relatively small sample numbers (<350) have found none (Oliveira et al., 2010; Roth et al., 2018). Similarly, analysis of 911 vegetable samples available at retail in Italy failed to detect any *Salmonella* spp. or *Listeria monocytogenes* (Terio et al., 2017). One of the largest such studies took 4 years and collected 10 070 samples only to find no *Salmonella*, *E. coli* O157:H7, *Shigella* or *Campylobacter*. Thirty-seven samples were found to contain *L. monocytogenes* for a positivity rate of 0.37% (Zhang et al., 2020). It would appear that attempting to find bacterial pathogens in produce is somewhat like looking for a 'needle in a haystack' and that any such attempts would need to be through a longitudinal ongoing surveillance study rather than cross-sectional analysis.

Attempts to isolate norovirus in fresh produce, however, have resulted in higher rates of the pathogen detectable in produce. Fresh produce sold at retail in the UK was examined for the presence of norovirus with a sample size of 1152. Five percent of lettuce samples were positive, as well as 2.3% of fresh raspberries and 3.6% of frozen raspberries (Cook et al., 2019). The presence of norovirus in berries in Argentina was only 0.5%, but this was detectable with the relatively small sample size of 184 (Oteiza et al., 2022). Loutreul et al. (2014) sampled 210 lettuce samples and 200 frozen berry samples provided by industrial producers in multiple European countries. Norovirus was isolated from 11% and 15.5% of lettuce and berry samples, respectively (Loutreul et al., 2014). Analysis of 275 point-of-sale fresh lettuces from Greece, Serbia and Poland, resulted in only three samples (1.1%) positive for norovirus, while 275 samples of leafy greens from Ontario, Canada, resulted in a 16 (6%) positivity rate (Kokkinos et al., 2012; Mattison et al., 2010). Even a recent analysis of only

70 produce samples collected in Sicily resulted in a 2.9% of samples positive for norovirus (Purpari et al., 2019). This higher rate of norovirus detectability, as compared to bacterial pathogen detection, in fresh produce, is consistent with norovirus being a leading cause of illness and outbreaks associated with produce (Li et al., 2018; Machado-Moriera et al., 2019).

4 Outbreaks and illnesses caused by produce

While isolating pathogens from produce can be somewhat elusive, outbreak and illness data confirm that pathogens are associated with fresh produce available at the consumer level. Foodborne illness outbreaks associated with fresh produce increased both in number and as a proportion of all reported foodborne outbreaks from 1973 through 1997, with a total of 190 produce-associated outbreaks during this time period. Outbreaks with a known food item being produce during the same time period grew from 0.7% to 6% (Sivapalasingam et al., 2004).

Machado-Moreira et al. (2019) reviewed outbreaks and illnesses associated with fresh fruits and vegetables from 1980 to 2016 in developed countries as identified in peer-reviewed scientific publications. The top five produce products associated with the largest number of foodborne illness cases were strawberries (11553), radish (10126), salad (9627), raspberries (6970) and lettuce (6000). The top five commodities associated with the greatest number of outbreaks were salad (202), lettuce (84), raspberries (75), fruit (37) and alfalfa (33). Additional commodities associated with illness and outbreaks included tomatoes, cantaloupe, beans sprouts and onions (Machado-Moriera et al., 2019). Pathogens most associated with both illnesses and outbreaks included norovirus, *E. coli*, *Salmonella*, *Cyclospora* and hepatitis A, with norovirus being the most commonly associated with both cases (27495) and outbreaks (304). The pathogen associated with the most deaths, however, during the same time period, was *Listeria monocytogenes* (Machado-Moriera et al., 2019).

Li et al. (2018) examined produce-associated outbreak and case data from 2010 through 2015 in different regions and countries using outbreak surveillance systems. These included the US (Centers for Disease Control), European Union (European Food Safety Authority and European CDC), Australia (Dept. of Health), New Zealand (Institute of Environmental Science and Research Limited), Japan (Ministry of Health, Labor and Welfare) and Canada (Public Health Agency of Canada). There were a total of 988 outbreaks and 45723 cases over this period with the most outbreaks being in the United States (472) and the European Union (401), and fewer outbreaks in Japan (38), Australia (32), Canada (25) and New Zealand (20). The top four pathogens associated with outbreaks were norovirus (42.4%), *Salmonella enterica* (19.9%) and Shiga toxin-producing *E. coli* (STEC) (12.6%) (Li et al., 2018).

Over the past 5 years (2017-2021) in the United States, there have been 32 produce related outbreaks (CDC, 2022b). Thirteen of these have been associated with prepackaged salads (8), romaine lettuce (3) and leafy greens (2). The pathogens associated with these products included *E. coli* O157:H7 (8), *L. monocytogenes* (2), *Cyclospora* (2) and *Salmonella* (1). Eight outbreaks over this same time period that were associated with fruit included papaya (5) and cut fruit/melon (3). Two outbreaks were associated with sprouts and either *Salmonella* or *E.coli* O157:H7, and two outbreaks were associated with onions and *Salmonella*. These more recent data from CDC compared to data for the time period of 2010-2015 (Li et al., 2018) may indicate that producers are making some progress in reducing the incidence of pathogens associated with final produce products.

Data from detection of pathogens in actual produce samples and extensive data from outbreaks and illnesses indicate that norovirus is the leading cause of illnesses caused by produce and that fruits and leafy greens/salads/lettuces appear to be leading produce products associated with foodborne illness. With respect to consumers, there is a need to clearly identify when consumers can mitigate risk in their preparation and handling of produce, and when they cannot. In the event that risk reduction at the consumer level is not possible, there is also a need to clearly identify what produce products might be best avoided by vulnerable populations.

5 Organic produce

Organic food purchasing continues to increase with consumers being increasingly concerned about pesticides and fertilizers used on conventionally grown foods. Consumers, therefore, tend to perceive organically grown food as safer than produce that is conventionally grown. The most commonly purchased organic foods are produce items. The FDA Food Safety and Nutrition survey found that in the United States, 65% of consumer's report buying organic produce either some or most of the time with 7% always purchasing organic produce (Lando et al., 2021). Twenty-eight percent of consumers think organically grown produce is less likely to have harmful bacteria associated with it compared to conventionally grown, while 47% think they are about the same with respect to harmful bacteria (Lando et al., 2021). Given that manure is needed for most organic crops, there are reasons to believe that organic produce may pose a greater microbial risk than conventionally grown.

5.1 Microbial quality of conventionally grown versus organic produce

A limited number of studies have attempted to compare the microbiological quality of conventional and organic produce. Oliveira et al. (2010) sampled conventional and organic lettuce at the farm level. Seventy-two samples of each type of lettuce were collected and analysed for levels of a range of indicator organisms, as well as the presence of the pathogens *Salmonella* spp., *L. monocytogenes* and *E. coli* O15:H7. Organically grow lettuce generally had a higher bacterial load; however, no pathogens were isolated from any of the samples (Oliveira et al., 2010). Antimicrobial resistance of bacteria on organic versus conventionally grown samples was assessed by collection of 218 conventionally grown produce samples and 181 organically grown produce samples (Ruimy et al., 2010). Samples represented a range of different fruit and vegetable products and were tested for the prevalence of gram-negative bacteria that would grow on antibiotics. It was determined that all raw fruits and vegetables sampled were frequently associated with antibiotic resistant gram-negative bacteria and that there were no significant differences between organic and conventionally grown produce. Similar to other studies, Maffei et al. (2013) found that while organically grown produce generally had higher bacterial loads of some indicator organisms than conventionally grown produce, no *Salmonella* were found on any of the samples (Maffei et al., 2013). More recently, Srisamran et al. (2022) sampled conventional and organic produce from supermarkets and, in contrast to previous studies, found lower concentration of *E. coli* in conventional versus organic produce samples. A limitation of all of these studies is the relatively low sample numbers (130-349) given the sporadic nature of pathogens in produce of any type. There is a need for a large, longitudinal study of organically versus conventionally grown produce to determine whether there may be a significant difference in microbial risks with one type of produce or the other. The data to date seem to indicate that at the least, the two have an equivalent risk, with the possibility that organically grown may have a slightly greater risk.

5.2 Foodborne outbreaks associated with organic foods

Another indication of risk that can be measured is the number of illnesses and outbreaks associated with organically grown produce. Harvey et al. (2016) identified 18 outbreaks caused by organic foods from 1992 to 2014. Nine of these outbreaks were attributed to produce items and their seeds. Together these outbreaks accounted for 636 illnesses, 232 hospitalizations and 3 deaths. This is far more than the 146 illness and 21 hospitalizations caused by non-produce organic foods involved in the other nine outbreaks due to organic dairy or meat products. Outbreaks increased in the 2000s with the rising increase in the availability and popularity of organically regulated products (Harvey et al., 2016). Traditionally, outbreaks have not been tracked by production method; however, perhaps this is an area of need going forward which could further help to elucidate whether there is a difference in safety between conventionally and organically grown produce.

6 Farmers markets

Farmers markets have grown significantly since 2000 when the estimated number of farmers markets was 2863 (USDA, ERS, 2022). The National Farmers Market Managers Survey, administered in spring of 2020, estimated that approximately 8140 farmer's markets were in operation in the lower 48 states of the United States (USDA, 2020). Produce was the most common food category sold at markets, with 99.6% of markets reporting selling fruits and vegetables. Specifically, 99.3% of markets sold fresh vegetables, 95.3% sold fresh fruit, and 79.6% sold fresh or dried herbs. Only 53% of Farmers Market Managers were paid employees, earning an average of $18.40/hour, while 39% of managers were volunteers. Managers worked approximately 20 hours/week. The nature of employment (i.e. paid vs. volunteer or part vs. full-time) could be relevant when considering food safety training and food safety practices employed at farmers markets.

6.1 Microbial quality of produce items offered at traditional retail versus farmers markets

A limited number of studies have looked at the microbial quality of produce available at farmer's markets with a comparison to products available from a traditional retail vendor. Arthur et al. collected 1183 produce samples in Ontario, CA, from retail stores (82%), farmer's markets (15%) and organic farms (2%). Only two samples were found to contain *Salmonella*, a tomato from a retail outlet and organic lettuce from an organic farm. No other pathogens were found and the overall prevalence of *E. coli* did not significantly differ by location type (Arthur et al., 2007). Pan et al. (2015) examined the presence of indicator organisms and pathogens on vegetables sold from Asian vendors at farmer's markets in California. All 242 samples had detectable faecal coliforms and 20% of samples had detectable *E. coli*. Sixteen (6.6%) of samples were positive for *Salmonella* and there was a twofold probability of *Salmonella* contamination in samples from growers reporting using organic farming practices compared to those using conventional farming practices (Pan et al., 2015).

More recently, assessment of microbial quality of fresh produce sold at farmer's markets versus supermarkets in Florida found farmer's market produce to be slightly less safe than that found in supermarkets (Roth et al., 2018). One hundred samples were obtained from supermarkets and 301 samples from farmer's markets. *E. coli* was identified in seven farmer's market samples and *L. monocytogenes* in four, while neither pathogen was found in any of the supermarket samples. *Salmonella* was found in one each of the farmer's market and supermarket samples (Roth et al., 2018). Pires et al. (2020) found that of 128 produce samples collected and tested from farmers markets in California, all were negative for *Salmonella*, while 31.3% were positive for *E. coli* (Pires et al., 2020). The information that is available about the quality of produce from farmers markets indicates that it is no safer than produce sold from traditional retail outlets, and may be less safe.

6.2 Foodborne illnesses due to farmers markets

Identifying foodborne illness and outbreaks associated with farmer's markets is particularly challenging because retail origin of a food associated with illness or outbreak is not always indicated. Bellemare and Nguyen (2018) examined the relationship between farmers markets and foodborne illnesses over 7 years between 2004 and 2013. There was a positive relationship between the number of farmers markets per million individuals and the number of total outbreaks and cases of foodborne illness. Additionally, there was a positive relationship between number of farmers markets and outbreaks and cases of norovirus, and outbreaks of *Campylobacter*. Since farmer's markets sell more than just produce, not all of this correlation may be due to produce. Additionally, the authors point out that this research represented correlation only, not causation (Bellemare and Nguyen, 2018). It provides evidence, however, that there is a need to further examine the relationship between foodborne illness and farmer's markets to inform whether interventions are needed. Young et al. (2017c) identified nine outbreaks and two case reports associated with foods sold at farmer's markets between 2000 and 2016. Six of these were associated with produce or produce products, the largest of which involved raw peas and *C. jejuni* found to be linked to Sandhill crane faeces on the farm. Another outbreak with an identified source involved strawberries and *E. coli* O157:H7 with the identified contamination caused by deer faeces on the farm (Young et al., 2017c). There is a need, therefore, to continue to monitor foodborne illnesses and outbreaks associated with farmer's markets.

6.3 Perceptions versus observations of food safety at farmers markets

The perceptions of the safety of produce sold at eight farmers markets in Kentucky was assessed using self-administered surveys to 239 shoppers at the markets. Researchers found that 71.8% of respondents said that fewer microbial safety concerns of produce purchased at the markets was very or extremely important to them. It should be noted that 78.5% reported the same about their belief that the produce was free from chemicals. The majority of respondents (84.7%) reported washing produce purchased at the markets (Khouryieh et al., 2019). Conversely, in research conducted with 45 farm market managers, 42% reported having no food safety standards in place for the market. Less than 25% reported sanitizing market surfaces and only 11% reported always cleaning market containers between uses. Finally, the majority (75%) offered no sanitation training to workers or vendors (Harrison et al., 2013). Multiple studies have used observation methods to assess the safe food handling practices and conditions at farmers markets in the United States and Canada (Behnke et al., 2012; McIntyre et al., 2014; Pollard et al., 2016). Common unsafe practices identified included insufficient handwashing and improper handwashing as well as handling money and produce without proper handwashing. Common unsafe conditions associated with farmers markets included lack of hand sanitizer and lack of handwashing stations as well as lack of temperature control for pre-cut produce. Given the volunteer and part-time nature of Farmers Market Managers (USDA, 2020), as well as the seasonality of such markets, there is a need to develop targeted education modules related to handwashing and handwashing facilities in farmers markets to increase the safety of produce sold through these markets. While these markets represent a relatively small proportion of all produce sold, their increasing popularity warrants finding ways to increase the safety of conditions and produce sold in these facilities before they become an even larger proportion of produce sold directly to consumers.

7 Consumer knowledge and behaviours around safe produce handling and consumption

7.1 Safe produce handling knowledge

US consumers may be unaware of the risks that produce may present with regard to food safety. The most recently available FDA Food and Nutrition Survey found that a large majority of US respondents believed that raw fruit (78%) and raw vegetables (75%) were 'not at all or very unlikely' to be contaminated with harmful bacteria (Lando et al., 2021). This indicates that a large number of Americans are not aware that unwashed raw produce may contain harmful bacteria, or that some RTE produce products need to be handled carefully with respect to refrigeration, in order to ensure their safety. Regarding knowledge about specific produce products, 72% of respondents thought that whole lettuce was somewhat to very likely to contain harmful bacteria while 55% thought that pre-cut lettuce or salads were not at all or very unlikely to contain harmful bacteria (Lando et al., 2021). Twenty-five percent of respondents thought that pre-cut fruit such as melon was not at all likely to contain harmful bacteria and 40% of respondents thought the same about frozen vegetables. Consumers recognized the foodborne pathogens *Salmonella* (97%) and *E. coli* (88%), but only 44%, 21% and 7% recognized *Listeria*, norovirus and *Campylobacter*, respectively (Lando et al., 2021). While norovirus is a leading cause of foodborne illness associated with produce, 57.5% of consumers in a nationwide survey did not know that any food may become contaminated with norovirus, and 74.3% were not aware that norovirus outbreaks have been commonly linked to contaminated green, leafy vegetables, such as lettuce and spinach (Cates et al., 2015).

Most US consumers do not believe that frozen produce is a potential vector for foodborne pathogen exposure (Lando et al., 2021); however, frozen produce has been identified as a source of both norovirus and hepatitis A outbreaks throughout North America and Europe (Nasheri et al., 2019). Both viruses can be inactivated by cooking (CDC, 2016, 2019), but 5-12% of US consumers reported eating frozen vegetables without heating, depending upon the vegetable (Lando et al., 2021). Therefore, it may be important to balance the public health goal of encouraging frozen produce consumption for its nutritional benefits, while at the same time encouraging safe handling practices of frozen produce. Scott et al. (2009) provided evidence that targeted produce safety education modules can improve consumers' knowledge around how to properly wash, refrigerate and prevent cross-contamination of fresh produce. Researchers used a 1-hour education module administered to 2651 individuals in 69 Texas counties and saw significant knowledge increase around the practices of washing raw poultry and preventing cross-contamination (Scott et al., 2009). In general, consumers lack knowledge around the risks for foodborne illness and how to properly wash and prevent cross-contamination of fresh produce, as well as the risks for RTE produce products if not properly chilled.

7.2 Produce handling behaviours

While produce contaminated with pathogens in production represents a risk consumers need to be aware of, it is much more difficult to identify the level of illness caused by produce contaminated with pathogens by consumers in the home, via cross-contamination with raw meat, poultry or fish. Anderson et al. (2004) videotaped 99 consumers preparing a meal containing both meat or poultry and raw produce. They found that both handwashing and produce washing were inadequate, and that nearly all subjects cross-contaminated raw meat, poultry, seafood, eggs and/or unwashed produce with RTE foods multiple times during food preparation (Anderson et al., 2004). Similarly, Møretrø et al. (2021) observed and interviewed 87 households in six European countries, while they made a chicken and vegetable meal. They found that rinsing chicken in sinks was common and washing vegetables in the same sink was also common. Conversely, using the same board and knife for vegetables after using it for chicken and without washing with detergent was common in some countries but not others (Møretrø et al., 2021). A review of popular recipe blogs found that among 304 recipes containing fresh produce, only 10 (3.3%)

suggested washing produce to be consumed raw (Morrison and Young, 2019). There is a need, therefore, to continue to educate consumers about the risk of cross-contamination of fresh produce in the home and the importance of separating raw meat and poultry and unwashed produce from produce and other foods that will be consumed uncooked.

Limited research indicates that produce washing increased at the beginning of the COVID-19 pandemic. Thomas and Feng (2021) used surveys and focus groups with consumers from April to August 2021 and found that for each of the 5 months, reported produce washing increased both with water and with water plus soap. Therefore, while the COVID-19 pandemic may have increased the correct behaviour to wash produce, it also increased the incorrect behaviour of utilizing soap on produce (Thomas and Feng, 2021). Research by the International Food Information Council found that produce washing was up to 30% in 2020 and fell back down to approximately 22% in 2021 (IFIC, 2021). Given these results and assuming that the motivation to increase washing produce was likely to remove the novel coronavirus, it is unlikely that the pandemic will have any lasting positive effects on safe produce handling.

8 Produce safety and vulnerables and immunocompromised consumers

The pathogens most often associated with fresh and RTE produce, *Salmonella*, *E. coli* O157:H7 and *L. monocytogenes*, are those which are most likely to result in complications, hospitalizations and even death in vulnerable populations (Scallan et al., 2015). Vulnerable populations include young children, older adults, pregnant women, cancer and HIV/AIDs patients, as well as transplant recipients. While there is extensive research around these groups and foodborne illness in general, there is a paucity of research specific to produce safety and these consumers. Examination of the literature around safe food handling knowledge and practices of these populations finds that in most cases the only produce product specifically addressed has been raw sprouts (Lindup et al., 2020; Hoffman et al., 2005; Medeiros et al., 2008; Kendall et al., 2017).

Evans and Redmond (2018) assessed the knowledge of cancer patients and caregivers and found that very few were aware that unpeeled or unwashed fruit and vegetables and salad should be avoided during chemotherapy (Evans and Redmond, 2018). Similarly, in a study with older adults only 3% were aware of the possible association of prepacked salads with *L. monocytogenes*, and 64% and 41% reported consuming prepackaged salads and pre-cut fruits, respectively, after the use-by date (Evans and Redmond, 2016). Research with pregnant women found that pathogen-specific food safety education led to increased knowledge and safe food handling practices, so this may be a potential strategy to educate vulnerable populations (Kendall et al., 2017).

Dworkin et al. (2013) found that a comic book food safety intervention for AIDS patients significantly increased their knowledge not to consume raw carrots, bean sprouts, lettuce and spinach (Dworkin et al., 2013). Kavanaugh et al. (2021) conducted focus groups with older adults in order to explore their food handling and consumption practices in connection with several potentially hazardous foods. Approximately half of the participants reported eating bagged or boxed salads; however, some participants avoided them due to food safety concerns. Similarly, about a quarter of participants reported occasionally eating sprouts, but other participants reported that they had stopped eating sprouts out of food safety concerns. In general, participants in the focus groups reported avoiding certain potentially hazardous foods when they had heard about recalls and outbreaks associated with those foods (Kavanaugh et al., 2021).

The literature indicates that vulnerable populations are receptive to targeted food safety messages, particularly when they are pathogen-specific. These consumers in general, however, seem to be unaware of the risks associated with fresh and RTE produce and continue to report consuming produce products that may put them at risk for illness and complications from *Salmonella*, *L. monocytogenes* or STEC. Since there is great diversity in the individuals included in these groups, there is a need to better understand the consumption patterns and handling practices of all categories of vulnerable consumers with respect to raw and prepackaged produce consumption. This could help inform new produce safety education materials targeted to these vulnerable populations.

9 Fresh produce food safety handling recommendations and resources

Consumer education around safe handling of produce is both prolific and yet at the same time somewhat hidden. In many cases, safe produce handling is embedded within general food safety education resources. These may be at the international, national or state/community level.

The World Health Organization promotes the Five Keys to Safer Food (WHO, 2022). These include 1) Keep clean, 2) Separate raw and cooked, 3) Cook thoroughly, 4) Keep food at safe temperatures and 5) Use safe water and raw materials. Within these are included the messages to wash fruits and vegetables, especially if eaten raw, as well as to refrigerate perishable food below 5°C (WHO, 2022).

In the United States, General Food Safety education materials that also mention produce safety include those from US Food and Drug Administration (FDA, 2022), the Centers for Disease Control (CDC, 2022a) and the Partnership for Food Safety Education (Partnership, 2022). The general food safety materials promoted revolve around the 'Core Four', that is, 1) Clean, 2) Separate, 3) Cook and 4) Chill (Partnership, 2022). Within these core four, messages related to produce include Clean - wash/rinse raw produce. Additionally, Separate includes the message to separate raw and RTE produce from raw meat or fish products. Finally, Chill focuses on refrigerating any perishable food to 40°F or below within 2 hours (Partnership, 2022).

Health Canada also provides general Food Safety Tips (Government of Canada, 2022). Finally, in addition to these International and Country-wide food safety education resources, a number of states, local public health agencies and nutrition and food educators have developed food safety education materials that are now available on websites (CO State, 2010; NC State, 2021; Government of Canada, 2020a). These are all helpful resources, but a challenge many of them present is that they must be sought out by the consumer, or used to teach the consumer in a more structured educational setting.

There is a need, with the growing increase in both produce consumption, and illnesses and outbreaks due to produce, to have more targeted education materials to inform consumers about specific guidelines for produce safety. Some of these types of materials are beginning to emerge. In the United States, the Partnership for Food Safety Education's 'Produce Pro' focuses on produce and instructs consumers on what produce should be washed and how, as well as what produce (prewashed/prepackaged) should not be washed (Partnership, 2022) (Fig. 1).

Produce safety education is further complicated by the fact that in some cases the final safety steps are out of the hands of consumers. That is, in the sporadic occasions that are known to occur of RTE prepackaged salads, sprouts or leafy greens being contaminated with a pathogen, consumers do not have a safety step with the exception of avoiding the food product. In some cases, for consumers at high risks for complications from foodborne illness, there is a need to inform them about what produce products should be avoided and what are the alternative, safer choices. Only in the case of raw sprouts are high-risk consumers consistently told to avoid the product. There is a need for produce

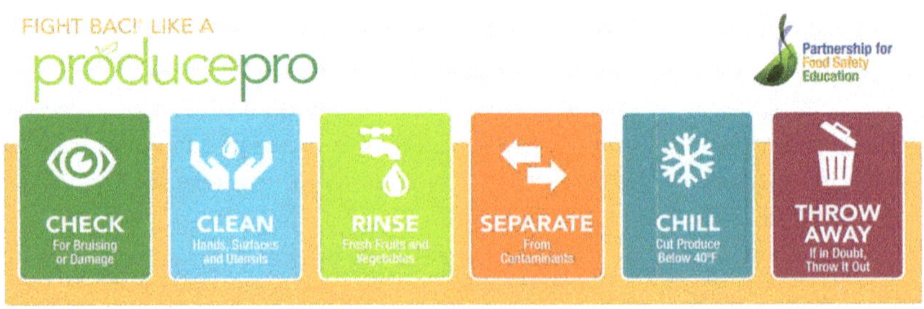

Figure 1 Fresh produce specific food safety education materials developed by the Partnership for Food Safety Education, available with supporting materials at fightbac.org.

safety messaging that clearly identifies potentially hazardous produce products that should be avoided by vulnerable populations due to their increased likelihood of experiencing severe consequences from foodborne illness.

10 Development of new and targeted education materials around produce safety: the need to include behavioural theories and formative research

There is a need for the development of new and targeted food safety education campaigns and education modules around safe produce handling and consumption. There is also an opportunity to develop such education resources based on appropriate formative research and behavioural theories. At least 82 behavioural theories are available to guide behavioural or social science research; however, relatively few of those theories have been applied in the context of food safety research (Davis et al., 2015; Sivaramalingam et al., 2015; Young et al., 2017a).

Incorporating a behavioural theory is recommended for all future consumer-based food safety research because doing so provides a framework for the intervention and also offers insight into how various psychosocial constructs might be leveraged to promote behaviour change (Glanz et al., 2008; Sivaramalingam et al., 2015; Young et al., 2017a). While 27% of research studies have incorporated a behavioural theory into a food safety education intervention (Sivaramalingam et al., 2015), very few of the interventions supported by a behavioural theory included produce safety (Mayer and Harrison, 2012; Medeiros et al., 2008; Quick et al., 2013). Moreover, few food safety research studies incorporating a behavioural theory have targeted vulnerable populations (Sivaramalingam et al., 2015), whom are more likely to suffer severe illness as the result of foodborne infection.

Future produce safety education campaigns targeting high-risk groups would benefit from incorporating a theory of behaviour change; however, there is evidence that existing theories will need to be modified to fit the unique needs of the target population (Cates et al., 2006; Chen et al., 2010; Medeiros et al., 2008; Dickinson et al., 2014; Young et al., 2017a). Therefore, formative research is recommended as a preliminary step for all future food safety education interventions, especially when targeting vulnerable populations (Young et al., 2017a).

Formative research is a broad term encompassing a number of activities, but the overarching principle is to conduct preliminary research with members of the target population in order to better understand motivations, attitudes, behaviours, etc., in order to guide later phases of research or programme development (Kimball and Loya, 2017). In practice, formative research has only been used to guide approximately 41% of food safety education interventions (Sivaramalingam et al., 2015), whereas the majority of studies evaluating associations between psychosocial constructs and food handling behaviours have employed some type of formative research (Young et al., 2017a,b). Future food safety education interventions around safe produce handling would benefit from conducting formative research with the target audience in order to modify the selected behavioural theory and ultimately enhance the potential effectiveness of the related food safety message. Moreover, since the populations at greatest risk for foodborne illness are underrepresented in the literature, research is needed to determine how to best apply behaviour change theories when designing interventions for vulnerable populations, such as older adults or immunocompromised individuals (Sivaramalingam et al., 2015; Young et al., 2017a).

11 Improving consumer's handling of fresh produce: current needs and next steps

Safe produce handling at the consumer level presents a unique challenge because of the continued sporadic presence of pathogens in fresh, frozen and RTE produce. Increased consumer consumption of leafy greens and RTE produce is consistent with recommendations for a healthy diet. At the same time, these are some of the produce items (types) most commonly associated with outbreaks and illnesses due to both bacterial and viral pathogens. For a healthy individual, the benefits of consuming these products likely outweigh the very small risk for foodborne illness; however, this may not be true for vulnerable populations. Further complicating messaging is that fresh produce needs to be washed, while some of the same produce in RTE form should not be washed prior to consumption. Finally, a subset of consumers seem to perceive organic and locally grown produce as preferred safer choices, but data indicate that from a microbiological perspective these produce are no safer, and are potentially less safe, than traditionally grown produce.

Overall, the complexities of messages needed to guide consumers around safe handling of produce require careful formative research. Education materials based on formative research grounded in behavioural theory models are the recommended best practices for increasing the efficacy of food safety education materials. Including such formative research will allow researchers and educators to better understand consumers' intentions, influencers and barriers to adopting safe produce handling. There is a great need, therefore, for interdisciplinary teams to work together to identify key messages and outlets/venues to deliver appropriate safe produce handling messages to both vulnerable populations and the general population.

12 References

Anderson, J. B., Shuster, T. A., Hansen, K. E., Levy, A. S. and Volk, A. (2004). A camera's view of consumer food-handling behaviors. *Journal of the American Dietetic Association* 104(2), 186–191. https://doi.org/10.1016/j.jada.2003.11.010.

Arthur, L., Jones, S., Fabri, M. and Odumeru, J. (2007). Microbial survey of selected Ontario-grown fresh fruits and vegetables. *Journal of Food Protection* 70(12), 2864–2867. https://doi.org/10.4315/0362-028X-70.12.2864.

Behnke, C., Seo, S. and Miller, K. (2012). Assessing food safety practices in farmers' markets. *Food Protection Trends* 32(5), 232–239.

Bellemare, M. F. and Nguyen, N. (2018). Farmers markets and food-borne illness. *American Journal of Agricultural Economics* 100(3), 676–690. https://doi.org/10.1093/ajae/aay011.

Bowman, S. A., Clemens, J. C. and Friday, J. E. (2021a). Food pattern group and macronutrient intake of adults: WWEIA, NHANES 2003–2004 to 2017–2018, food surveys research group dietary data Brief No. 35. United States Department of Agriculture. Available at: https://www.ars.usda.gov/northeast-area/beltsville-md-bhnrc/beltsville-human-nutrition-research-center/food-surveys-research-group/docs/wweia-data-briefs/.

Bowman, S. A., Clemens, J. C., Friday, J. E. and Anand, J. (2021b). Changes in total fruit and fruit juice intakes of individuals: WWEIA, NHANES 2005–2006 to 2017–2018, food surveys research group dietary data brief no. 41. United States Department of Agriculture. Available at: https://www.ars.usda.gov/northeast-area/beltsville-md-bhnrc/beltsville-human-nutrition-research-center/food-surveys-research-group/docs/wweia-data-briefs/.

Cates, S. C., Kosa, K. M., Brophy, J. E., Hall, A. J. and Fraser, A. (2015). Consumer education needed on Norovirus prevention and control: Findings from a nationally representative survey of US adults. *Journal of Food Protection* 78(3), 484–490. https://doi.org/10.4315/0362-028X.JFP-14-313.

Cates, S. C., Kosa, K. M., Karns, S., Godwin, S. L., Speller-Henderson, L., Harrison, R. and Ann Draughon, F. (2009). Food safety knowledge and practices among older adults: Identifying causes and solutions for risky behaviors. *Journal of Nutrition for the Elderly* 28(2), 112–126. https://doi.org/10.1080/01639360902949986.

Cates, S. C., Kosa, K. M., Teneyck, T., Jaykus, L. A., Moore, C. and Cowen, P. (2006). Older adults' knowledge, attitudes, and practices regarding listeriosis prevention. *Food Protection Trends* 26, 774–785.

Centers for Disease Control and Prevention (CDC) (2016). Hepatitis A prevention. Retrieved March 24, 2022. Available at: https://wonder.cdc.gov/wonder/prevguid/p0000433/p0000433.asp#head007000000000000.

Centers for Disease Control and Prevention (CDC) (2019). Preventing Norovirus outbreaks. Retrieved March 24, 2022. Available at: https://www.cdc.gov/vitalsigns/norovirus/index.html.

Centers for Disease Control and Prevention (CDC) (2022a). Fruit and vegetable safety. Retrieved March 24, 2022. Available at: https://www.cdc.gov/foodsafety/communication/steps-healthy-fruits-veggies.html.

Centers for Disease Control and Prevention (CDC) (2022b). List of selected multistate foodborne outbreak investigations. Retrieved March 24, 2022. Available at: https://www.cdc.gov/foodsafety/outbreaks/multistate-outbreaks/outbreaks-list.html.

Chen, G., Kendall, P. A., Hillers, V. N. and Medeiros, L. C. (2010). Qualitative studies of the food safety knowledge and perceptions of transplant patients. *Journal of Food Protection* 73(2), 327–335. https://doi.org/10.4315/0362-028X-73.2.327.

Colorado State University (2010, March). Extension (CO state). *Guide to Washing Fresh Produce, Fact Sheet No. 9.380*. Retrieved March 13, 2022. Available at: https://nifa.usda.gov/sites/default/files/resource/Guide%20to%20Washing%20Fresh%20Produce508.pdf.

Cook, N., Williams, L. and D'Agostino, M. (2019). Prevalence of Norovirus in produce sold at retail in the United Kingdom. *Food Microbiology* 79, 85–89. https://doi.org/10.1016/j.fm.2018.12.003.

Davis, R., Campbell, R., Hildon, Z., Hobbs, L. and Michie, S. (2015). Theories of behaviour and behaviour change across the social and behavioural sciences: A scoping review. *Health Psychology Review* 9(3), 323–344. https://doi.org/10.1080/17437199.2014.941722.

Dickinson, A., Wills, W., Meah, A. and Short, F. (2014). Food safety and older people: The Kitchen Life study. *British Journal of Community Nursing* 19(5), 226–232. https://doi.org/10.12968/bjcn.2014.19.5.226.

Dworkin, M. S., Peterson, C. E., Gao, W., Mayor, A., Hunter, R., Negron, E., Fleury, A. and Besch, C. L. (2013). Efficacy of a food safety comic book on knowledge and self-reported behavior for persons living with AIDS. *PLoS ONE* 8(10), e72874. https://doi.org/10.1371/journal.pone.0072874.

Eurostat (2018). Fruit and vegetable consumption statistics. Retrieved March 22, 2022. Available at: https://ec.europa.eu/eurostat/statistics-explained/index.php?title=Fruit_and_vegetable_consumption_statistics#Fruit_consumption.

Evans, E. W. and Redmond, E. C. (2016). Older adult consumer knowledge, attitudes, and self-reported storage practices of ready-to-eat food products and risks associated with listeriosis. *Journal of Food Protection* 79(2), 263–272. https://doi.org/10.4315/0362-028X.JFP-15-312.

Evans, E. W. and Redmond, E. C. (2018). Food safety knowledge and self-reported food-handling practices in cancer treatment. *Oncology Nursing Forum* 45(5), E98–E110. https://doi.org/10.1188/18.ONF.E98-E110.

Glanz, K., Rimer, B. K. and Viswanath, K. (2008). Theory, research and practice in health behavior and health education. In: Glanz, K., Rimer, B. K. and Viswanath, K. (Eds.), *Health Behavior and Health Education: Theory, Research, and Practice* (4th edn., pp. 26–27). San Francisco, CA: Jossey-Bass.

Godwin, S. L. and Stone, R. (2012). Food safety because you care! An educational program for older adults. *Cooperative Extension Faculty Research*. Available at: http://digitalscholarship.tnstate.edu/extension/44, Paper 44.

Government of Canada (2018). 2010–2011 Salmonella in Cantaloupes. Available at: https://inspection.canada.ca/food-safety-for-industry/food-chemistry-and-microbiology/food-safety-testing-bulletin-and-reports/salmonella-in-cantaloupes/eng/1396554421068/1396554422178.

Government of Canada (2019). Canada's dietary guidelines. Retrieved March 23, 2022. Available at: https://food-guide.canada.ca/en/guidelines/.

Government of Canada (2020a). Fruits and vegetables - fact sheets. Retrieved March 23, 2022. Available at: https://inspection.canada.ca/food-safety-for-consumers/fact-sheets/specific-products-and-risks/fruits-and-vegetables/eng/1363562389536/1363562475361.

Government of Canada (2020b). Updated food recall warning - Sprouts alive brand and sunsprout brand micro-greens alfalfa recalled due to Salmonella. Available at: https://inspection.canada.ca/food-recall-warnings-and-allergy-alerts/2020-09-19/eng/1600566035595/1600566042478.

Government of Canada (2022). General food safety tips. Retrieved March 23, 2022. Available at: https://www.canada.ca/en/health-canada/services/general-food-safety-tips.html.

Harrison, J. A., Gaskin, J. W., Harrison, M. A., Cannon, J. L., Boyer, R. R. and Zehnder, G. W. (2013). Survey of food safety practices on small to medium-sized farms and in farmers markets. *Journal of Food Protection* 76(11), 1989–1993. https://doi.org/10.4315/0362-028X.JFP-13-158.

Harvey, R. R., Zakhour, C. M. and Gould, L. H. (2016). Foodborne disease outbreaks associated with organic foods in the United States. *Journal of Food Protection* 79(11), 1953–1958. https://doi.org/10.4315/0362-028X.JFP-16-204.

Health Canada (2011). It's your health: Risks associated with sprouts. Available at: https://publications.gc.ca/collections/collection_2012/sc-hc/H13-7-30-2011-eng.pdf.

Hoelzer, K., Pouillot, R., Egan, K. and Dennis, S. (2012). Produce consumption in the United States: An analysis of consumption frequencies, serving sizes, processing forms, and high-consuming population subgroups for microbial risk assessments. *Journal of Food Protection* 75(2), 328–340. https://doi.org/10.4315/0362-028X.JFP-11-313.

Hoffman, E. W., Bergmann, V., Shultz, J. A., Kendall, P., Medeiros, L. C. and Hillers, V. N. (2005). Application of a five-step message development model for food safety education materials targeting people with HIV/AIDS. *Journal of the American Dietetic Association* 105(10), 1597–1604. https://doi.org/10.1016/j.jada.2005.07.035.

Hoy, M. K., Clemens, J. C. and Moshfegh, A. J. (2021a). Intake of fruits by adults, food surveys research group dietary data brief no. 37. United States Department of Agriculture. Available at: https://www.ars.usda.gov/northeast-area/beltsville-md-bhnrc/beltsville-human-nutrition-research-center/food-surveys-research-group/docs/wweia-data-briefs/.

Hoy, M. K., Clemens, J. C. and Moshfegh, A. J. (2021b). Intake of vegetables by adults, food surveys research group dietary data brief no. 39. United States Department of Agriculture. Available at: https://www.ars.usda.gov/northeast-area/beltsville-md-bhnrc/beltsville-human-nutrition-research-center/food-surveys-research-group/docs/wweia-data-briefs/.

International Food Information Council (IFIC) (2021). *Food and Health Survey*. Retrieved March 1, 2022. Available at: https://foodinsight.org/2021-food-health-survey/ (Vol. 2021).

Jacxsens, L., Ibañez, I. C., Gomez-Lopez, V. M., Fernandes, J. A., Allende, A., Uyttendaele, M. and Huybrechts, I. (2015). Belgian and Spanish consumption data and consumer handling practices for fresh fruits and vegetables useful for further microbiological and chemical exposure assessment. *Journal of Food Protection* 78(4), 784–795. https://doi.org/10.4315/0362-028X.JFP-14-376.

Kavanaugh, M., Fisher, K. and Quinlan, J. J. (2021). Use of focus groups to identify food safety risks for older adults in the US. *Foods* 11(1), 37. https://doi.org/10.3390/foods11010037.

Kendall, P., Scharff, R., Baker, S., LeJeune, J., Sofos, J. and Medeiros, L. (2017). Food safety instruction improves knowledge and behavior risk and protection factors for foodborne illnesses in pregnant populations. *Maternal and Child Health Journal* 21(8), 1686–1698. https://doi.org/ 10.1007/s10995-017-2291-2.

Khouryieh, M., Khouryieh, H., Daday, J. K. and Shen, C. (2019). Consumers' perceptions of the safety of fresh produce sold at farmers' markets. *Food Control* 105, 242–247. https://doi.org/10.1016/j.foodcont.2019.06.003.

Kim, S. A., Moore, L. V., Galuska, D., Wright, A. P., Harris, D., Grummer-Strawn, L. M., Merlo, C. L., Nihiser, A. J. and Rhodes, D. G. (2014). Vital signs: Fruit and vegetable intake among children–United States. *MMWR. Morbidity and Mortality Weekly Report* 63(31), 2003–2010.

Kimball, E. and Loya, K. I. (Eds.) (2017). *Using Qualitative Research to Promote Organizational Intelligence: New Directions for Institutional Research, Number 174*. John Wiley & Sons.

Kokkinos, P., Kozyra, I., Lazic, S., Bouwknegt, M., Rutjes, S., Willems, K., Moloney, R., de Roda Husman, A. M., Kaupke, A., Legaki, E., D'Agostino, M., Cook, N., Rzezutka, A., Petrovic, T. and Vantarakis, A. (2012). Harmonised investigation of the occurrence of human enteric viruses in the leafy green vegetable supply chain in three European countries. *Food and Environmental Virology* 4(4), 179–191. https://doi.org/ 10.1007/s12560-012-9087-8.

Kosa, K. M., Cates, S. C., Godwin, S. L., Ball, M. and Harrison, R. E. (2011). Effectiveness of educational interventions to improve food safety practices among older adults. *Journal of Nutrition in Gerontology and Geriatrics* 30(4), 369–383. https://doi.org/10.1080/21551197.2011.623943.

Lando, A., Verrill, L. and Wu, F. (2021). *FDA's Food and Nutrition Survey 2019, Survey*. Food and Drug Administration. Available at: https://www.fda.gov/media/146532/download.

Li, M., Baker, C. A., Danyluk, M. D., Belanger, P., Boelaert, F., Cressey, P., Gheorghe, M., Polkinghorne, B., Toyofuku, H. and Havelaar, A. H. (2018). Identification of biological hazards in produce consumed in industrialized countries: A review. *Journal of Food Protection* 81(7), 1171–1186. https://doi.org/10.4315/0362-028X.JFP-17-465.

Lindup, M., van den Bogaart, L., Golshayan, D., Aubert, J. D., Vionnet, J., Regamey, J., Pascual, M., Manuel, O. and Mombelli, M. (2020). Real-life food-safety behavior and incidence of foodborne infections in solid organ transplant recipients. *American Journal of Transplantation* 20(5), 1424–1430. https://doi.org/10.1111/ajt.15725.

Loutreul, J., Cazeaux, C., Levert, D., Nicolas, A., Vautier, S., Le Sauvage, A. L., Perelle, S. and Morin, T. (2014). Prevalence of human noroviruses in frozen marketed shellfish, red fruits and fresh vegetables. *Food and Environmental Virology* 6(3), 157–168. https://doi.org/10.1007/s12560-014-9150-8.

Machado-Moreira, B., Richards, K., Brennan, F., Abram, F. and Burgess, C. M. (2019). Microbial contamination of fresh produce: What, where, and how? *Comprehensive Reviews in Food Science and Food Safety* 18(6), 1727–1750. https://doi.org/10.1111/1541-4337.12487.

Maffei, D. F., de Arruda Silveira, N. F. and Catanozi, M. D. P. L. M. (2013). Microbiological quality of organic and conventional vegetables sold in Brazil. *Food Control* 29(1), 226–230. https://doi.org/10.1016/j.foodcont.2012.06.013.

Mattison, K., Harlow, J., Morton, V., Cook, A., Pollari, F., Bidawid, S. and Farber, J. M. (2010). Enteric viruses in ready-to-eat packaged leafy greens. *Emerging Infectious*

Diseases 16(11), 1815–1817; discussion 1817. https://doi.org/10.3201/eid1611.100877.

Mayer, A. B. and Harrison, J. A. (2012). Safe eats: An evaluation of the use of social media for food safety education. *Journal of Food Protection* 75(8), 1453–1463. https://doi.org/10.4315/0362-028X.11-551.

McIntyre, L., Karden, L., Shyng, S. and Allen, K. (2014). Survey of observed vendor food-handling practices at farmers' markets in British Columbia, Canada. *Food Protection Trends* 34, 397–408.

Medeiros, L. C., Chen, G., Hillers, V. N. and Kendall, P. A. (2008). Discovery and development of educational strategies to encourage safe food handling behaviors in cancer patients. *Journal of Food Protection* 71(8), 1666–1672. https://doi.org/10.4315/0362-028X-71.8.1666.

Møretrø, T., Nguyen-The, C., Didier, P., Maître, I., Izsó, T., Kasza, G., Skuland, S. E., Cardoso, M. J., Ferreira, V. B., Teixeira, P., Borda, D., Dumitrascu, L., Neagu, C., Nicolau, A. I., Anfruns-Estrada, E., Foden, M., Voysey, P. and Langsrud, S. (2021). Consumer practices and prevalence of Campylobacter, Salmonella and Norovirus in kitchens from six European countries. *International Journal of Food Microbiology* 347, 109172. https://doi.org/10.1016/j.ijfoodmicro.2021.109172.

Morrison, E. and Young, I. (2019). The missing ingredient: Food safety messages on popular recipe blogs. *Food Protection Trends* 39(1), 28–39.

Nasheri, N., Vester, A. and Petronella, N. (2019). Foodborne viral outbreaks associated with frozen produce. *Epidemiology and Infection* 147, e291. https://doi.org/10.1017/S0950268819001791.

National Health Service (NHS) (2019). The Eatwell guide. Retrieved March 24, 2022. Available at: https://www.nhs.uk/live-well/eat-well/food-guidelines-and-food-labels/the-eatwell-guide/.

Oliveira, M., Usall, J., Vinas, I., Anguera, M., Gatius, F. and Abadias, M. (2010). Microbiological quality of fresh lettuce from organic and conventional production. *Food Microbiology* 27(5), 679–684. https://doi.org/10.1016/j.fm.2010.03.008.

Oteiza, J. M., Prez, V. E., Pereyra, D., Jaureguiberry, M. V., Sánchez, G., Sant'Ana, A. S. and Barril, P. A. (2022). Occurrence of norovirus, rotavirus, hepatitis A virus, and enterovirus in berries in Argentina. *Food and Environmental Virology* 14(2), 170–177.

Pan, F., Li, X., Carabez, J., Ragosta, G., Fernandez, K. L., Wang, E., Thiptara, A., Antaki, E. and Atwill, E. R. (2015). Cross-sectional survey of indicator and pathogenic bacteria on vegetables sold from Asian vendors at farmers' markets in Northern California. *Journal of Food Protection* 78(3), 602–608. https://doi.org/10.4315/0362-028X.JFP-14-095.

Partnership for Food Safety Education (Partnership) (2022). *Safe Produce*. Retrieved March 8, 2022. Available at: https://www.fightbac.org/food-safety-education/safe-produce/.

Pires, A. F. A., Stover, J., Kukielka, E., Haghani, V., Aminabadi, P., De Melo Ramos, T. and Jay-Russell, M. T. (2020). Salmonella and *Escherichia coli* prevalence in meat and produce sold at farmers' markets in Northern California. *Journal of Food Protection* 83(11), 1934–1940. https://doi.org/10.4315/JFP-20-079.

Pollard, S., Boyer, R., Chapman, B., di Stefano, J., Archibald, T., Ponder, M. A. and Rideout, S. (2016). Identification of risky food safety practices at southwest Virginia farmers' markets. *Food Protection Trends* 36(3), 168–175.

Polsky, J. Y. and Garriguet, D. (2020). Change in vegetable and fruit consumption in Canada between 2004 and 2015. *Health Reports* 31(4), 3–12.

Purpari, G., Macaluso, G., Di Bella, S., Gucciardi, F., Mira, F., Di Marco, P., Lastra, A., Petersen, E., La Rosa, G. and Guercio, A. (2019). Molecular characterization of human enteric viruses in food, water samples, and surface swabs in Sicily. *International Journal of Infectious Diseases* 80, 66–72. https://doi.org/10.1016/j.ijid.2018.12.011.

Quick, V., Corda, K. W., Chamberlin, B., Schaffner, D. W. and Byrd-Bredbenner, C. (2013). Ninja kitchen to the rescue: Evaluation of a food safety education game for middle school youth. *British Food Journal* 115(5), 686–699. https://doi.org/10.1108/00070701311331481.

Rehm, C. D., Peñalvo, J. L., Afshin, A. and Mozaffarian, D. (2016). Dietary intake among US adults, 1999–2012. *JAMA* 315(23), 2542–2553. https://doi.org/10.1001/jama.2016.7491.

Roth, L., Simonne, A., House, L. and Ahn, S. (2018). Microbiological analysis of fresh produce sold at Florida farmers' markets. *Food Control* 92, 444–449. https://doi.org/10.1016/j.foodcont.2018.05.030.

Ruimy, R., Brisabois, A., Bernede, C., Skurnik, D., Barnat, S., Arlet, G., Momcilovic, S., Elbaz, S., Moury, F., Vibet, M. A., Courvalin, P., Guillemot, D. and Andremont, A. (2010). Organic and conventional fruits and vegetables contain equivalent counts of Gram-negative bacteria expressing resistance to antibacterial agents. *Environmental Microbiology* 12(3), 608–615. https://doi.org/10.1111/j.1462-2920.2009.02100.x.

Santé publique France (2019). *Recommendations Concerning Diet, Physical Activity and Sedentary Behaviour for Adults.* Available at: https://www.santepubliquefrance.fr/content/download/186842/document_file/42466_2019-rapport-nutrition-english-bat-ok.PDF.

Scallan, E., Crim, S. M., Runkle, A., Henao, O. L., Mahon, B. E., Hoekstra, R. M. and Griffin, P. M. (2015). Bacterial enteric infections among older adults in the United States: Foodborne diseases active surveillance network, 1996–2012. *Foodborne Pathogens and Disease* 12(6), 492–499. https://doi.org/10.1089/fpd.2014.1915.

Scott, A. R., Pope, P. E. and Thompson, B. M. (2009). Consumer's fresh produce food safety practices: Outcomes of a fresh produce safety education program. *Journal of Food Science Education* 8(1), 8–12. https://doi.org/10.1111/j.1541-4329.2008.00062.x.

Sebastian, R. S., Enns, C. W., Goldman, J. D., Hoy, M. K. and Moshfegh, A. J. (2018). Salad consumption in the U.S. What we eat in America, NHANES 2011–2014, food surveys research group dietary data Brief No. 19. United States Department of Agriculture. Available at: https://www.ars.usda.gov/northeast-area/beltsville-md-bhnrc/beltsville-human-nutrition-research-center/food-surveys-research-group/docs/wweia-data-briefs/.

Sivaramalingam, B., Young, I., Pham, M. T., Waddell, L., Greig, J., Mascarenhas, M. and Papadopoulos, A. (2015). Scoping review of research on the effectiveness of food-safety education interventions directed at consumers. *Foodborne Pathogens and Disease* 12(7), 561–570. https://doi.org/10.1089/fpd.2014.1927.

Sivapalasingam, S., Friedman, C. R., Cohen, L. and Tauxe, R. V. (2004). Fresh produce: A growing cause of outbreaks of foodborne illness in the United States, 1973 through 1997. *Journal of Food Protection* 67(10), 2342–2353. https://doi.org/10.4315/0362-028X-67.10.2342.

Srisamran, J., Atwill, E. R., Chuanchuen, R. and Jeamsripong, S. (2022). Detection and analysis of indicator and pathogenic bacteria in conventional and organic fruits and vegetables sold in retail markets. *Food Quality and Safety* 6.

State, N. C. (2021). Extension (NC state). *N.C. Fresh Produce Safety*. Retrieved March 13, 2022. Available at: https://ncfreshproducesafety.ces.ncsu.edu/ncfreshproducesafety-consumers-foodborne-illness/.

Storey, M. and Anderson, P. (2018). Total fruit and vegetable consumption increases among consumers of frozen fruit and vegetables. *Nutrition* 46, 115–121.

Terio, V., Bottaro, M., Pavoni, E., Losio, M. N., Serraino, A., Giacometti, F., Martella, V., Mottola, A., Di Pinto, A. and Tantillo, G. (2017). Occurrence of hepatitis A and E and Norovirus GI and GII in ready-to-eat vegetables in Italy. *International Journal of Food Microbiology* 249, 61–65. https://doi.org/10.1016/j.ijfoodmicro.2017.03.008.

Thomas, M. S. and Feng, Y. (2021). Food handling practices in the era of COVID-19: A mixed-method longitudinal needs assessment of consumers in the United States. *Journal of Food Protection* 84(7), 1176–1187. https://doi.org/10.4315/JFP-21-006.

Tooby, M., Morton, V., Nesbitt, A., Ciampa, N. and Thomas, M. K. (2021). Consumption of high-risk foods in the Canadian population, Foodbook study, 2014 to 2015. *Journal of Food Protection* 84(11), 1925–1936. https://doi.org/10.4315/JFP-21-101.

U.S. Department of Agriculture (USDA) (2020). AMS national farmers market managers survey report. Available at: https://www.nass.usda.gov/Publications/Todays_Reports/reports/nfar0820.pdf.

U.S. Department of Agriculture and Economic Research Service (USDA, ERS) (2022). ERS charts of note, local foods. Retrieved April 12, 2022. Available at: https://www.ers.usda.gov/data-products/charts-of-note/charts-of-note/?topicId=f5a7d42d-5209-47db-abbb-2e2cc3634cde#:~:text=A%20farmers'%20market%20is%20a,to%20USDA's%20Agricultural%20Marketing%20Service.

U.S. Department of Agriculture and U.S. Department of Health and Human Services (USDA & HHS) (2020). *Dietary Guidelines for Americans, 2020-2025* (9th edn.). Available at: https://www.dietaryguidelines.gov.

U.S. Food and Drug Administration (FDA) (2022). Safe food handling. Retrieved April 11, 2022. Available at: https://www.fda.gov/food/buy-store-serve-safe-food/safe-food-handling.

World Health Organization (WHO) (2022). Five keys to safer food manual. Retrieved March 8, 2022. Available at: https://www.who.int/publications/i/item/9789241594639.

Young, I., Reimer, D., Greig, J., Meldrum, R., Turgeon, P. and Waddell, L. (2017a). Explaining consumer safe food handling through behavior-change theories: A systematic review. *Foodborne Pathogens and Disease* 14(11), 609–622. https://doi.org/10.1089/fpd.2017.2288.

Young, I., Reimer, D., Greig, J., Turgeon, P., Meldrum, R. and Waddell, L. (2017b). Psychosocial and health-status determinants of safe food handling among consumers: A systematic review and meta-analysis. *Food Control* 78, 401–411. https://doi.org/10.1016/j.foodcont.2017.03.013.

Young, I., Thaivalappil, A., Reimer, D. and Greig, J. (2017c). Food safety at farmers' markets: A knowledge synthesis of published research. *Journal of Food Protection* 80(12), 2033–2047. https://doi.org/10.4315/0362-028X.JFP-17-193.

Zhang, H., Yamamoto, E., Murphy, J. and Locas, A. (2020). Microbiological safety of ready-to-eat fresh-cut fruits and vegetables sold on the Canadian retail market. *International Journal of Food Microbiology* 335, 108855. https://doi.org/10.1016/j.ijfoodmicro.2020.108855.

Index

Milton Keynes UK
Ingram Content Group UK Ltd.
UKHW021145180823
427088UK00001B/6

9 781801 462686